BIBLIOGRAPHIC RECORD TARGET

Graduate Library
University of Michigan

Preservation Office

Storage Number: ______________________

ABR0993
UL FMT B RT a BL m T/C DT 07/18/88 R/DT 07/18/88 CC STAT mm E/L 1
035/1: : |a (RLIN)MIUG86-B49708
035/2: : |a (CaOTULAS)160122320
040: : |a RPB |c RPB |d NIC |d MiU
100:1 : |a Casey, John, |d 1820-1891.
245:02: |a A treatise on the analytical geometry of the point, line, circle, and conic sections, |b containing an account of its most recent extensions, with numerous examples.
260: : |a Dublin, |b Hodges, Figgis, and Co.,; |a London, |b Longmans, Green, and Co., |c 1885.
300/1: : |a xvi, [2], 331 p. |c 20 cm.
650/1: 0: |a Geometry, Analytic
998: : |c WFA |s 9124

DUBLIN UNIVERSITY PRESS SERIES.

A TREATISE

ON THE

ANALYTICAL GEOMETRY

OF THE

POINT, LINE, CIRCLE, AND CONIC SECTIONS,

CONTAINING

An Account of its most recent Extensions,

WITH NUMEROUS EXAMPLES.

BY

JOHN CASEY, LL.D., F.R.S.,

Fellow of the Royal University of Ireland;
Member of the Council of the Royal Irish Academy;
Member of the Mathematical Societies of London and France; and
Professor of the Higher Mathematics and Mathematical Physics
in the Catholic University of Ireland.

DUBLIN: HODGES, FIGGIS, & CO., GRAFTON-STREET.
LONDON: LONGMANS, GREEN, & CO., PATERNOSTER-ROW.

1885

DUBLIN:

PRINTED AT THE UNIVERSITY PRESS,

BY PONSONBY AND WELDRICK.

PREFACE.

In the present Work I have endeavoured, without exceeding the usual size of an Elementary Treatise, to give a comprehensive account of the Analytical Geometry of the Conic Sections, including the most recent additions to the Science.

For several years Analytical Geometry has been my special study, and some of the investigations in the more advanced portions of this Treatise were first published in Papers written by myself. These include: finding the Equation of a Circle touching Three Circles; of a Conic touching Three Conics; extending the Equations of Circles inscribed and circumscribed to Triangles to Circles inscribed and circumscribed to Polygons of any number of sides; the extension to Conics of the properties of circles cutting orthogonally; proving that the Tact-invariant of two Conics is the product of Six Anharmonic Ratios; and some others.

Of the Propositions in the other parts of the Treatise, the proofs given will be found to be not only simple and elementary, but in some instances original.

In compiling my Work I have consulted the writings of various authors. Those to whom I am most indebted are: SALMON, CHASLES, and CLEBSCH, from the last of whom I have taken the comparison of Point and Line and Line Co-ordinates (Chapter II., Section III.); and Aronhold's notation (Chapter VIII., Section III.), now published for the first time in an English Treatise on Conic Sections. For Recent Geometry, the writings of BROCARD, NEUBERG, LEMOINE, M'CAY, and TUCKER.

The exercises are very numerous. Those placed after the Propositions are for the most part of an elementary character, and are intended as applications of the propositions to which they are appended. The exercises at the ends of the chapters are more difficult. Some have been selected from the Examination Papers set at the Universities, from Roberts' examples on Analytic Geometry, and Wolstenholme's Mathematical Problems. Some are original; and for a very large number I am indebted to my Mathematical friends Professors NEUBERG, R. CURTIS, S.J., CROFTON, and the Messrs. J. and F. PURSER.

The work was read in manuscript by my lamented and esteemed friend, the late Rev. Professor TOWNSEND, F.R.S.; by Dr. HART, Vice-Provost of Trinity College, Dublin; and

Professor B. Williamson, F.R.S. Their valuable suggestions have been incorporated.

In conclusion, I have to return my best thanks to the last-named gentleman for his kindness in reading the proof sheets, and to the Committee of the 'Dublin University Press Series' for defraying the expense of publication.

JOHN CASEY.

86, South Circular Road, Dublin,
October 5, 1885.

[The following Course, omitting the Articles marked with asterisks, is recommended for Junior readers: Chapter I., Sections I., II., III.; Chapter II., Section I.; Chapter III., Section I.; Chapters V., VI., VII.]

CONTENTS.

CHAPTER I.

THE POINT.

SECTION I.—CARTESIAN CO-ORDINATES.

SECTION II.

SECTION III.

SECTION IV.—COMPLEX VARIABLES

CHAPTER II.

THE RIGHT LINE.

SECTION I.—CARTESIAN CO-ORDINATES.

SECTION II.—TRILINEAR CO-ORDINATES.

Section III.—Comparison of Point and Line Co-ordinates.

CHAPTER III.

THE CIRCLE.

Section I.—Cartesian Co-ordinates.

Section II.—A System of Tangential Circles.

Section III.—Trilinear Co-ordinates.

Section IV.—Tangential Equations.

CHAPTER IV.

THE GENERAL EQUATION OF THE SECOND DEGREE.

Cartesian Co-ordinates.

CHAPTER V.

THE PARABOLA.

CHAPTER VI.

THE ELLIPSE.

CHAPTER VII.

THE HYPERBOLA.

CHAPTER VIII.

MISCELLANEOUS INVESTIGATIONS.

SECTION I.—CONTACT OF CONIC SECTIONS.

SECTION II.—SIMILAR FIGURES.

SECTION III.—THE GENERAL EQUATION—TRILINEAR CO-ORDINATES.

Section IV.—Theory of Envelopes.

Section V.—Theory of Projection.

Section VI.—Sections of a Cone.

Section VII.—Theory of Homographic Division.

ERRATA.

Page 26, line 3, *for* (see 18, 2°), *read* (see 17, 2°).
,, 33, Ex. 5, ,, x', x'', x''', ,, y', y'', y'''.
,, ,, ,, y', y'', y''', ,, x', x'', x'''.
,, ,, ,, ρ', ρ'', ρ''', ,, $\rho'^2, \rho''^2, \rho'''^2$.
,, 57, line 2 from bottom, *for* λ'', *read* α''.
,, 183, Ex. 16, *after* "between," *insert* the tangents to.
,, 313, line 9, *for* B_1C_1, *read* $B'C'$.
,, 319, line 1, *for* PQ, *read* QT.

ERRATA.

The Author is indebted to the REV. SEBASTIAN SIRCOM, S.J., Stonyhurst, for the greater number of the following corrections:—

Page 44, line 13, *for* ω, *read* 0.

,, 58, line 10 from bottom, *for* $2(\beta_1-\beta_2)(\gamma_1-\gamma_2)$ *read* $2bc(\beta_1-\beta_2)(\gamma_1-\gamma_2)$.

,, 61, last line, *for* $b\sin^2\alpha$ *read* $b\cos^2\alpha$.

,, 65, line 5, *for* A *read* B; line 7, *for* 1 *read* -1; line 16, *for* β *read* B.

,, 69, line 10 from bottom, *dele* $-$ *before* $\dfrac{\cos\frac{1}{2}C+\cos\frac{1}{2}A\sin\frac{1}{2}B}{\cos\frac{1}{2}C}$.

,, 75, line 2 from bottom, *for* $(x-x')$ *read* $(x-x')(x-x'')$.

,, 77, line 6 from bottom, *for* r' *read* r^2.

,, 83, line 15, *for* f''' *read* f''''; *for* $r'''\cos\phi'''$ *read* $r''''\cos\phi''''$; *for* g''', *read* g''''; line 23, *for* g''' *read* g''''.

,, 91, line 11, *for* 31, 12 *read* $\overline{31}$, $\overline{12}$.

,, 92, line 4 from bottom, *for* $\cos\frac{1}{2}c$ *read* $\cos^2\frac{1}{2}C$; line 2 from bottom, *for* S' *read* S_1; and *for* $\dfrac{S_3}{3}$, *read* $\dfrac{S_3}{r_3}$.

,, 94, last line, *for* $\dfrac{l1}{lx\,.\,1x}$ *read* $\dfrac{\overline{l_1}}{\overline{lx}\,.\,\overline{1x}}$.

,, 95, line 13, *for* 14, 12 *read* $\overline{14}$, $\overline{12}$.

,, 96, line 4 from bottom, *for* A, B, C, *read* A_1, B_1, C_1; last line, *for* $r_4\sin\frac{1}{2}A'\sin\frac{1}{2}B'\sin\frac{1}{2}C$ *read* $r_4\sin\frac{1}{2}A_1\sin\frac{1}{2}B_1\sin\frac{1}{2}C_1$.

,, 97, line 4, *for* A', *read* A_1.

,, 102, line 2 from bottom, *for* $\sin\frac{1}{2}B$ *read* $\cos\frac{1}{2}B$.

,, 107, line 12, *for* F', D', E, *read* F', D', E'.

,, 115, line 3 from bottom, *for* y''' *read* y''.

,, 116, line 16, *for* $\sin(C-A)\,.\,B$ *read* $\sin(C-A)\,.\,\beta$.

,, 117, line 8, *for* $\overline{13}'$ *read* $\overline{13}$.

,, 119, line 3, *for* d *read* δ.

,, 120, line 5, *for* $2gx'+2fy'$, *read* $2g'x+2f'y$.

,, 121, line 8 from bottom, *for* $b\sin\theta\cos\theta$, *read* $2h\sin\theta\cos\theta$.

,, 126, line 20, *for* $(ax+hx+g)$ *read* $(ax+hy+g)$.

,, 127, line 8, *for* $ax+by$ *read* $ax+hy$.

,, 131, 4°. The general proposition, Art. 100, *Cor.*, does not extend to this case; and the conclusion here stated does not hold for the hyperbola, unless the point P' is on the line AB, which is supposed to be parallel to an asymptote.

Errata.

Page 132, line 18, *for* $-$ *read* $=$.

,, 133, line 5 from bottom, *for* $\dfrac{\Delta}{C^2}$ *read* $\dfrac{(a+b)\Delta}{C^2}$.

,, 140, line 2, *for* $(a + x^2)$ *read* $(a + x)^2$.

,, 148, last line, and last but one, *for* $\tan\phi', \tan\phi''$ *read* $\tan\phi' . \tan\phi''$.

,, 161, Ex. 51, *for* $a^{\frac{1}{3}} + b^{\frac{1}{3}}$ *read* $a^{\frac{2}{3}} + b^{\frac{2}{3}}$.

,, 170, last line, *for* yy *read* yy'.

,, 176, line 10, *for* MG^2 *read* MG.

,, 189, line 10, *put* $-$ *before* $\left(\dfrac{xx'}{a^2} + \dfrac{yy'}{b^2} - 1\right)^2$.

,, 192, line 7, *for* $-$ *read* $=$.

,, 201, Ex. 63, *for* $\beta\gamma$ *read* $2\beta\gamma$.

,, 222, line 21, *for* 2ϕ *read* $2\phi'$.

,, 240, line 7, *for* $\dfrac{b'^3}{ab}$ *read* $\left(\dfrac{b'^3}{ab}\right)^2$.

,, 241, line 8, *for* $Ax^2 + By^2$ *read* $A^2x^2 + B^2y^2$.

,, 249, line 11 from bottom, *for* C, *read* c; and line 10 from bottom, *for* $(\cos C + \theta)$ *read* $\cos(C + \theta)$.

,, 252, line 4 from bottom, *for* y *read* ay; and *for* $ac \sin B$ *read* $\frac{1}{2} ac \sin B$.

,, 253, line 15, and p. 255, line 13 from bottom, *for* $\dfrac{a \cot \omega}{3}$ *read* $\dfrac{a \cot \omega}{3} y$.

,, 252 and 253, last line but one, *for* A *read* A'.

,, 257, line 18, *for* C *read* c; p. 259, line 9, *for* ϕ_2 *read* $2\phi_2$.

,, 268, line 6, *for* $+$ *read* $-$; p. 269, line 9, and p. 271, line 5, *for* $+$ *read* $=$.

,, 271, line 21, *for* $(bc + ca - ab)^2$ *read* $(bc + ca - ab)^2 \gamma^2$.

,, 278, line 8 from bottom, *dele* 'sin.'

,, 282, line 10, *for* $HR . RL$ *read* $HR . RK$.

,, 299, line 4 from bottom, *for* k_3 *read* k^3.

,, 301, line 13, *for* $2(fg - ch)g'$ *read* $2(hf - bg)g'$.

,, 302, line 7, *for* S *read* S_1; p. 316, line 12, *for* e *read* c.

,, 317, line 7, *for* L_2 *read* λ_2; and *for* L_3 *read* λ_3.

,, 318, line 15, insert $+$ *after* $\cos\theta \cos\theta''$; and in line 16, *for* θ' *read* θ.

,, 319, line 16, *for* $2f_1 y$ *read* $2f_1 yz$; and line 11 from bottom, *for* m'' *read* m^4.

,, 320, line 5, *for* circle's *read* circles of.

,, 322, line 1, *for* 1 *read* 1,; and in line 3, *for* Z *read* z; line 12 from bottom, *for* QQ *read* QQ'.

,, 324, line 13 from bottom, *for* f *read* q; and lines 6 and 2 from bottom, *for* g *read* q.

,, 325, line 6, *for* gx^2 *read* qv^2; line 19, *for* $\dfrac{ds}{dx}$, $\dfrac{ds}{dy}$, *read* $\dfrac{dS}{dx}$, $\dfrac{dS}{dy}$; line 8 from bottom, *before* 'the parabola', *insert* the parallel to; line 6 from bottom, *for* $k^2x^2 - a^2y^2$, *read* $k^2x - a^2y$; line 3 from bottom, *for* $\{(a^2 - r^2)x^2 + \&c.\}$ *read* $\{(a^2 - r^2)x^2 + \&c.\}^2$.

A

TREATISE ON ANALYTIC GEOMETRY.

CHAPTER I.

THE POINT.

SECTION I.—CARTESIAN CO-ORDINATES.

DEFINITION I.—Two fixed fundamental lines XX', YY' in a plane, which are used for the purpose of defining the positions of all figures that may be drawn in the plane, are called *axes*. When these are at right angles to each other they are called *rectangular axes*, otherwise they are called *oblique axes*.

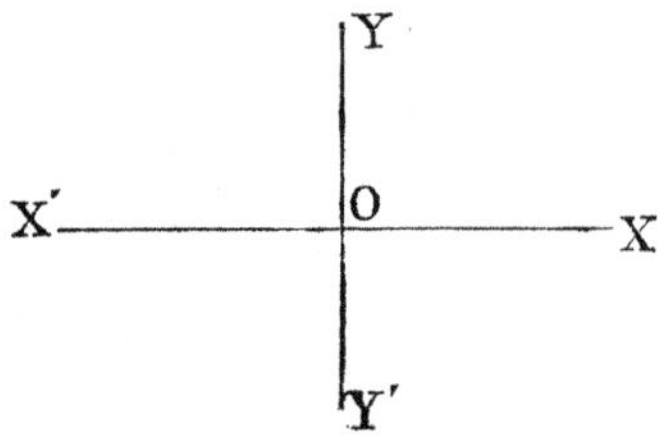

DEF. II.—The lines XX', YY' are called respectively the axis of *abscissæ*, and the axis of *ordinates*. XX' is also called, for reasons that will appear further on, the axis of x, and YY' the axis of y.

DEF. III.—The point O, the intersection of the axes, is called the origin.

DEF. IV.—The origin divides each axis into two parts, one *positive*, the other *negative*.* Thus, $X'X$ is divided into the

* A little consideration will show that the distinction of positive and negative in connexion with the position of a point is absolutely necessary, and not merely a convention, as stated by some writers. All that is conventional is the direction which we fix upon as positive; but whatever that be, the opposite *must* be negative.

parts OX, OX', of which OX measured to the right is usually considered positive, and OX' negative, because it is measured in the opposite direction. Similarly the upward direction, OY, is regarded as positive, and the downward, OY', negative. When the axes are oblique the angle XOY between their positive directions is denoted by ω. The axes will be rectangular unless the contrary is stated.

DEF. V.—Any quantities serving to define the position of a point in a plane are called its *co-ordinates*. Three different systems of co-ordinates are in use, namely, *Cartesian* (called after Descartes, the founder of Analytic Geometry), *Polar*, and *Trilinear co-ordinates*.

DEF. VI.—The Cartesian co-ordinates of a point P are found thus :—Through P draw PM parallel to OY; then the lines OM, MP are the co-ordinates of P; and since OM is measured along OX it is positive, and MP parallel to OY is also positive. Thus both co-ordinates of P are positive. Similarly the co-ordinates of R, viz., ON', $N'R$ are both negative; and lastly, the points Q, S have each one co-ordinate positive and the other negative.

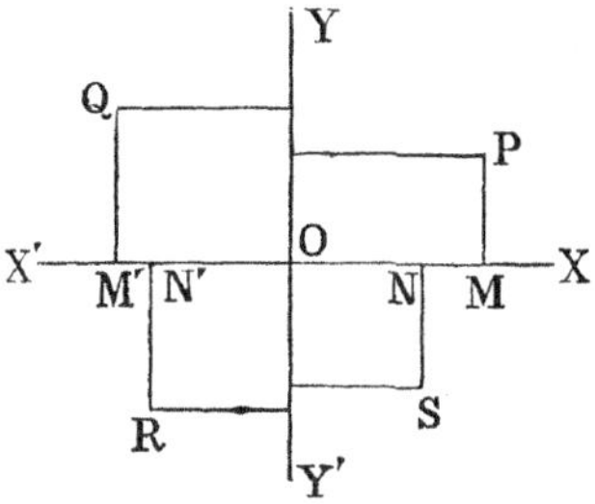

DEF. VII.—The Cartesian co-ordinates of a known or *fixed* point are usually denoted by the initial letters of the alphabet, such as a, b. They are also denoted by the letters x, y, with accents or suffixes, thus : x', y'; x'', y'', &c.; x_1, y_1; x_2, y_2, &c. The co-ordinates of an unknown or of a *variable* point are denoted by the final letters, such as x, y, without either accents or suffixes, and sometimes by the Greek letters α, β; but these are more frequently employed in trilinear co-ordinates, which will be explained further on.

1. *To find the distance δ between two points in terms of their co-ordinates.*

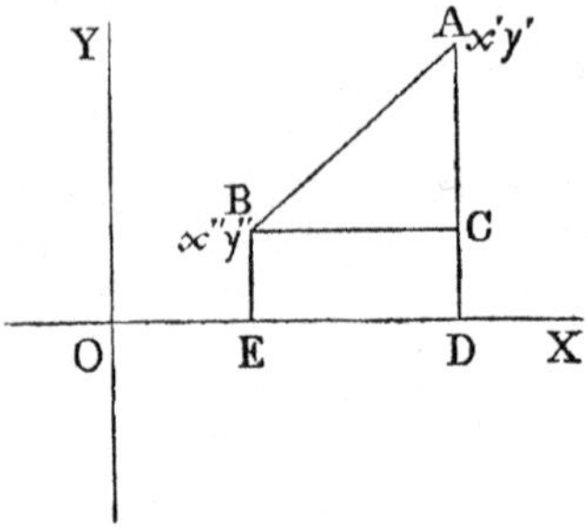

1°. *Let the axes be rectangular.*

Let A, B be the points, $x'y'$, $x''y''$ their co-ordinates. Draw BC parallel to OX; AD, BE parallel to OY. Then, since the co-ordinates of A are $x'y'$, we have

$$OD = x', \qquad DA = y'.$$

Similarly $$OE = x'', \qquad EB = y''.$$

Hence $$BC = x' - x'', \quad CA = y' - y'';$$

but $$AB^2 = BC^2 + CA^2;$$

therefore $$\delta^2 = (x' - x'')^2 + (y' - y'')^2. \qquad (1)$$

Hence we have the following rule:—*Subtract the x of one point from the x of the other, also the y of one point from the y of the other; then the sum of the squares of the remainders is equal to the square of the required distance.*

2°. *Let the axes be oblique.*

Since the angle ACB is the supplement of XOY, we have

$$ACB = 180^\circ - \omega.$$

Hence $$AB^2 = BC^2 + CA^2 + 2BC \cdot CA \cos\omega,$$

that is, $$\delta^2 = (x' - x'')^2 + (y' - y'')^2 + 2(x' - x'')(y' - y'')\cos\omega.$$

In practice, oblique axes are seldom employed; but as they sometimes are, we shall give the principal formulæ in both forms.

EXAMPLES.

1. Find the distance of the point $x'y'$ from the origin—

1°. When the axes are rectangular. *Ans.* $\delta^2 = x'^2 + y'^2$. (2)

2°. When they are oblique. *Ans.* $\delta^2 = x'^2 + y'^2 + 2x'y' \cos\omega$. (3)

2. Find the distance between the points $(r\cos\theta', r\sin\theta')$, $(r\cos\theta'', r\sin\theta'')$.

Ans. $\delta = 2r \sin\frac{1}{2}(\theta' - \theta'')$. (4)

3. Find the distance between the points $\left(-\frac{C}{A}, 0\right)$, $\left(0, -\frac{C}{B}\right)$.

1°. When the axes are rectangular. *Ans.* $\delta = \frac{C}{AB}\sqrt{A^2 + B^2}$. (5)

2°. When oblique. *Ans.* $\delta = \frac{C}{AB}\sqrt{A^2 + B^2 + 2AB\cos\omega}$. (6)

4. Find the distance between the points $\{a\cos(\alpha+\beta),\ b\sin(\alpha+\beta)\}$, $\{a\cos(\alpha-\beta),\ b\sin(\alpha-\beta)\}$.

Ans. $\delta = 2\sin\beta\,\{a^2\sin^2\alpha + b^2\cos^2\alpha\}^{\frac{1}{2}}$. (7)

DEF.—*The line joining two points will for shortness be called the join of the two points.*

5. Find the condition that the join of the points $x'y'$, $x''y''$ may subtend a right angle at xy. Since the triangle formed by the three points is right-angled, the square on one side is equal to the sum of the squares on the other two. Hence

$$(x'-x'')^2 + (y'-y'')^2 = (x-x')^2 + (y-y')^2 + (x-x'')^2 + (y-y'')^2;$$

and reducing, we get

$$(x-x')(x-x'') + (y-y')(y-y'') = 0. \tag{8}$$

If the axes be oblique, the condition is

$$(x-x')(x-x'') + (y-y')(y-y'')$$
$$+ 2\{(x-x')(y-y'') + (x-x'')(y-y')\}\cos\omega = 0. \tag{9}$$

2. *To find the condition that three points $x'y'$, $x''y''$, $x'''y'''$ shall be collinear.*

Let A, B, C be the points: drawing parallels we have, from similar triangles, $BD : AD :: CE : EB$.

Hence
$$\frac{x' - x''}{y' - y''} = \frac{x'' - x'''}{y'' - y'''}, \tag{10}$$

or $$(x'y'' - x''y') + (x''y''' - x'''y'') + (x'''y' - x'y''') = 0. \tag{11}$$

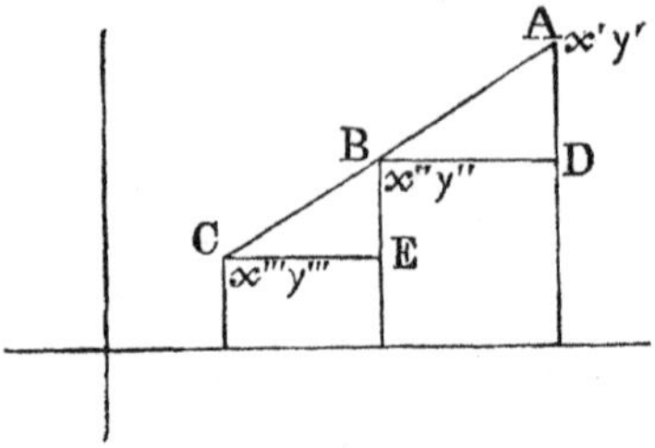

This may be written in the form of the determinant

$$\begin{vmatrix} x', & y', & 1, \\ x'', & y'', & 1, \\ x''', & y''', & 1, \end{vmatrix} = 0. \tag{12}$$

3. This proposition may be proved otherwise, and by a method which will connect it with another of equal importance.

Lemma.—*The area of the triangle whose angular points are* $x'y'$, $x''y''$, *and the origin, is* $\frac{1}{2}(x'y'' - x''y')\sin\omega$.

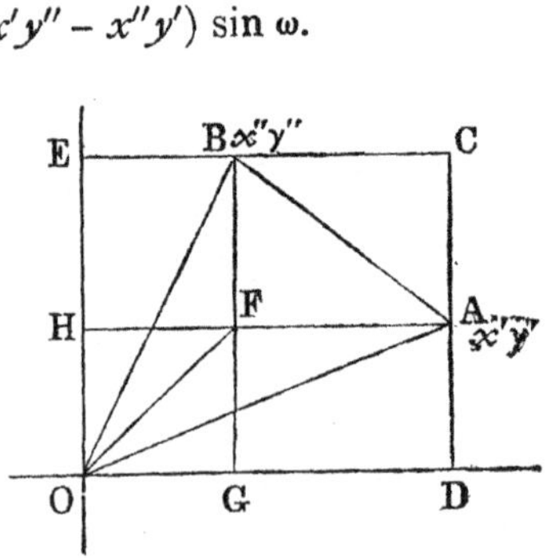

Dem.—Through the points $x'y'$, $x''y''$ draw parallels to the axes; then the parallelograms *ODCE*, *OGFH* are respectively equal to $x'y''\sin\omega$, $x''y'\sin\omega$. Hence the triangle *OAB*, which is evidently equal to half the difference of these parallelograms, is

$$\tfrac{1}{2}(x'y'' - x''y')\sin\omega. \tag{13}$$

Cor. 1.—If the axes be rectangular, the triangle

$$AOB = \tfrac{1}{2}(x'y'' - x''y'). \qquad (14)$$

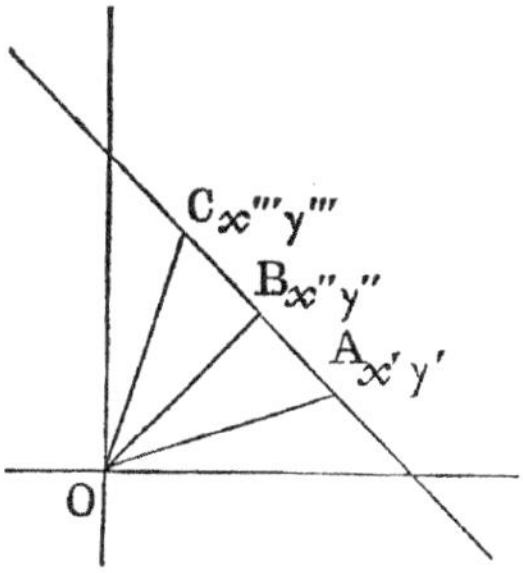

To apply this, let A, B, C be three collinear points. Join OA, OB, OC; then we have

$$\triangle\, OAB + \triangle\, OBC = \triangle\, OAC;$$

therefore $x'y'' - x''y' + x''y''' - x'''y'' = x'y''' - x'''y'$,

or $(x'y'' - x''y') + (x''y''' - x'''y'') + (x'''y' - x'y''') = 0$.

Cor. 2.—The $\triangle\, OBA = - \triangle\, OAB$. For $OBA = x''y' - x'y''$, and $OAB = x'y'' - x''y'$.

4. The Lemma of Art. 3 enables us *to find the area of a triangle in terms of the co-ordinates of its vertices.*

For, if any point O within the triangle be taken as the origin of rectangular axes, and the co-ordinates of the vertices be $x'y'$, $x''y''$, $x'''y'''$, then join OA, OB, OC. Since the triangle

$$ABC = OAB + OBC + OCA,$$

we have

$$\triangle\, ABC = \tfrac{1}{2}\{x'y'' - x''y' + x''y''' - x'''y'' + x'''y' - x'y'''\}, \qquad (15)$$

or

$$= \tfrac{1}{2}\begin{vmatrix} x', & y', & 1, \\ x'', & y'', & 1, \\ x''', & y''', & 1 \end{vmatrix}. \qquad (16)$$

It is evident that we get the same result if we take the origin outside the triangle by attending to the signs of the areas (see *Cor.* 2, Art. 3).

From this proposition it follows that the geometrical interpretation of the condition that three points should be collinear is, that the area of the triangle formed by them is zero.

5. In the same manner it follows that the area of any polygon, in terms of the co-ordinates of its vertices, is

$$\tfrac{1}{2}\{(x_1y_2 - x_2y_1) + (x_2y_3 - x_3y_2) + \ldots (x_ny_1 - x_1y_n)\}. \quad (17)$$

EXAMPLES.

Find the areas of the triangles whose vertices are—

1. (1, 2); (3, 4); (5, 2).
2. (3, 4); (5, 3); (6, 2).
3. (− 5, 4); (− 6, 5); (6, 2).
4. (2, 1); (3, − 2); (− 4, − 1).
5. $$(x'y'),\ \left(\frac{-C}{A}, 0\right),\ \left(0, \frac{-C}{B}\right). \quad (18)$$

   Ans. $\dfrac{C}{2AB}(Ax' + By' + C.)$

6. $$(at'^2, 2at'),\ (at''^2, 2at''),\ (at'''^2, 2at''').$$

   $$Ans.\ a^2 \begin{vmatrix} 1, & t', & t'^2, \\ 1, & t'', & t''^2, \\ 1, & t''', & t'''^2 \end{vmatrix}. \quad (19)$$

7. $\{at't'', a(t' + t'')\};\ \{at''t''', a(t'' + t''')\};\ \{at'''t', a(t''' + t')\}.$

   Ans. Half the area of Ex. 2.

8. $(a \cos\phi', b \sin\phi'),\ (a \cos\phi'', b \sin\phi''),\ (a \cos\phi''', b \sin\phi''').$

   $$Ans.\ 2ab \sin\tfrac{1}{2}(\phi' - \phi'') \sin\tfrac{1}{2}(\phi'' - \phi''') \sin\tfrac{1}{2}(\phi''' - \phi'). \quad (20)$$

*9. $$\left\{\frac{a\cos\frac{1}{2}(\alpha+\beta)}{\cos\frac{1}{2}(\alpha-\beta)}, \frac{b\sin\frac{1}{2}(\alpha+\beta)}{\cos\frac{1}{2}(\alpha-\beta)}\right\};\ \left\{\frac{a\cos\frac{1}{2}(\beta+\gamma)}{\cos\frac{1}{2}(\beta-\gamma)}, \frac{b\sin\frac{1}{2}(\beta+\gamma)}{\cos\frac{1}{2}(\beta-\gamma)}\right\};$$

$$\left\{\frac{a\cos\frac{1}{2}(\gamma+\alpha)}{\cos\frac{1}{2}(\gamma-\alpha)}, \frac{b\sin\frac{1}{2}(\gamma+\alpha)}{\cos\frac{1}{2}(\gamma-\alpha)}\right\}.$$

$$Ans.\ ab \tan\tfrac{1}{2}(\alpha-\beta) \tan\tfrac{1}{2}(\beta-\gamma) \tan\tfrac{1}{2}(\gamma-\alpha). \quad (21)$$

10. $(k \tan\phi, k \cot\phi),\ (k \tan\phi', k \cot\phi'),\ (k \tan\phi'', k \cot\phi'').$

$$Ans. \frac{k^2}{\tan\phi \,.\, \tan\phi' \,.\, \tan\phi''} \begin{vmatrix} 1, & \tan\phi, & \tan^2\phi, \\ 1, & \tan\phi', & \tan^2\phi', \\ 1, & \tan\phi'', & \tan^2\phi'' \end{vmatrix}. \quad (22)$$

6. *To find the co-ordinates of the point which divides in a given ratio $l : m$ the join of two points, $x'y'$, $x''y''$.* If A, B be the given points, let C be the point of division, xy its co-ordinates; then, drawing parallels, we have

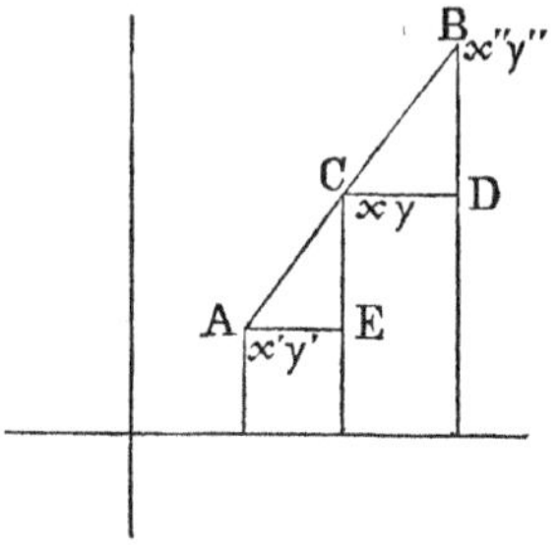

$$\frac{AC}{CB} = \frac{AE}{CD};$$

but $$\frac{AC}{CB} = \frac{l}{m}; \therefore \frac{AE}{CD} = \frac{l}{m}.$$

Hence $$\frac{x - x'}{x'' - x} = \frac{l}{m};$$

therefore $$x = \frac{lx'' + mx'}{l + m}$$

Similarly, $$y = \frac{ly'' + my'}{l + m}$$ (23)

If the join of the two points be cut externally, we get

$$\frac{l}{m} = \frac{x - x'}{x - x''}.$$

Hence $$x = \frac{lx'' - mx'}{l - m}$$

and $$y = \frac{ly'' - my'}{l - m}$$ (24)

Hence the formulæ for external division can be obtained from those for internal section by changing the sign of the ratio. This is evident *a priori* if we consider that for internal section the segments AC, CB are measured in the same direction, and therefore have a *positive* ratio; but for external section, being measured in opposite directions, they have contrary signs, and hence a *negative* ratio.

Cor. 1.—If the ratio $\frac{l}{m}$ be denoted by λ, we have

$$x = \frac{x' + \lambda x''}{1 + \lambda}, \quad y = \frac{y' + \lambda y''}{1 + \lambda}. \qquad (25)$$

Hence, by varying λ we get the co-ordinates of any point in the line AB, in terms of a single parameter λ.

Cor. 2.—If λ be equal to unity, we get

$$x = \frac{x' + x''}{2}, \quad y = \frac{y' + y''}{2}. \tag{26}$$

Hence we have the following rule:—

The co-ordinates of the middle point of the join of two given points are respectively half the sums of the corresponding co-ordinates of these points.

EXAMPLES.

1. If the segment AB be divided in the points L, M, N in the ratios λ, μ, ν respectively, find the ratios of the segments—

1°. Into which AL is divided in M. *Ans.* $\dfrac{\mu(1 + \lambda)}{\mu - \lambda}$; (27)

2°. Into which LM is divided in N. *Ans.* $\dfrac{(\lambda - \nu)}{(1 + \lambda)} \div \dfrac{(\mu - \nu)}{(1 + \mu)}$. (28)

2. The joins of the middle points of opposite sides, and the join of the middle points of the diagonals of a quadrilateral, are concurrent. For, if $x_1 y_1$, $x_2 y_2$, $x_3 y_3$, $x_4 y_4$ be the co-ordinates of its angular points, then the co-ordinates of the point of bisection of the join of the middle points of its diagonals, or of either pair of opposite sides, are

$$\tfrac{1}{4}(x_1 + x_2 + x_3 + x_4), \quad \tfrac{1}{4}(y_1 + y_2 + y_3 + y_4).$$

DEF.—*The point whose abscissa and ordinate are respectively the arithmetic means of the abscissæ and ordinates of any system of points is called the mean centre of that system of points. Thus the point whose co-ordinates are those found in Ex.* 2 *is the mean centre of the angular points of the quadrilateral.*

3. If O be the mean centre of a system of m points, O' the mean centre of another system of n points; prove that the mean centre of the system composed of both divides the line OO' inversely in the ratio of $m : n$.

4. The medians of a triangle are concurrent (each passes through the mean centre of its angular points).

5. Find the co-ordinates of the mean centre of the points

$(a \cos \alpha,\ b \sin \alpha),\quad (a \cos \beta,\ b \sin \beta),\quad (a \cos \gamma,\ b \sin \gamma),$

$(a \cos (\alpha + \beta + \gamma),\quad - b \sin (\alpha + \beta + \gamma)).$

Ans.
$$\left.\begin{aligned}\bar{x} &= a \cos \tfrac{1}{2}(\alpha + \beta) \cos \tfrac{1}{2}(\beta + \gamma) \cos \tfrac{1}{2}(\gamma + \alpha), \\ \bar{y} &= b \sin \tfrac{1}{2}(\alpha + \beta) \sin \tfrac{1}{2}(\beta + \gamma) \sin \tfrac{1}{2}(\gamma + \alpha)\end{aligned}\right\}. \tag{29}$$

It is usual to put a horizontal line over the co-ordinates of the mean centre of a system of points.

7. The definition of mean centre may be extended as follows :—

If $A, B, C \ldots L$ be any system of points $x_1 y_1, x_2 y_2 \ldots x_n y_n$, $a, b, c \ldots l$, a corresponding system of multiples, then the point whose coordinates are

$$\left.\begin{aligned}\bar{x} &= \frac{ax_1 + bx_2 \ldots lx_n}{a + b + \ldots t} \\ \bar{y} &= \frac{ay_1 + by_2 + \ldots ly_n}{a + b \ldots l}\end{aligned}\right\}, \tag{30}$$

is called the mean centre of the points $A, B, C \ldots L$ for the system of multiples $a, b, c \ldots l$.—(Sequel to Euclid, p. 13).

The equations (30) are, for shortness, usually written

$$\bar{x} = \frac{\Sigma(ax_1)}{\Sigma(a)}, \quad \bar{y} = \frac{\Sigma(ay_1)}{\Sigma(a)}. \tag{31}$$

Section II.—Polar Co-ordinates.

8. The polar co-ordinates of a point P are—

1°. *Its distance OP from a fixed point O, called the origin. OP is usually denoted by ρ, and is called the radius vector of the point P.*

2°. *The angle θ, which OP makes with a fixed line (called the initial line), passing through the origin.*

From these definitions it is evident that any equation in Cartesian co-ordinates will be transformed into polar co-ordinates if the initial line coincide with the axis of x, by the substitution $x = \rho \cos\theta$, $y = \rho \sin\theta$; or by the substitution $x = \rho\cos(\theta - \alpha)$, $y = \rho\sin(\theta - \alpha)$, if it make an angle α with the axis of x.

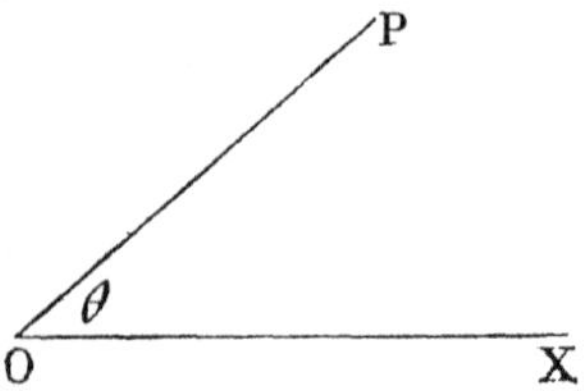

EXAMPLES.

1. Change the following equations to polar co-ordinates:—

   1°. $x^2 + y^2 = 2ax$. 3°. $x^3 = y^2(2a - x)$.

   2°. $x^2 - y^2 = 2ax$. 4° $y^3 = \dfrac{x^2(a + x)}{a - x}$.

2. Change the following equations to rectangular co-ordinates:—

   1°. $\rho^2 = a^2 \cos 2\theta$. 3°. $\rho^2 \sin 2\theta = a^2$.

   2°. $\rho^{\frac{1}{2}} \cos \frac{1}{2}\theta = a^{\frac{1}{2}}$. 4°. $\rho^{\frac{1}{2}} = a^{\frac{1}{2}} \cos\frac{1}{2}\theta$.

3. What is the condition that the points $\rho_1\theta_1$; $\rho_2\theta_2$; $\rho_3\theta_3$ may be collinear? *Ans.* $\rho_1\rho_2\sin(\theta_1 - \theta_2) + \rho_2\rho_3\sin(\theta_2 - \theta_3) + \rho_3\rho_2\sin(\theta_3 - \theta_1) = 0$.

4. Express the area of any rectilineal figure in terms of the polar co-ordinates of its angular points.

SECTION III.—TRANSFORMATION OF CO-ORDINATES.

9. *The co-ordinates of any point P with respect to one system of axes being known, to find its co-ordinates with respect to a parallel system.*

Let Ox, Oy be the old axes, $O'X$, $O'Y$ the new, so that O' is the new origin; then let the co-ordinates of O', with respect to Ox, Oy, be x', y'—that is, let $OL = x'$, $LO' = y'$. Again, let x, y be the old co-ordinates of P, that is, let $OM = x$, $MP = y$.

Lastly, let X, Y be the co-ordinates with respect to the new

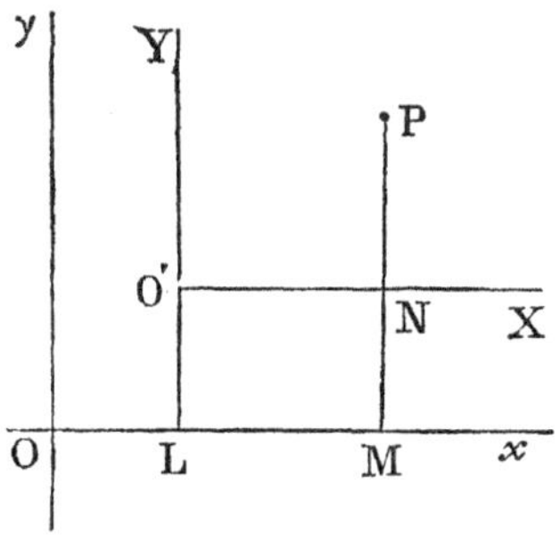

axes; then we have

$$O'N = X, \quad NP = Y;$$

therefore, since

$$OM = OL + O'N, \quad \text{and} \quad MP = LO' + NP,$$

we have

$$x = x' + X, \quad \text{and} \quad y = y' + Y. \qquad (32)$$

Hence, if in any equation we replace x, y by $x' + X$, $y' + Y$, we have it referred to parallel axes through the point $x'y'$.

Examples.

1. Refer the following equations to parallel axes:—

1°, $x^2 + y^2 - 12x - 16y - 44 = 0$. New origin 6, 8.

Ans. $x^2 + y^2 - 144 = 0$.

2°. $3x^2 - 4xy + 2y^2 + 7x - 5y - 3 = 0$. New origin, 1, 1.

2. Find the co-ordinates of a point, so that when the following equations are referred to parallel axes passing through it they may be deprived of terms of the first degree:—

1°. $3x^2 + 5xy + y^2 - 3x + 2y + 21 = 0$. *Ans.* $-\frac{16}{13}$, $\frac{27}{13}$.

2°. $5x^2 + 2xy + y^2 - 10x + 2y + 10 = 0$. *Ans.* $\frac{3}{2}$, $-\frac{5}{2}$.

3°. $4x^2 + 4xy + y^2 - 8x - 6y - 10 = 0$. *Ans.* ∞, ∞.

10. *The co-ordinates of a point P with respect to a rectangular system Ox, Oy of axes being known, to find its co-ordinates with respect to another rectangular system OX, OY, having the same origin, but making an angle θ with the former.*

Let OM, MP, the co-ordinates with respect to the old axes,

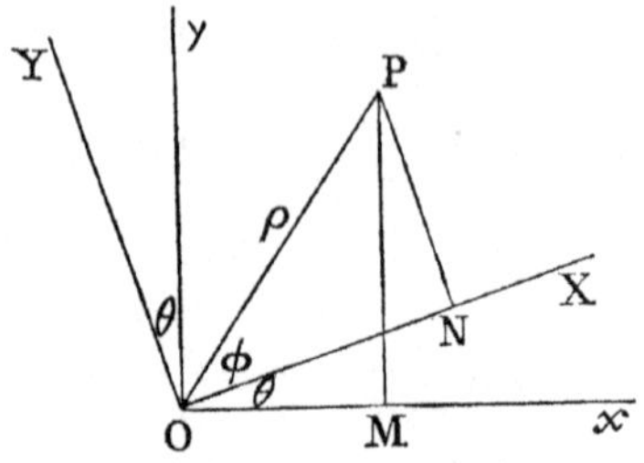

be denoted by x, y; and ON, NP the new co-ordinates, by X, Y.

Let OP be denoted by ρ, and the angle PON by ϕ. Now since

$$\cos(\theta + \phi) = \cos\theta \cos\phi - \sin\theta \sin\phi,$$

and
$$\sin(\theta + \phi) = \sin\theta \cos\phi + \cos\theta \sin\phi,$$

multiplying each by ρ, and substituting, we get

$$\left.\begin{aligned} x &= X\cos\theta - Y\sin\theta, \\ y &= X\sin\theta + Y\cos\theta \end{aligned}\right\}. \qquad (33)$$

Cor.—If the equations (33) be solved, we get

$$\left.\begin{aligned} X &= x\cos\theta + y\sin\theta, \\ Y &= y\cos\theta - x\sin\theta \end{aligned}\right\}. \qquad (34)$$

Observation.—Those who are acquainted with the Differential Calculus will see that

$$x = \frac{dy}{d\theta}, \quad \text{and} \quad y = -\frac{dx}{d\theta}.$$

Examples.

1. If we transform from oblique co-ordinates to rectangular, retaining the old axis of x; prove $Y = y \sin\omega$, $X = x + y\cos\omega$.

2. If in transforming from one set of oblique axes to another, retaining the old origin, α, β denote the angles which the new axes make with the

old axis of x; $\alpha'\beta'$ those which they make with the old axis of y; prove

$$x \sin\omega = X \sin\alpha' + Y \sin\beta',$$
$$y \sin\omega = X \sin\alpha + Y \sin\beta.$$

3. Show that both transformations are included in the formulæ—

$$x = \lambda x + \mu y + \nu,$$
$$y = \lambda' x + \mu' y + \nu',$$

by giving suitable values to the constants λ, μ, &c.

*4. If the old axes be inclined at an angle ω, and the new at an angle ω', and if the quantic $ax^2 + 2hxy + by^2$, referred to the old axes, be transformed to $a'X^2 + 2h'XY + b'Y^2$, referred to the new; prove—

$$1^\circ. \quad \frac{ab - h^2}{\sin^2\omega} = \frac{a'b' - h'^2}{\sin^2\omega'}. \tag{35}$$

$$2^\circ. \quad \frac{a + b - 2h\cos\omega}{\sin^2\omega} = \frac{a' + b' - 2h'\cos\omega'}{\sin^2\omega'}. \tag{36}$$

*Section IV.—Complex Variables.

11. *An expression $x + iy$, in which x, y are the rectangular Cartesian co-ordinates of a point P, and i the imaginary radical, $\sqrt{-1}$ is called a complex magnitude. If $\rho = \sqrt{x^2 + y^2} = OP$, ρ is called the modulus, and the angle θ, made by OP with the axis of x, the inclination or argument.*

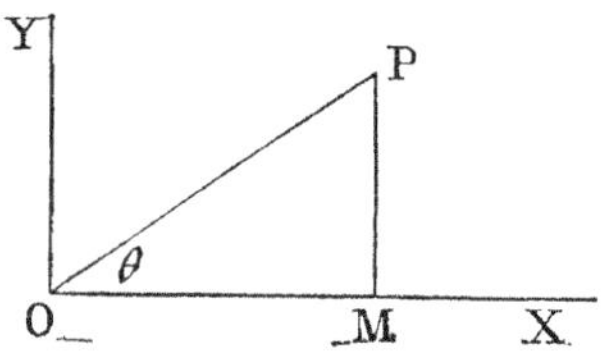

Complex magnitudes were introduced by Cauchy in 1825, in a memoir, "*Sur les integrales définies prises entre des limites imaginaires:*" the method of representing them geometrically is due to Gauss. The introduction of these variables is one of the greatest strides ever made in Mathematics. The whole of the modern theory of functions depends on them, and they are so connected with modern Mathematics, that some knowledge of them is essential to the student. We shall give only their most elementary principles.

12. *Being given two points A, B, which are the geometric representations of two complex magnitudes z_1, z_2, it is required to find the point which represents their sum or their difference.*

1°. *Their sum.*—Let $z_1 = x_1 + iy_1$; $z_2 = x_2 + iy_2$; then $z_1 + z_2 = (x_1 + x_2) + i(y_1 + y_2)$. Now if C represent $z_1 + z_2$, the co-ordinates of C are $x_1 + x_2$, $y_1 + y_2$. Hence the co-ordinates of C are the doubles of the co-ordinates of D (Art. 6, *Cor.* 2), the middle point of AB. Therefore C will be the fourth angular point of the parallelogram which has OA, OB as two adjacent sides. *Hence the vector, from the origin to the point which represents the sum of two complex variables, is the diagonal of the parallelogram which has the vectors of the two components as adjacent sides.*

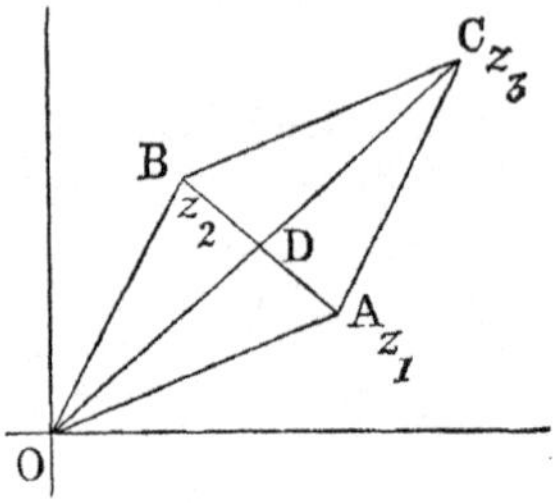

2°. *Their difference.*—If we put $z_1 + z_2 = z_3$, we have $z_2 = z_3 - z_1$. Hence we have the following construction for finding the vector and the point which represent the difference of two complex magnitudes. *Draw from the origin a line OB equal and parallel to the line AC, joining the representative points A, C of z_1, z_3; then OB will be the vector, and B the point required.*

13. *Being given the points which represent two complex magnitudes, to find the points which represent their product and their quotient.*

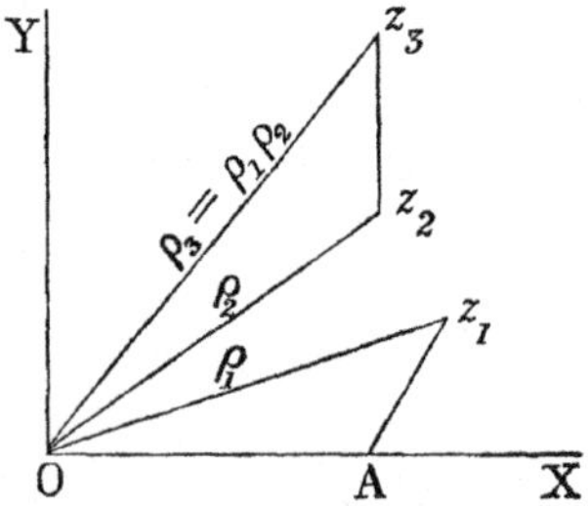

1°. *Their product.*—Let z_1, z_2 be the given points, ρ_1, ρ_2 their moduli, and θ_1, θ_2 their arguments; then we have

$$z_1 = \rho_1(\cos\theta_1 + i\sin\theta_1),$$

$$z_2 = \rho_2(\cos\theta_2 + i\sin\theta_2);$$

therefore $$z_1 z_2 = \rho_1\rho_2\{\cos(\theta_1 + \theta_2) + i\sin(\theta_1 + \theta_2)\}$$

$$= \rho_3(\cos\theta_3 + i\sin\theta_3).$$

Hence, if z_3 be the point required, ρ_3 its modulus, and θ_3 its argument, we see *that the product of two complex magnitudes is a complex magnitude, whose modulus is equal to the product of their moduli, and argument equal to the sum of their arguments.* Hence, if we make OA equal to the linear unit, the triangle AOz_1 is similar to z_2Oz_3, and the method of constructing the point z_3 is known.

2°. *Their quotient.*—This follows from 1°. For we have

$$\frac{z_3}{z_2} = \frac{z_1}{1}.$$

Hence the quotient $z_3 \div z_2$ *makes with axis of* x *an angle equal to that which* z_3 *makes with* z_2, *and the modulus is a fourth proportional to* ρ_2, ρ_3, and 1.

EXAMPLES.

1. Transform $x + iy$ to polar co-ordinates. *Ans.* $\rho e^{i\theta}$.

2. Find the point which represents—

   1°. The square of the magnitude $\alpha + i\beta$.

   2°. Its square root. 3°. Its nth power. 4°. Its nth root.

3. If z_1, z_2, z_3 be three coinitial complex variables, prove that if three multiples l, m, n can be found satisfying the two equations

$$lz_1 + mz_2 + nz_3 = 0, \qquad l + m + n = 0,$$

the corresponding points are collinear.

4. If O be the origin, α, β, γ complex magnitudes representing the angular points of the triangle ABC; prove that if $l\alpha + m\beta + n\gamma = 0$, the points A', B', C', in which the lines AO, BO, CO meet the sides of the triangle, are denoted by either of the systems

$$\frac{-l\alpha}{m+n}, \quad \frac{-m\beta}{n+l}, \quad \frac{-n\gamma}{l+m}; \quad \frac{m\beta + n\gamma}{m+n}, \quad \frac{n\gamma + l\alpha}{n+l}, \quad \frac{l\alpha + m\beta}{l+m}.$$

5. If α, β, γ, δ represent any four coplanar points A, B, C, D, and if the multiples l, m, n, p satisfy the two equations $l\alpha + m\beta + n\gamma + p\delta = 0$, $l + m + n + p = 0$, prove that the point of intersection of AB and CD is $\dfrac{l\alpha + m\beta}{l+m}$, of BC, AD is $\dfrac{m\beta + n\gamma}{m+n}$, and of AC, BD is $\dfrac{l\alpha + n\gamma}{l+n}$.

6. If $\bar{z}$ be the complex magnitude which represents the mean centre of the points z_1, $z_2 \ldots z_n$, &c., for the system of multiples a, b, $c \ldots l$, prove

$$\bar{z} = \frac{\Sigma(az_1)}{\Sigma(a)}.$$

Miscellaneous Exercises.

1. Show that the polar co-ordinates (ρ, θ); $(-\rho, \pi + \theta)$; $(\rho, \theta - \pi)$, all represent the same point.

2. Prove that the three points

$$(a, b); \ (a + 28\sqrt{2}, \ b + 28\sqrt{2}); \ \left(a + \frac{33}{\sqrt{2}}, \ b - \frac{33}{\sqrt{2}}\right),$$

form a right-angled triangle.

3. Find the perimeter of the quadrilateral whose vertices, taken in order, are

$$(a, \ a\sqrt{3}); \quad (-b\sqrt{3}, \ b); \quad (-c, \ -c\sqrt{3}); \quad (d\sqrt{3}, \ -d).$$

4. If the three sides of a triangle, taken in order, be divided in the ratios $l : -m$, $m : -n$, $n : -l$, prove that the three points of section are collinear.

5. If (x, y) (x', y') be the co-ordinates of a point referred respectively to rectangular and oblique axes having a common origin, prove that if the axes of the first system bisect the angles between those of the second,

$$x = (x' + y') \cos \frac{\omega}{2}, \quad y = (x' - y') \sin \frac{\omega}{2}.$$

6. If the points (ab), $(a'b')$, $(a - a', \ b - b')$ be collinear, prove $ab' = a'b$.

7. If the co-ordinates $(x'y')$, $(x''y'')$, $(x'''y''')$ of three variable points satisfy the relations

$$(x' - x'') = \lambda(x'' - x''') - \mu(y'' - y'''),$$
$$(y' - y'') = \lambda(y'' - y''') + \mu(x'' - x'''),$$

where λ and μ are constants, p ove that the triangle of which these points are vertices is given in species.

8. If two systems of co-ordinates have the same origin and the same axis of x, prove that

$$x = x' + y' \frac{\sin(\omega - \omega')}{\sin \omega}, \quad y = y' \frac{\sin \omega'}{\sin \omega}.$$

9. Prove that the orthocentre of a triangle is the mean centre of its angular points for the multiples $\tan A$, $\tan B$, $\tan C$.

10. For what system of multiples is the circumcentre of a triangle the mean centre of its angular points?

11. If $x'y'$, $x''y''$, $x'''y'''$ be the vertices of a triangle, a, b, c the lengths of its sides, prove that the co-ordinates of the centre of its inscribed circle are

$$\frac{ax' + bx'' + cx'''}{a + b + c}, \quad \frac{ay' + by'' + cy'''}{a + b + c}.$$

12. If O be the mean centre of three points A, B, C for the system of multiples p, q, r; prove $p : q : r :: \triangle \, OBC : OCA : OAB$.

13. Prove that the degree of any equation cannot be altered by transformation of co-ordinates.

14. If A, B, C, D be four collinear points, prove that

$$AB \,.\, CD + BC \,.\, AD + CA \,.\, BD = 0.$$

15. Prove the following formulæ of transformation from oblique axes to polar co-ordinates :—

$$x = \rho\,\frac{\sin(\omega - \theta)}{\sin\omega}, \quad y = \rho\,\frac{\sin\theta}{\sin\omega}.$$

16. Prove that the diameter of the circle passing through the two points $\rho'\theta'$, $\rho''\theta''$, and the origin, is

$$\frac{\sqrt{\rho'^2 + \rho''^2 - 2\rho'\rho''\cos(\theta' - \theta'')}}{\sin(\theta' - \theta'')}.$$

17. Find the area of the triangle whose vertices are the three points

$$(a, \theta), \quad \left(2a, \theta + \frac{\pi}{3}\right), \quad \left(3a, \theta + \frac{2\pi}{3}\right).$$

*18. If O be the mean centre of the system of points A, B, C, &c., for the system of multiples a, b, c, &c., prove, for any point P,

$$\Sigma(a \,.\, AP^2) = \Sigma(a \,.\, AO^2) + \Sigma a \,.\, OP^2.$$

*19. In the same case prove

$$\Sigma(a) \,.\, \Sigma(a \,.\, AO^2) = \Sigma ab \,.\, AB^2.$$

*20. If A, B, C, D be four coplanar points, and if we denote

$$\begin{array}{llll} BC^2, & AD^2, & \text{by} & a, \ f, \\ CA^2, & BD^2, & ,, & b, \ g, \\ AB^2, & CD^2, & ,, & c, \ h, \end{array}$$

prove the determinant equation

$$\begin{vmatrix} 0, & c, & b, & f, & 1, \\ c, & 0, & a, & g, & 1, \\ b, & a, & 0, & h, & 1, \\ f, & g, & h, & 0, & 1, \\ 1, & 1, & 1, & 1, & 0, \end{vmatrix} = 0.$$

Multiply together the two matrices

$$\begin{vmatrix} 1, & 0, & 0, & 0, \\ x'^2 + y'^2, & -2x', & -2y', & 1, \\ & \text{\&c.} & & \end{vmatrix} \times \begin{vmatrix} 0, & 0, & 0, & 1, \\ 1, & x', & y', & x'^2 + y'^2, \\ & & \text{\&c.} & \end{vmatrix},$$

each consisting of five rows and four columns.—(CAYLEY).

CHAPTER II.

THE RIGHT LINE.

SECTION I.—CARTESIAN CO-ORDINATES.

14. *To represent a right line by an equation*, there are three cases to be considered.

1°. *When the line intersects both axes, but not at the origin.*

First method.—Let the line be SQ, and let it cut the axes in the points A, B; then OA, OB are called the intercepts on the axes, and are usually denoted by a, b. Also when the axes are rectangular, the tangent of the angle which the line makes with the axis of x on the positive direction (viz. the angle PAX) is denoted by m.

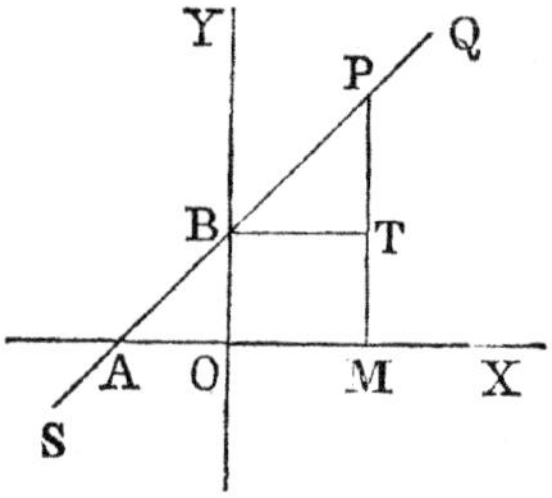

Now take any point P in SQ, and draw PM parallel to OY; then OM, MP are the co-ordinates of P; and if the axes be rectangular, we have, drawing BT parallel to OX, since $PT = PM - OB = y - b$,

$$\frac{PT}{BT} = \tan PAX,$$

or $$\frac{y-b}{x} = m;$$

therefore $$y = mx + b. \qquad (37)$$

If we had taken any other point in SQ, and called its co-ordinates x and y, we should have obtained the same equation. On this account $y = mx + b$ is called *the equation of the line.* If the axes were not rectangular, the equation would still be of the same form. For in that case $PT \div BT = OB \div OA$ $= \sin OAB \div \sin OBA = \sin A \div \sin (\omega - A)$,

or $$\frac{y-b}{x} = \sin A \div \sin (\omega - A) = m;$$

therefore $$y = mx + b,$$

and the only thing changed is the quantity represented by m. Since x, y denote the co-ordinates of any point along the line, they are called *current co-ordinates.* They are also called *variables*, because they vary as the point which they represent moves along the line.

The quantities m, b are called *constants;* because they retain the same value, while the line remains in the same position, and vary only when the position of the line varies. Hence we have the following definition:—

The equation of a line is such a relation between the co-ordinates of a variable point, which, if fulfilled, the point must be on the line.

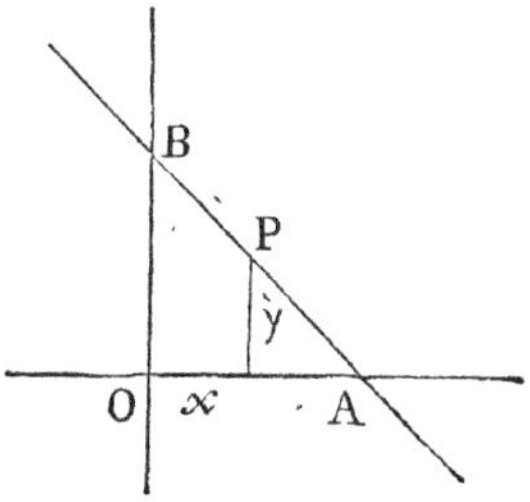

Second method.—Let AB be the line; and denoting the co-ordinates of any point P in it by x, y, and the intercepts (see *first method*) OA, OB by a, b, we have, from similar triangles,

$$\frac{x}{a} = \frac{PB}{AB}, \text{ and } \frac{y}{b} = \frac{AP}{AB};$$

therefore $$\frac{x}{a} + \frac{y}{b} = 1. \qquad (38).$$

Third method.—Let AB be the line. Let fall the perpendicular OP from the origin; and denoting OP by p, and the angles AOP, POB by α, β, respectively, we, from (38), have

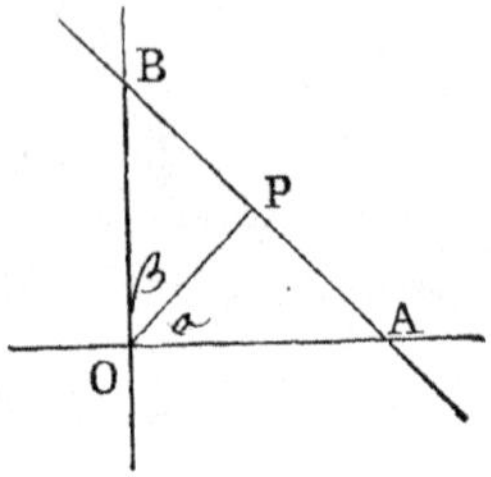

$$\frac{x}{OA} + \frac{y}{OB} = 1;$$

hence
$$\frac{p}{OA}x + \frac{p}{OB}y = p,$$

or
$$x\cos\alpha + y\cos\beta = p. \tag{39}$$

Hence, if the axes be rectangular,
$$x\cos\alpha + y\sin\alpha = p. \tag{40}$$

This form of equation, which in many investigations is more manageable than any other, has been called the *standard form*. See HESSE, *Vorlesungen Analytische Geometrie.*

Fourth method.—The general equation $Ax + By + C = 0$, *of the first degree, represents a right line.*

Dem.—By transposition, and dividing by B, we get
$$y = -\frac{A}{B}x - \frac{C}{B};$$

and this (see *first method*), being of the form $y = mx + b$, represents a right line.

15. 2°. *When the line passes through the origin.*

Let OA be the line. Take any point P in it, and draw PM parallel to OY; then, if the angle POM be denoted by α, we have

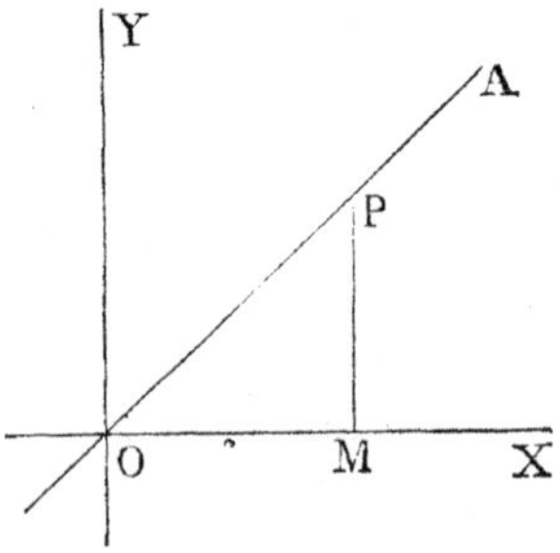

$$PM : OM :: \sin\alpha : \sin(\omega - \alpha),$$

or
$$y : x :: \sin\alpha : \sin(\omega - \alpha);$$

therefore
$$y = \frac{\sin\alpha}{\sin(\omega - \alpha)}x.$$

Hence, putting $\dfrac{\sin\alpha}{\sin(\omega-\alpha)} = m$, we get $y = mx$. (41)

This equation may be inferred from (37) by putting $b = 0$. Hence—*If the equation of a line contain no absolute term, the line passes through the origin.*

16. 3°. *When the line is parallel to one of the axes.*

Let the line AB be parallel to the axis of x, and make an intercept b on the axis of y. Now take any point P in AB, and draw the ordinate PM, which is equal to b [Euc. I. xxxiv.]. Hence the ordinate of any point P in the line AB is equal to b, and this statement is expressed algebraically by the equation $y = b$, which is therefore the equation of the line AB.

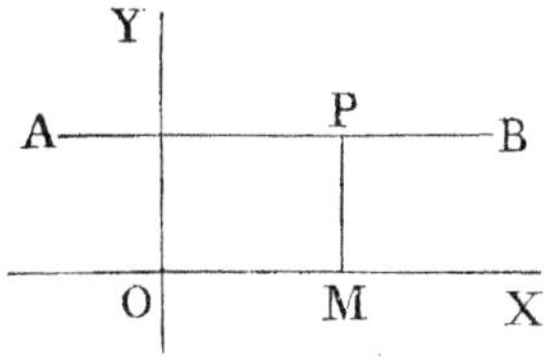

This result can be obtained differently, and in a way that will connect it with a fundamental theorem of Modern Geometry.

From equation (38) we have $\dfrac{x}{a} + \dfrac{y}{b} = 1$, where a and b are the intercepts on the axes. Now if the intercept a be infinite, that is, if the line meet the axis of x at infinity, the term $\dfrac{x}{a}$ will vanish, and we get $\dfrac{y}{b} = 1$, or $y = b$; but $y = b$ denotes a line parallel to the axis of x. Hence a line which meets the axis of x at infinity is parallel to it; and we have the general theorem, *that lines which meet at infinity are parallel.* In a similar manner $x = a$ denotes a line parallel to the axis of y at the distance a. Hence we have the following general proposition:—*If the equation of a line contains no x, it is parallel to the axis of x; and if it contains no y, it is parallel to the axis of y.*

EXAMPLES.

1. What line is represented by the equation $y = 0$?

Ans. The axis of x. For if $b = 0$ in the equation $y = b$, we get $y = 0$.

2. Prove that if the equations of two lines differ only in their absolute terms, the lines are parallel.

3. Find the intercepts which the line $Ax + By + C = 0$ makes on the axes.

$$Ans.\quad -\frac{C}{A},\ -\frac{C}{B}.$$

4. If the equation of a line be multiplied by any constant it still represents the same line; for the intercepts made by $\lambda Ax + \lambda By + \lambda C = 0$ on the axes are the same as those made by $Ax + By + C = 0$.

5. Prove that the line which divides two sides of a triangle proportionally is parallel to the third side.

6. Find the locus of a point which is equally distant from the origin and the point $(2x', 2y')$.

If (xy) be equally distant from $(0, 0)$ $(2x', 2y')$, we have

$$x^2 + y^2 = (x - 2x')^2 + (y - 2y')^2.$$

Hence

$$xx' + yy' = x'^2 + y'^2. \tag{42}$$

And since this contains x and y in the first degree, the locus is a right line.

7. Find the loci of points equally distant from the following pairs of points:—

1°. $(a \cos\phi,\ b \sin\phi)$; $(a \cos\phi',\ b \sin\phi')$.

$$Ans.\quad \frac{ax}{\cos\frac{1}{2}(\phi+\phi')} - \frac{by}{\sin\frac{1}{2}(\phi+\phi')} = (a^2 - b^2)\cos\tfrac{1}{2}(\phi-\phi'). \tag{43}$$

2°. $\{(a\cos(\alpha+\beta),\ b\sin(\alpha+\beta)\}$; $\{a\cos(\alpha-\beta),\ b\sin(\alpha-\beta)\}$.

$$Ans.\quad \frac{ax}{\cos\alpha} - \frac{by}{\sin\alpha} = (a^2 - b^2)\cos\beta. \tag{44}$$

3°. $\left(kt,\ \frac{k}{t}\right)$; $\left(kt',\ \frac{k}{t'}\right)$.

$$Ans.\quad 2x - \frac{2y}{tt'} = k\left(1 - \frac{1}{t^2 t'^2}\right)(t + t'). \tag{45}$$

4°. $(at^2,\ 2at)$; $(at'^2,\ 2at')$.

$$Ans.\quad 2(t+t)x + 4y = a(t+t')(t^2 + t'^2 + 4). \tag{46}$$

5°. $(a\sec\phi,\ b\tan\phi)$; $(a\sec\phi',\ b\tan\phi')$.

$$Ans.\quad \frac{2ax}{\cos\phi+\cos\phi'} + \frac{2by}{\sin(\phi+\phi')} = \frac{a^2+b^2}{\cos\phi\cos\phi'}. \tag{47}$$

17. *If the equations* $Ax+By+C=0$; $x\cos\alpha+y\sin\alpha-p=0$, *represent the same line, it is required to find the relations between their coefficients.*

1°. *When the axes are rectangular.*

Dividing the first equation by R, and equating with the second, we get

$$\frac{A}{R}=\cos\alpha,\quad \frac{B}{R}=\sin\alpha.$$

Square, and add, and we get

$$\frac{A^2+B^2}{R^2}=1;\text{ therefore } R=\sqrt{A^2+B^2}.$$

Hence $$\cos\alpha=\frac{A}{\sqrt{A^2+B^2}},\quad \sin\alpha=\frac{B}{\sqrt{A^2+B^2}}. \tag{48}$$

2°. *When the axes are oblique. It is required to compare the equations*

$$Ax+By+C=0,$$

and $$x\cos\alpha+y\cos\beta-p=0.$$

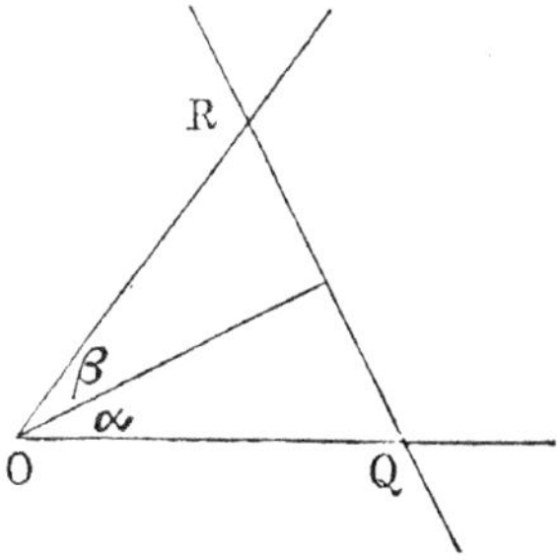

Let OQ, OR be the intercepts; then we have

$$OQ=-\frac{C}{A},\quad OR=-\frac{C}{B}.$$

Hence $$QR=\frac{C}{AB}\sqrt{A^2+B^2-2AB\cos\omega};$$

but $$QR:OR::\sin\omega:\sin Q \text{ or } \cos\alpha.$$

Hence $$\cos\alpha=\frac{A\sin\omega}{\sqrt{A^2+B^2-2AB\cos\omega}}.$$

In like manner, $$\cos\beta=\frac{B\sin\omega}{\sqrt{A^2+B^2-2AB\cos\omega}}.$$

Cor. 1.—

$$\sin\alpha=\frac{B-A\cos\omega}{\sqrt{A^2+B^2-2AB\cos\omega}},\quad \sin\beta=\frac{A-B\cos\omega}{\sqrt{A^2+B^2-2AB\cos\omega}}. \tag{49}$$

Cor. 2.— $$\tan\alpha=\frac{B-A\cos\omega}{A\sin\omega},\quad \tan\beta=\frac{A-B\cos\omega}{B\sin\omega}. \tag{50}$$

18. *To find the angle between the lines* $Ax + By + C = 0$ (1); *and* $A'x + B'y + C' = 0$ (2).

1°. *Let the axes be rectangular.* Then, if ϕ be the angle between (1) and (2), it is equal to the difference of their inclinations to the axis of x; but the tangents of these inclinations are (see Art. 14, *fourth method*),

$$-\frac{A}{B}, \text{ and } -\frac{A'}{B'}.$$

Hence $$\tan\phi = \left(\frac{A'}{B'} - \frac{A}{B}\right) \div \left(1 + \frac{AA'}{BB'}\right) = \frac{A'B - AB'}{AA' + BB'}. \tag{51}$$

Cor. 1.—If the lines (1) and (2) be parallel, they make equal angles with the axis of x; therefore

$$-\frac{A}{B} = -\frac{A'}{B'}.$$

Hence the condition of parallelism is

$$AB' - A'B = 0. \tag{52}$$

Cor. 2.—If $\phi = \frac{\pi}{2}$, $\tan\phi$ is infinite, and the condition of the lines, being at right angles to each other, is

$$AA' + BB' = 0: \tag{53}$$

That is, *if two lines whose equations are given be perpendicular to each other, the sum of the products of the coefficients of like variables is zero.*

Cor. 3.—If the lines $y = mx + b$, $y = m'x + b'$ be perpendicular to each other,

$$mm' + 1 = 0. \tag{54}$$

Cor. 4.—The angle between the lines $y = mx + b$, $y = m'x + b'$ is given by the formula

$$\tan\phi = \frac{m - m'}{1 + mm'}. \tag{55}$$

Cor. 5.—If the equations of the given lines be in the standard form,

$$x\cos\alpha + y\sin\alpha - p = 0, \quad x\cos\beta + y\sin\beta - p' = 0,$$

we have $$\phi = \alpha - \beta. \tag{56}$$

2°. *Let the axes be oblique.*

If θ, θ' denote the angles which the given lines make with the axis of x; then (*see* 18, 2°) we have $\theta = \alpha + 90$; therefore

$$\tan\theta = -\cot\alpha = \frac{A\sin\omega}{A\cos\omega - B}. \text{ (See equation (50).)}$$

Similarly,

$$\tan\theta = \frac{A'\sin\omega}{A'\cos\omega - B}.$$

Hence

$$\tan\phi = \tan(\theta - \theta') = \frac{(A'B - AB')\sin\omega}{AA' + BB' - (AB' + A'B)\cos\omega}. \quad (57)$$

Cor.—If the lines be perpendicular to each other

$$AA' + BB' - (AB' + A'B)\cos\omega = 0. \quad (58)$$

EXAMPLES.

1. Find the angle between the lines

$$\frac{x\cos\beta}{a} + \frac{y\sin\beta}{b} - 1 = 0, \quad \frac{x\cos\gamma}{a} + \frac{y\sin\gamma}{b} - 1 = 0.$$

$$\textit{Ans.}\ \sin\phi = \frac{ab\sin(\beta - \gamma)}{\sqrt{a^2\sin^2\beta + b^2\cos^2\beta}\sqrt{a^2\sin^2\gamma + b^2\cos^2\gamma}}. \quad (59)$$

2. Find the angle between the lines $x - y = 0$ and

$$\frac{x}{\tan\phi' + \tan\phi''} + \frac{y}{\cot\phi' + \cot\phi''} = k.$$

$$\textit{Ans.}\ \tan^{-1}\left\{\frac{1 + \tan\phi'\tan\phi''}{1 - \tan\phi'\tan\phi''}\right\}. \quad (60)$$

DEF.—*The result of substituting the co-ordinates of any point in the equation of any line or curve is called the* POWER *of that point with respect to the line or curve.*

[This definition, first given by STEINER, is now employed by all the French and German writers.]

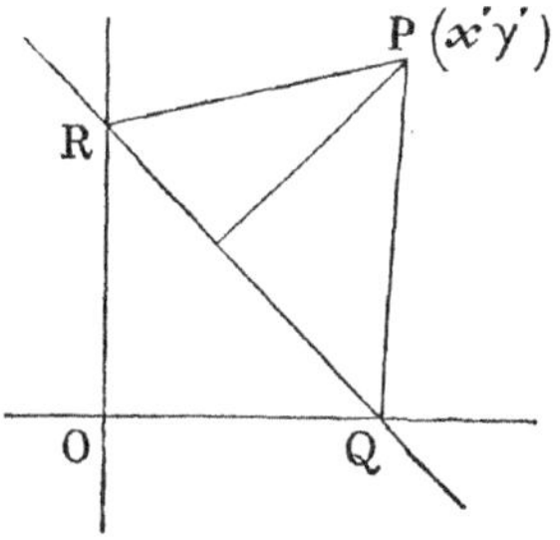

19. *To find the length of the perpendicular from the point $x'y'$ on the line $Ax + By + C = 0$.*

1°. *Let the axes be rectangular.*

Let the line intersect the axes in the points Q, R, then the perpendicular from P is equal to twice the area of the triangle PQR divided by the base QR; but the area of

$$PQR = \frac{C}{2AB}(Ax' + By' + C), \text{ (Equation (18).)}$$

and

$$QR = \frac{C}{AB}\sqrt{A^2 + B^2}. \qquad \text{(Equation (5).)}$$

Therefore the length of the perpendicular is

$$\frac{Ax' + By' + C}{\sqrt{A^2 + B^2}}. \qquad (61)$$

Hence we have the following rule for finding the length of the perpendicular from a given point on a given line :—

Divide the power of the given point with respect to the given line by the square root of the sum of the squares of the coefficients of the variables, and the quotient will be the length required.

Cor. 1.—If the equation of the given line be in the standard form $x \cos \alpha + y \sin \alpha - p = 0$, the length of the perpendicular on it from any given point $x'y'$ is equal to the power of that point with respect to the line, for the sum of the squares of the coefficients of the variables is unity.

This result being a very important one, we shall give another proof of it. Let MN be the line $x \cos \alpha + y \sin \alpha - p = 0$; R the given point $x'y'$. Through R draw the line RQ parallel to MN. Draw OQ perpendicular to RQ, and let it cut MN in V; then OV is equal to p, and denoting OQ by p', the equation of RQ is $x \cos \alpha + y \sin \alpha = p'$; and since this line passes through the point $x'y'$, these co-ordinates must satisfy its equation.

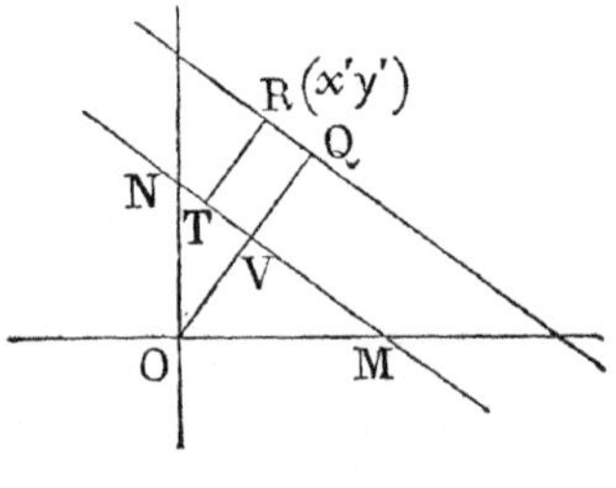

Hence $\qquad x' \cos \alpha + y' \sin \alpha = p';$

therefore $\qquad x' \cos \alpha + y' \sin \alpha - p = p' - p;$

but $p' - p = OQ - OV = VQ = TR$ = perpendicular from R on the line $x \cos \alpha + y \sin \alpha - p$; therefore

$$\text{perpendicular} = x' \cos \alpha + y' \sin \alpha - p. \tag{62}$$

Cor. 2.—From the figure it follows that if p' be greater than p, the point R is on the side of the line remote from the origin.

Hence the power of any point with respect to a line which passes between it and the origin is positive, and in the contrary case negative.

2°. *Let the axes be oblique.*

Since the axes are oblique, the area of the triangle PQR is

$$\frac{C(Ax' + By' + C) \sin \omega}{2AB};$$

and the length of QR is

$$\frac{C\sqrt{A^2 + B^2 - 2AB \cos \omega}}{AB}. \quad \text{(Equation (6).)}$$

Therefore the perpendicular is

$$\frac{(Ax' + By' + C) \sin \omega}{\sqrt{A^2 + B^2 - 2AB \cos \omega}}. \tag{63}$$

Cor. 1.—The power of any point on a line with respect to the line is zero; and, conversely, if the power of a point with respect to a line be zero, the point must be on the line.

Cor. 2.—If $S \equiv Ax + By + C = 0$, $S' \equiv A'x + B'y + C' = 0$, be the equations of any two lines, and l, m any two multiples (including unity), either positive or negative, then

$$lS + mS' = 0 \tag{64}$$

is the equation of some line passing through the intersection of the lines S and S'.

For, since S and S' are of the first degree with respect to x and y, $lS + mS' = 0$ will also be of the first degree, and therefore will be the equation of some line. Again, if P be the point of intersection of S and S', the powers of P (*Cor.* 1) with respect to S, S' are respectively zero. Hence the power of P with respect to $lS + mS' = 0$ is zero, and therefore the line $lS + mS' = 0$ must pass through P.

Cor. 3.—The line $y - y' - m(x - x') = 0$ passes through the point $x'y'$; for the power of $x'y'$ with respect to it is zero.

Or thus: $y - y' = 0$ denotes (Art. 16) a line parallel to the axis of x at the distance y'; and $x - x' = 0$ a line parallel to the axis of y at the distance x'. Hence, *Cor.* 2,

$$y - y' - m(x - x') = 0 \tag{65}$$

denotes a line passing through their intersection, that is, through the point $x'y'$.

Cor. 4.—In the same manner it may be shown that if $S = 0$, $S' = 0$, be the equations of any two loci (such as a line and a circle, or two circles, &c.), $lS + mS' = 0$ will denote some curve passing through all the points of intersection of S and S'.

20. *To find the equation of a line passing through two points* $x'y'$, $x''y''$.

Take any variable point xy on the line, then the three points xy, $x'y'$, $x''y''$ are collinear. Hence, equation (12),

$$\begin{vmatrix} x, & y, & 1, \\ x', & y', & 1, \\ x'', & y'', & 1, \end{vmatrix} = 0, \tag{66}$$

which is the required equation.

It may be otherwise seen that this is the equation of a line passing through the two given points. 1°. It contains x and y in the first degree; hence it is the equation of a right line. 2°. If we substitute $x'y'$ for xy the determinant will have two rows alike, and therefore will vanish; hence the co-ordinates $x'y'$ satisfy it, and the line passes through $x'y'$. Similarly it passes through $x''y''$. The determinant (66) expanded gives

$$(y'-y'')x-(x'-x'')y+x'y''-x''y'=0; \qquad (67)$$

from which we infer the following practical rule for writing down the equation of a line passing through two given points $x'y'$, $x''y''$:—

Place the co-ordinates of one of the given points under those of the other, as in the margin; then the difference of the ordinates of the given points will give the coefficient of x*: the corresponding difference of the abscissæ with sign changed will be the coefficient of* y*. Lastly, the determinant, with two rows formed by the given co-ordinates, will be the absolute term.*

$$\begin{matrix} x', & y', \\ x'', & y'', \end{matrix}$$

Cor. 1.—If the equation of the line joining $x'y'$, $x''y''$ be written in the form $Ax+By+C=0$, we have

$$y'-y''=A, \quad (x'-x'')=-B, \quad x'y''-x''y'=C.$$

Cor. 2.—Hence may be inferred the condition that the points $x''y''$, $x'''y'''$ may subtend a right angle at $x'y'$.

For, let the join of the points

$$x'y',\ x''y'' \text{ be } Ax+By+C=0,$$

and the join of the points

$$x'y',\ x'''y''' \text{ be } A'x+B'y+C'=0;$$

and, since these are at right angles to each other,

$$AA'+BB'=0;$$

and, substituting, we get

$$(x'-x'')(x'-x''')+(y'-y'')(y'-y''')=0. \quad \text{(Comp. (8).)}$$

Examples.

1. Find the equation of the join of $(2, -4)$, $(3, -5)$.

Ans. $x + y + 2 = 0$.

2. Find the medians of the triangle whose vertices are $x'y'$, $x''y''$, $x'''y'''$.

$$\text{Ans. } (y'' + y''' - 2y')\,x - (x'' + x''' - 2x')\,y + (x'' + x''')\,y' - (y'' + y''')\,x' = 0, \text{ \&c.} \quad (68)$$

3. Find the equations of the joins of the pairs of points—

1°. $(r \cos \phi', r \sin \phi')$; $(r \cos \phi'', r \sin \phi'')$.

$$\text{Ans. } \cos \tfrac{1}{2}(\phi' + \phi'')\,x + \sin \tfrac{1}{2}(\phi + \phi'')\,y = r \cos \tfrac{1}{2}(\phi' - \phi''). \quad (69)$$

2°. $(a \cos \phi', b \sin \phi')$; $(a \cos \phi'', b \sin \phi'')$.

$$\text{Ans. } \cos \tfrac{1}{2}(\phi' + \phi'')\frac{x}{a} + \sin \tfrac{1}{2}(\phi' + \phi'')\frac{y}{b} = \cos \tfrac{1}{2}(\phi' - \phi''). \quad (70)$$

3°. $\{a \cos(\alpha + \beta), b \sin(\alpha + \beta)\}$; $\{a \cos(\alpha - \beta), b \sin(\alpha - \beta)\}$.

$$\text{Ans. } \cos \alpha \frac{x}{a} + \sin \alpha \frac{y}{b} = \cos \beta. \quad (71)$$

4°. $(at^2, 2at)$; $(at'^2, 2at')$. *Ans.* $2x - (t + t')\,y + 2att' = 0$. (72)

5°. $(a \sec \phi, b \tan \phi)$; $(a \sec \phi', b \tan \phi')$.

$$\text{Ans. } \cos \tfrac{1}{2}(\phi - \phi')\frac{x}{a} - \sin \tfrac{1}{2}(\phi + \phi')\frac{y}{b} = \cos \tfrac{1}{2}(\phi + \phi'). \quad (73)$$

6°. $(k \tan \phi, k \cot \phi)$; $(k \tan \phi', k \cot \phi')$.

$$\text{Ans. } \frac{x}{\tan \phi + \tan \phi'} + \frac{y}{\cot \phi + \cot \phi'} = k. \quad (74)$$

4. Find the equations of the joins of the middle points of the opposite sides, and also of the joins of the middle points of the diagonals of the quadrilateral whose vertices are $x'y'$, $x''y''$, $x'''y'''$, $x''''y''''$, and show that the three lines thus found are concurrent.

21. *To find the co-ordinates of the point of intersection of two lines whose equations are given.*

Since the co-ordinates of the point of intersection must satisfy the equation of each line, this problem is identical with the algebraic one of solving two simultaneous equations of the first degree. Thus the co-ordinates of the point ot intersection of the lines

$$\frac{x}{m} + \frac{y}{n} = 1, \quad \frac{x}{n} + \frac{y}{m} = 1, \quad \text{are} \quad \frac{mn}{m+n}, \quad \frac{mn}{m+n}.$$

Examples.

1. Find the co-ordinates of the points of intersection of the following pairs of lines :—

1°. $x\cos\phi + y\sin\phi = r,\quad x\cos\phi' + y\sin\phi' = r.$

$$Ans.\ x = \frac{r\cos\frac{1}{2}(\phi+\phi')}{\cos\frac{1}{2}(\phi-\phi')},\quad y = \frac{r\sin\frac{1}{2}(\phi+\phi')}{\cos\frac{1}{2}(\phi-\phi')}. \tag{75}$$

2°. $\frac{x}{a}\cos\phi + \frac{y}{b}\sin\phi = 1,\quad \frac{x}{a}\cos\phi' + \frac{y}{b}\sin\phi' = 1.$

$$Ans.\ x = \frac{a\cos\frac{1}{2}(\phi+\phi')}{\cos\frac{1}{2}(\phi-\phi')},\quad y = \frac{b\sin\frac{1}{2}(\phi+\phi')}{\cos\frac{1}{2}(\phi-\phi')}. \tag{76}$$

3°. $x - ty + at^2 = 0,\quad x - t'y + at'^2 = 0.$

$$Ans.\ x = att',\quad y = a(t+t'). \tag{77}$$

2. If $\frac{x}{2a} + \frac{y}{2b} = 1$, $\frac{x}{2a'} + \frac{y}{2b'} = 1$, be one pair of opposite sides of a quadrilateral, and the co-ordinate axes the other pair; find the co-ordinates of the middle points of its three diagonals, and prove that they are collinear.

3. Find the co-ordinates of a point equally distant from the three points

$$(a\cos\phi,\ b\sin\phi);\ (a\cos\phi',\ b\sin\phi');\ (a\cos\phi'',\ b\sin\phi'').$$

The locus of a point equally distant from

$$(a\cos\phi,\ b\sin\phi);\ \text{and}\ (a\cos\phi',\ b\sin\phi'),$$

is the line
$$\frac{ax}{\cos\frac{1}{2}(\phi+\phi')} - \frac{by}{\sin\frac{1}{2}(\phi+\phi')} = (a^2-b^2)\cos\frac{1}{2}(\phi-\phi').$$

Similarly,
$$\frac{ax}{\cos\frac{1}{2}(\phi'+\phi'')} - \frac{by}{\sin\frac{1}{2}(\phi'+\phi'')} = (a^2-b^2)\cos\frac{1}{2}(\phi'-\phi'')$$

is the locus of a point equally distant from

$$(a\cos\phi',\ b\sin\phi');\ \text{and}\ (a\cos\phi'',\ b\sin\phi'').$$

Hence, solving from these equations, we get

$$\left.\begin{aligned} x &= \frac{a^2-b^2}{a}\cos\tfrac{1}{2}(\phi+\phi')\cos\tfrac{1}{2}(\phi'+\phi'')\cos\tfrac{1}{2}(\phi''+\phi),\\ y &= \frac{b^2-a^2}{b}\sin\tfrac{1}{2}(\phi+\phi')\sin\tfrac{1}{2}(\phi'+\phi'')\sin\tfrac{1}{2}(\phi''+\phi)\end{aligned}\right\}. \tag{78}$$

*4. Find the co-ordinates of a point equally distant from—

1°. $(at^2, 2at)$; $(at'^2, 2at')$; $(at''^2, 2at'')$.

$$Ans.\ x = \frac{a}{2}(t^2 + t'^2 + t''^2 + tt' + t't'' + t''t + 4),$$

$$y = -\frac{a}{4}(t + t')(t' + t'')(t'' + t). \qquad (79)$$

*2°. $(a \sec\phi, b \tan\phi)$; $(a \sec\phi', b \tan\phi')$; $(a \sec\phi'', b \tan\phi'')$.

$$\left.\begin{aligned} Ans.\ x &= \frac{a^2+b^2}{a}\,\frac{\cos\frac{1}{2}(\phi-\phi')\cos\frac{1}{2}(\phi'-\phi'')\cos\frac{1}{2}(\phi''-\phi)}{\cos\phi\cos\phi'\cos\phi''}, \\ y &= \frac{a^2+b^2}{b}\,\frac{\sin\frac{1}{2}(\phi+\phi')\sin\frac{1}{2}(\phi'+\phi'')\sin\frac{1}{2}(\phi''+\phi)}{\cos\phi\cos\phi'\cos\phi''} \end{aligned}\right\}. \qquad (80)$$

*3°. $(k \tan\phi, k \cot\phi)$; $(k \tan\phi', k \cot\phi')$; $(k \tan\phi'')$.

$$\left.\begin{aligned} Ans.\ x &= \frac{k}{2}(\tan\phi\tan\phi'\tan\phi'' + \cot\phi + \cot\phi' + \cot\phi''), \\ y &= \frac{k}{2}(\cot\phi\cot\phi'\cot\phi'' + \tan\phi + \tan\phi' + \tan\phi'') \end{aligned}\right\}. \qquad (81)$$

*4°. $(a\cos\alpha, b\sin\alpha)$; $(a\cos(\alpha+\beta), b\sin(\alpha+\beta))$; $(a\cos(\alpha-\beta), b\sin(\alpha-\beta))$.

$$\left.\begin{aligned} Ans.\ x &= \frac{a^2-b^2}{a}\cos(\alpha-\tfrac{1}{2}\beta)\cos\alpha\cos(\alpha+\tfrac{1}{2}\beta), \\ y &= \frac{b^2-a^2}{b}\sin(\alpha-\tfrac{1}{2}\beta)\sin\alpha\sin(\alpha+\tfrac{1}{2}\beta) \end{aligned}\right\}. \qquad (82)$$

*5°. $(x'y')$, $(x''y'')$, $(x'''y''')$.

Ans. If ρ', ρ'', ρ''' denote the respective distances of the points from the origin, Δ the area of the triangle formed by joining them,

$$x = \begin{vmatrix} 1, & 1, & 1 \\ x', & x'', & x''' \\ \rho', & \rho'', & \rho''' \end{vmatrix} \div \Delta; \quad y = \begin{vmatrix} 1, & 1, & 1 \\ y', & y'', & y''' \\ \rho', & \rho''', & \rho''' \end{vmatrix} \div \Delta. \qquad (83)$$

22. *To find the equation of the line through $x'y'$, making an angle ϕ with $Ax + By + C = 0$.*

Let $A'x + B'y + C' = 0$ be the required line; and since

this passes through $x'y'$, we have $A'x' + B'y' + C' = 0$. Hence $A'(x - x') + B'(y - y') = 0$ is the form of the required equation.

Again, we have $$\tan\phi = \frac{A'B - AB'}{AA' + BB'}. \quad \text{(Equation (51).)}$$

Hence $$A'(B - A\tan\phi) = B'(A + B\tan\phi).$$

And the required equation is—

$$\frac{x - x'}{B - A\tan\phi} + \frac{y - y'}{A + B\tan\phi} = 0, \tag{84}$$

which may be written in either of the following forms:—

$$\frac{x - x'}{B\cos\phi - A\sin\phi} + \frac{y - y'}{A\cos\phi + B\sin\phi} = 0. \tag{85}$$

$$\begin{vmatrix} A\sin\phi - B\cos\phi, & A\cos\phi + B\sin\phi, & 0 \\ x, & y, & 1 \\ x', & y', & 1 \end{vmatrix} = 0. \tag{86}$$

23. If the angle ϕ be right, the equation (84) becomes

$$B(x - x') = A(y - y').$$

Hence the equation of the line through $x'y'$, perpendicular to $Ax + By + C$, is

$$B(x - x') = A(y - y'). \tag{87}$$

This may be otherwise proved as follows:—

The line $Bx - Ay + C'$ fulfils the condition (53) of being perpendicular to $Ax + By + C$; and if it pass through $x'y'$, we get $Bx' - Ay' + C' = 0$. Hence, subtracting, we get the equation just written.

24. The line through $x'y'$, making an angle ϕ with $y = mx + b$, is

$$\frac{x - x'}{1 + m\tan\phi} = \frac{y - y'}{m - \tan\phi}. \tag{88}$$

Cor.—The line through $x'y'$ perpendicular to $y = mx + b$ is

$$y - y' = -\frac{1}{m}(x - x'). \qquad (89)$$

EXAMPLES.

1. Find the line through (0, 1), making an angle of 30°, with $x + y = 2$.

2. Prove that the lines $x + y\sqrt{3} - 6 = 0$, $3x - y\sqrt{3} - 4 = 0$ are at right angles to each other.

3. Find the equations of the perpendiculars of the triangle whose angular points are $x'y'$, $x''y''$, $x'''y'''$.

4. Find the equation of the perpendicular to the line

$$\frac{x\cos\alpha}{a} + \frac{y\sin\alpha}{b} = 1 \text{ at the point } (a\cos\alpha,\ b\sin\alpha).$$

5. Find the perpendicular to

$$x - y\tan\phi + a\tan^2\phi = 0, \text{ at the point } (a\tan^2\phi,\ 2a\tan\phi).$$

*6. Show that the orthocentre of the triangle formed by the lines

$$x - ty + at^2 = 0; \quad x - t'y + at'^2 = 0; \quad x - t''y + at''^2 = 0$$

is the point $\qquad -a, \quad a(t + t' + t'' + tt't''). \qquad (90)$

25. *To find the equation of a line dividing either of the angles between the lines* $Ax + By + C = 0$, $A'x + B'y + C' = 0$, *into two parts whose sines have a given ratio* $a : b$.

Let LL', MM' be the given lines; ON the required line. From any point XY on ON let fall perpendiculars on the given lines: these perpendiculars will be to one another in the ratio of the sines of the angles, and will both be of the same sign (Art. 21, *Cor.* 2), if the origin

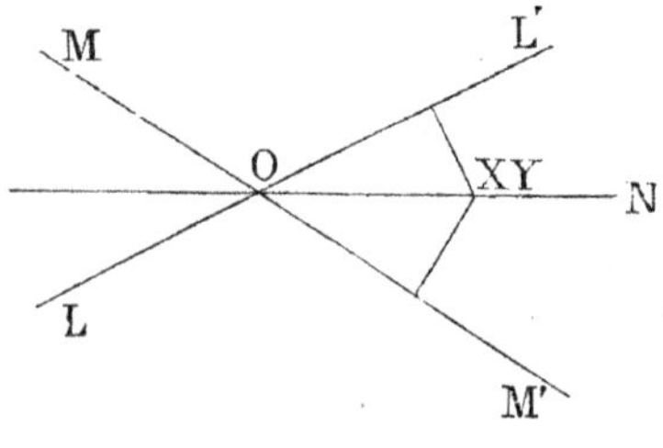

of co-ordinates lies in either of the angular spaces LOM,

$L'OM'$; and of different signs, if in either of the two remaining spaces. Hence

$$\frac{Ax + By + C}{\sqrt{A^2 + B^2}} \div \frac{A'x + B'y + C'}{\sqrt{A'^2 + B'^2}} = \pm \frac{a}{b},$$

the choice of sign depending on the position of the origin. Hence the equations of the lines dividing the angles between $Ax + By + C = 0$, $A'x + B'y + C' = 0$ in the ratio $a : b$, are

$$\frac{b(Ax + By + C)}{\sqrt{A^2 + B^2}} = \pm \frac{a(A'x + B'y + C')}{\sqrt{A'^2 + B'^2}}, \qquad (91)$$

the sign + being the proper one for one of them, and − for the other.

Cor. 1.—If we put

$$\frac{b}{\sqrt{A^2 + B^2}} = l, \text{ and } \frac{a}{\sqrt{A'^2 + B'^2}} = m,$$

the equations (91) are transformed into,

$$l(Ax + By + C) \pm m(A'x + B'y + C') = 0. \qquad (92)$$

Now if a and b are given, l and m will be given. Hence we have the following important theorem:—*If the equations of two given lines be multiplied respectively by given constants, and the products either added or subtracted, the result will be the equation of a line dividing one of their angles into parts whose sines have a given ratio.*

Cor. 2.—If in the equation

$$l(Ax + By + C) + m(A'x + B'y + C') = 0,$$

we put

$$\frac{m}{l} = \lambda, \text{ we get } Ax + By + C + \lambda(A'x + B'y + C') = 0;$$

and giving all possible values to λ, we get all possible lines through the intersection of

$$Ax + By + C = 0, \quad \text{and} \quad A'x + B'y + C' = 0;$$

Compare Art. 6, *Cor.* 1.

Cor. 3.—If the equations of the given lines be in the standard form, the ratio of the sines will be the same as the ratio of the multiples.

Cor. 4.—Since the line passing through a fixed point $x'y'$ and the intersection of the lines

$$Ax + By + C = 0, \quad A'x + B'y + C' = 0$$

divides the angle between the lines into parts whose sines are in the ratio of the perpendiculars on them from $x'y'$, we have

$$a = \frac{Ax' + By' + C}{\sqrt{A^2 + B^2}}, \quad b = \frac{A'x' + B'y' + C'}{\sqrt{A'^2 + B'^2}}.$$

Hence, substituting these values in (91), we get

$$(Ax + By + C)(A'x' + B'y' + C') \\ - (A'x + B'y + C')(Ax' + By' + C) = 0. \quad (93)$$

Cor. 5.—If three given lines be concurrent, viz.,

$$Ax + By + C = 0, \quad A'x + B'y + C' = 0, \quad A''x + B''y + C'' = 0,$$

we see (*Cor.* 2) that the third must be of the form

$$Ax + By + C + \lambda(A'x + B'y + C').$$

And, comparing coefficients, we get

$$A + \lambda A' - A'' = 0,$$

$$B + \lambda B' - B'' = 0,$$

$$C + \lambda C' - C'' = 0.$$

Hence, eliminating λ, the condition of concurrence is—

$$\begin{vmatrix} A, & A', & A'' \\ B, & B', & B'' \\ C, & C', & C'' \end{vmatrix} = 0. \quad (94)$$

Cor. 6.—If the coefficients in the equations of three lines be such that when the equations are multiplied by any suitable constants they vanish identically, the lines are concurrent.

For if

$$\lambda(Ax+By+C)+\mu(A'x+B'y+C')+\nu(A''x+B''y+C'')\equiv 0,$$

we have, comparing coefficients,

$$\lambda A + \mu A' + \nu A'' = 0,$$

$$\lambda B + \mu B' + \nu B'' = 0,$$

$$\lambda C + \mu C' + \nu C'' = 0;$$

and eliminating λ, μ, ν, we get the condition (94) of concurrence.

Examples.

1. Find the lines which divide the angles between

$$3x + 4y + 12 = 0, \quad 8x + 15y + 16 = 0,$$

into parts whose sines are in the ratio 2 : 3.

Ans. $51(3x + 4y + 12) \pm 10(8x + 15y + 16) = 0$.

2. Write the equations of the bisectors of the angles between

$$x \cos \alpha + y \sin \alpha - p = 0, \quad x \cos \beta + y \sin \beta - p' = 0,$$

in the standard forms.

3. Form the equations of the perpendiculars of the triangle whose sides are

$$A_1x+B_1y+C_1=0, \ (1) \quad A_2x+B_2y+C_2=0, \ (2) \quad A_3x+B_3y+C_3, \ (3)=0,$$

the perpendicular on (1) must be of the form (2) − k(3), and the condition of perpendicularity gives

$$k = (A_1A_2 + B_1B_2) \div (A_3A_1 + B_3B_1).$$

Hence the perpendicular is

$$(A_3A_1 + B_3B_1)(A_2x+B_2y+C_2) - (A_1A_2+B_1B_2)(A_3x+B_3y+C_3) = 0. \quad (95)$$

4. Find the equation of the line which passes through the intersection of

$$A_1x + B_1y + C_1 = 0, \quad A_2x + B_2y + C_2 = 0,$$

and is parallel to $A_3x + B_3y + C_3 = 0$.

5. Find the co-ordinates of a point equally distant from the three lines in Ex. 3.

6. If the distances of a certain point from the lines

$$x\cos\alpha + y\sin\alpha - p = 0,\quad x\cos\alpha' + y\sin\alpha' - p' = 0,\quad x\cos\alpha'' + y\sin\alpha'' - p'' = 0$$

be d, d', d'', respectively, and if

$$\lambda = p + d,\quad \lambda' = p' + d',\quad \lambda'' = p'' + d'';$$

prove $$\lambda\sin(\alpha' - \alpha'') + \lambda'\sin(\alpha'' - \alpha) + \lambda''\sin(\alpha - \alpha') = 0. \qquad (96)$$

*7. If $x\cos\alpha_1 + y\sin\alpha_1 - p_1 = 0$, $x\cos\alpha_2 + y\sin\alpha_2 - p_2 = 0$, &c., be the bisectors of the internal angles of any pentagon; prove

$$p_1\sin A_1 + p_2\sin A_2 + \ldots + p_5\sin A_5 = 0, \text{ where } A_1 = \alpha_2 - \alpha_3 + \alpha_4 - \alpha_5;$$
$$A_2 = \alpha_3 - \alpha_4 + \alpha_5 - \alpha_1, \text{ \&c.}$$

*8. The co-ordinates of the vertices of two triangles are

$$a_1b_1,\ a_2b_2,\ a_3b_3;\ \text{and } c_1d_1,\ c_2d_2,\ c_3d_3,$$

respectively; the joins of corresponding vertices are divided similarly in the points (D, D', D''): if perpendiculars from D, D', D'' on the sides of either triangle be concurrent, prove the relation

$$\begin{vmatrix} c_1, & a_1, & 1 \\ c_2, & a_3, & 1 \\ c_3, & a_3, & 1 \end{vmatrix} + \begin{vmatrix} d_1, & b_1, & 1 \\ d_2, & b_2, & 1 \\ d_3, & b_3, & 1 \end{vmatrix} = 0. \qquad (97)$$

BALTZER.

26. *To find when an equation of the second degree is the product of the equations of two lines.*

1°. Let the equation contain only one of the variables, such as

$$x^2 - (a + b)x + ab.$$

Since this is evidently the product of the equations

$$x - a = 0,\quad x - b = 0,$$

we see *that an equation of the second degree, containing only one of the variables, represents two lines parallel to the axis of the other variable.*

2°. *If the equation be homogeneous in both variables, it represents two lines passing through the origin.*

For example,

$x^2 - 5xy + 6y^2 = 0$ is the product of $(x - 2y) = 0$, $(x - 3y) = 0$.

3°. If the general equation

$$ax^2 + 2hxy + by^2 + 2gx + 2fy + c = 0$$

denotes two lines, throwing it into the form

$$(ax + hy + g)^2 - \{(h^2 - ab)y^2 + 2(gh - af)y + (g^2 - ac)\} = 0,$$

we see that the second member must be a perfect square.

Hence $$(h^2 - ab)(g^2 - ac) - (gh - af)^2 = 0,$$

or $$abc + 2fgh - af^2 - bg^2 - ch^2 = 0. \qquad (98)$$

This important function of the coefficients of the general equation of the second degree is called its *discriminant*. It may be written in determinant form thus:

$$\begin{vmatrix} a, & h, & g, \\ h, & b, & f, \\ g, & f, & c, \end{vmatrix} = 0. \qquad (99)$$

The student should carefully commit each of the formulæ (98) (99) to memory. The minors of the determinant (99) will be denoted by the corresponding capital letters. Thus,

$$A \equiv bc - f^2, \quad B \equiv ca - g^2, \quad C \equiv ab - h^2, \quad F \equiv gh - af,$$
$$G \equiv hf - bg, \quad H \equiv fg - ch.$$

27. *If the general equation represent two lines, it is required to find the co-ordinates of their point of intersection.*

Let

$$ax^2 + 2hxy + by^2 + 2gx + 2fy + c \equiv (lx + my + n)(l'x + m'y + n').$$

Hence, comparing coefficients, we get

$$a = ll', \quad b = mm', \quad c = nn',$$

$$2f = mn' + m'n, \quad 2g = nl' + n'l, \quad 2h = lm' + l'm;$$

and solving for x and y from the equations

$$lx + my + n = 0, \quad l'x + m'y + n' = 0,$$

we get $\quad x : y :: 1, :: mn' - m'n : nl' - n'l : lm' - l'm;$

Hence $\quad x : y : 1 :: A^{\frac{1}{2}} : B^{\frac{1}{2}} : C^{\frac{1}{2}},$

which are the required values.

Cor. 1.—If the general equation represent two perpendicular lines,

$$a + b = 0 \text{ for rectangular axes.} \qquad (100)$$

$$a + b - 2h \cos\omega = 0 \text{ for oblique axes.} \qquad (101)$$

Cor. 2.—If the general equation represent two lines making an angle ϕ, we have for oblique axes,

$$\tan\phi = \frac{2\sqrt{h^2 - ab}\,.\sin\omega}{a + b - 2h\cos\omega}. \qquad (102)$$

Hence, if $h^2 - ab = 0$, the lines are parallel.

Examples.

1. What lines are represented by $x^2 - y^2 = 0$?

2. What lines are represented by $x^2 - 2xy\sec\theta + y^2 = 0$?

3. Prove that the two lines $ax^2 + 2hxy + by^2 = 0$ are respectively at right angles to the lines $bx^2 - 2hxy + ay^2 = 0$.

4. Find the angle between the lines $ax^2 + 2hxy + by^2 = 0$. If the equation represent the two lines $y - mx = 0$, $y - m'x = 0$, we get

$$m = \frac{-h + \sqrt{h^2 - ab}}{b}, \quad m' = \frac{-h - \sqrt{h^2 - ab}}{b};$$

and since $\quad \tan\phi = \dfrac{m - m'}{1 + mm'},\;$ we have $\tan\phi = \dfrac{2\sqrt{h^2 - ab}}{a + b}. \qquad (103)$

5. The angle between the lines

$$(x^2 + y^2)(\cos^2\theta\sin^2\alpha + \sin^2\theta) - (x\tan\alpha - y\sin\theta)^2 \text{ is } \alpha.$$

6. The lines $x^2 + 2xy\sec 2\alpha + y^2 = 0$ are equally inclined to $x + y = 0$.

7. Find the bisectors of the angles made by the lines $ax^2 + 2hxy + by^2 = 0$. The bisectors of the angles between the lines $y - mx = 0$, $y - mx' = 0$, are—

$$\frac{y - mx}{\sqrt{1 + m^2}} + \frac{y - m'x}{\sqrt{1 + m'^2}} = 0, \quad \frac{y - mx}{\sqrt{1 + m^2}} - \frac{y - m'x}{\sqrt{1 + m'^2}} = 0.$$

Hence, multiplying and restoring values, we get

$$h(x^2 - y^2) - (a - b)xy = 0. \qquad (104)$$

8. The difference of the tangents which the lines

$$x^2(\tan^2\theta + \cos^2\theta) - 2xy\tan\theta + y^2\sin^2\theta = 0$$

make with the axis of x is 2.

9. If Δ denote the discriminant (98); prove the following relations—

$$a\Delta = BC - F^2, \quad b\Delta = CA - G^2, \quad c\Delta = AB - H^2. \qquad (105)$$

10. When $\Delta = 0$; prove $A : B : C :: \frac{1}{F^2} : \frac{1}{G^2} : \frac{1}{H^2}$. (106)

11. If $ax^2 + 2hxy + by^2 + 2gx + 2fy + c = 0$ represent two lines; prove that the lines $ax^2 + 2hxy + by^2 = 0$ are parallel to them.

12. Find the discriminant of

$$(ax^2 + 2hxy + by^2 + 2gx + 2fy + c) + \lambda(x^2 + y^2 + 2xy\cos\omega).$$

13. Prove that if in the result (104) we change x, y into

$$x + \frac{A^{\frac{1}{2}}}{C^{\frac{1}{2}}}, \quad y + \frac{B^{\frac{1}{2}}}{C^{\frac{1}{2}}},$$

we get the equations of the bisectors of the angles made by

$$(ax^2 + 2hxy + by^2 + 2gx + 2fy + c = 0),$$

when it denotes lines.

*14. If the sum of the angles $\phi, \phi', \phi'', \phi'''$ be 2π; prove that the points

$$(a\cos\phi,\ b\sin\phi);\ (a\cos\phi',\ b\sin\phi');\ (a\cos\phi'',\ b\sin\phi'');\ (a\cos\phi''',\ b\sin\phi''')$$

are concyclic.

*15. If $t + t' + t'' + t''' = 0$; prove that the points

$$(at^2,\ 2at);\ (at'^2,\ 2at');\ (at''^2,\ 2at'');\ (at'''^2,\ 2at''')$$

are concyclic.

*16. If $\bar{x}, \bar{y}$ denote the mean centre of the points in Ex. 14; prove that the co-ordinates of the circumcentre are

$$\frac{a^2 - b^2}{a^2}\bar{x}, \quad \frac{b^2 - a^2}{b^2}\bar{y}. \qquad (107)$$

*17. The points

$$(k\tan\phi,\ k\cot\phi);\ (k\tan\phi',\ k\cot\phi');\ (k\tan\phi'',\ k\cot\phi'');$$
$$(k\tan\phi\,.\tan\phi'\,.\tan\phi'',\quad k\cot\phi\,.\cot\phi'\,.\cot\phi''),$$

are concyclic.

SECTION II.—TRILINEAR CO-ORDINATES.

28. DEFINITIONS.—*Let ABC be a triangle given in position and magnitude; then if perpendiculars from any point P on the sides of ABC be denoted by* α, β, γ, α, β, γ *are called the* TRILINEAR CO-ORDINATES *of P.* If the point P be on the side BC, the perpendicular from it on BC will vanish. Hence, in this system of co-ordinates the equation of BC will be $\alpha = 0$. Similarly, the equations of CA, AB will be $\beta = 0$, $\gamma = 0$ respectively. *The triangle ABC is called the* TRIANGLE OF REFERENCE, and *its sides the* LINES OF REFERENCE. The lines of reference $\alpha = 0$, $\beta = 0$, $\gamma = 0$, may themselves be expressed in Cartesian co-ordinates. Thus we may take them as abridgments for three equations of the form $Lx + My + N = 0$, &c.; but it is more convenient to consider them as abridgments for three equations in the standard form. Thus, if the equations of BC, CA, AB be

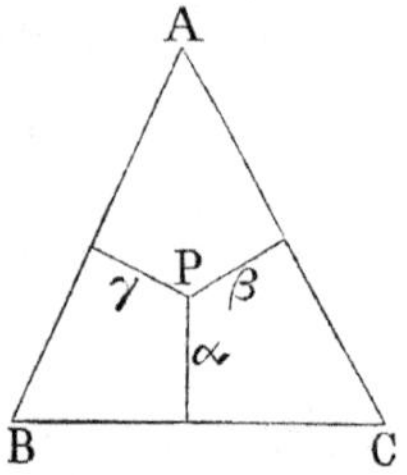

$$x \cos \alpha + y \sin \alpha - p = 0, \quad x \cos \beta + y \sin \beta - p' = 0,$$
$$x \cos \gamma + y \sin \gamma - p'' = 0,$$
$$\alpha = 0, \text{ and } x \cos \alpha + y \sin \alpha - p = 0$$

will be different modes of expressing the same thing. Again, if the Cartesian co-ordinates of P be X, Y, we see that

$$\alpha \equiv X \cos \alpha + Y \sin \alpha - p, \quad \beta \equiv X \cos \beta + Y \sin \beta - p',$$
$$\gamma \equiv X \cos \gamma + Y \sin \gamma - p'';$$

and, therefore, that any equation expressed in trilinear co-ordinates can be transformed into one in Cartesian co-ordinates.

Observation.—In these equations it will be seen that α, β, γ are used with different significations, but after a little practice this creates no confusion.

29. *If a line* (CD) *through the vertex* (C) *of a triangle* (ACB) *divide the base into segments* (BD, DA), *whose ratio is* λ; *and the vertical angle into segments, the ratio of whose sines is* k; *then the ratio of* $\lambda : k$ *is independent of the line* (CD).

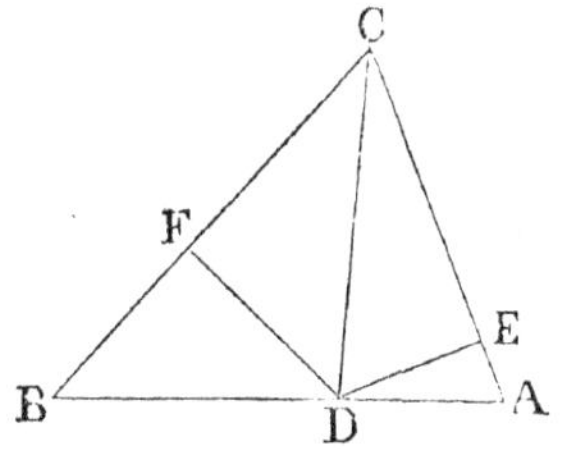

Dem.—From D let fall the perpendiculars DE, DF on AC, CB; then we have

$$\lambda = BD \div DA\ ;\ k = FD \div DE.$$

Hence $$\frac{k}{\lambda} = \frac{FD}{BD} \div \frac{DE}{DA} = \sin B \div \sin A. \qquad (108)$$

Examples.

1. Find the equations of the bisectors of the vertical angle. The equation of any line through C is of the form $\alpha - k\beta = 0$ where k is the ratio of the sines (Art. 25, *Cor.* 3). Hence the internal bisector is $\alpha - \beta = 0$, and the external, $\alpha + \beta = 0$. (109)

2. Find the equation of the median that bisects AB. Here the ratio of $BD : DA$ is unity. Hence $\lambda = 1$; therefore $k = \frac{\sin B}{\sin A}$, and the median is

$$\alpha - \frac{\sin B}{\sin A}\beta = 0, \quad \text{or} \quad \alpha \sin A - \beta \sin B = \omega. \qquad (110)$$

3. Find the equation of the perpendicular. Here $k = \cos B \div \cos A$. Therefore the perpendicular is

$$\alpha \cos A - \beta \cos B = 0. \qquad (111)$$

Observation.—We may write the equations of the internal bisectors of the three angles of the triangle of reference, viz.,

$$\alpha - \beta = 0, \quad \beta - \gamma = 0, \quad \gamma - \alpha = 0, \quad \text{in the form} \quad \alpha = \beta = \gamma\ ; \qquad (112)$$

where, by omitting each letter in succession, we have the bisector of the angle between the sides denoted by the remaining letters.

Similarly the three medians are

$$\alpha \sin A = \beta \sin B = \gamma \sin C, \qquad (113)$$

and the perpendiculars

$$\alpha \cos A = \beta \cos B = \gamma \cos C. \qquad (114)$$

4. Three lines whose equations are in the form $l\alpha = m\beta = n\gamma$ are concurrent, and the co-ordinates of their point of concurrence are

$$\frac{1}{l}, \quad \frac{1}{m}, \quad \frac{1}{n}. \qquad (115)$$

5. The lines $l\alpha = m\beta$, $\dfrac{\alpha}{l} = \dfrac{\beta}{m}$ make equal angles with $\alpha = \beta$ on opposite sides. Hence, if three lines through the vertices of a triangle be concurrent, the three lines equally inclined to the bisectors of its angles are concurrent.

DEF.—The three lines which make with the bisectors of the angles of a triangle, on the opposite sides angles equal to those which the medians make, are called the *symmedians* of the triangle, and their point of intersection its *symmedian point*.—M. D'OCAGNE.

6. The three symmedians of the triangle of reference are

$$\frac{\alpha}{\sin A} = \frac{\beta}{\sin B} = \frac{\gamma}{\sin C}. \qquad (116)$$

7. If the three lines $\dfrac{b}{c}\alpha = \dfrac{c}{a}\beta = \dfrac{a}{b}\gamma$ meet in O, and the three lines $\dfrac{c}{b}\alpha = \dfrac{a}{c}\beta = \dfrac{b}{a}\gamma$ meet in O', prove that the six angles OAB, OBC, OCA, $O'BA$, $O'CB$, $O'AC$, are all equal.—BROCARD.

DEF.—The points O, O' are called the *Brocard points*, and any of the six angles OAB, &c., the *Brocard angle* of the triangle.

8. Prove that the co-ordinates of the—

$$\left.\begin{array}{lll}
1^\circ.\ \text{Circumcentre} & \text{are} & \cos A,\ \cos B,\ \cos C, \\
2^\circ.\ \text{Orthocentre} & ,, & \sec A,\ \sec B,\ \sec C, \\
3^\circ.\ \text{Centroid} & ,, & \operatorname{cosec} A,\ \operatorname{cosec} B,\ \operatorname{cosec} C, \\
4^\circ.\ \text{Symmedian point} & ,, & \sin A,\ \sin B,\ \sin C, \\
5^\circ.\ \text{Point } O & ,, & \dfrac{c}{b},\ \dfrac{a}{c},\ \dfrac{b}{a}, \\
6^\circ.\ \text{Point } O' & ,, & \dfrac{b}{c},\ \dfrac{c}{a},\ \dfrac{a}{b} \\
7^\circ.\ \text{Centre of inscribed circle are } 1,\ 1,\ 1. & &
\end{array}\right\} . \qquad (117)$$

9. If the Brocard angle be denoted by α, prove

$$\cot \alpha = \cot A + \cot B + \cot C.$$

10. If the perpendicular erected to the base AB of a triangle ABC, at the foot of the symmedian line CS, meets in the points A', B' the perpendiculars at A, B to the sides AC, BC; prove $AA' : BB' :: AC^3 : BC^3$.—M. D'OCAGNE.

30. Def. i.—*If a line AB be divided in C into segments whose ratio is* λ, *and in D into segments whose ratio is* λ', *then the ratio of* $\lambda : \lambda'$ *is called the anharmonic ratio of the four points A, B, C, D.*

In the special case in which $\lambda = -\lambda'$, that is, when AB is divided internally and externally in the same ratio, AB is said to be divided *harmonically*, and the points C, D are called *harmonic conjugates* to A and B.

Def. ii.—*If an angle AOB be divided by a line OC into segments whose sine-ratio is k, and by a line OD into segments whose sine ratio is k', the ratio k : k' is called the anharmonic ratio of the pencil (O . ABCD), consisting of the rays OA, OB, OC, OD. The rays OC, OD are called conjugates to OA, OB.*

In the special case where $k = -k'$, $(O . ABCD)$ is called a harmonic pencil.

Observation.—The function of the segments of a line made by four points, which we have called their *anharmonic ratio*, has received different names from Geometers. Möbius calls it the *double ratio* (*doppelverhältniss*), Chasles, the *anharmonic ratio*, and the late Professor Clifford, the *cross ratio* of the four points. Chasles' nomenclatrue, although perhaps the least appropriate, is almost universally adopted.

31. *If a segment PQ be divided in the points A, B, C, D in the respective ratios* $a : 1$, $b : 1$, $c : 1$, $d : 1$,

the anharmonic ratio (ABCD) is independent of PQ.

Dem.—Since $PA : AQ :: a : 1$, we have $AQ = \dfrac{PQ}{a+1}$; similarly $BQ = \dfrac{PQ}{b+1}$. Hence $AB = \dfrac{(b-a)\,PQ}{(a+1)(b+1)}$, and we have corresponding values for the other segments; therefore

$$\frac{AB \, . \, CD}{AD \, . \, BC} = \frac{(a-b)(c-d)}{(a-d)(b-c)}.$$

32. *The anharmonic ratio of four collinear points A, B, C, D is equal to the anharmonic ratio of a pencil of rays (O . ABCD) passing through these points.*

Dem.—Let the ratio $AC : CB = \lambda$, $AD : DB = \lambda'$, $\sin AOC : \sin COB = k$, $\sin AOD : \sin DOB' = k'$; then Art. 29, $\lambda : \lambda' :: k : k'$. Hence the proposition is proved.

33. *If OP, OQ be two lines whose equations in the standard form are* $\alpha = 0$, $\beta = 0$, *and if OA, OB, OC, OD be four rays passing thoough O, whose equations are* $\alpha - k\beta = 0$, $\alpha - k'\beta = 0$, $\alpha - k''\beta = 0$, $\alpha - k'''\beta = 0$, *the anharmonic ratio of the pencil* $(O \, . \, ABCD)$ *is independent of* α *and* β.

Dem.—Draw any transversal cutting the pencil in the points A, B, C, D. Then, if PQ be divided in A, B, C, D in the ratios $a : 1$, $b : 1$, &c., we have $k = a\dfrac{\sin P}{\sin Q}$(Art. 29), &c. Hence $\dfrac{(k-k')(k''-k''')}{(k-k'')(k'-k''')} = \dfrac{(a-b)(c-d)}{(a-c)(b-d)}$ = the anharmonic ratio $(ABCD)$ = the pencil $(O \, . \, ABCD)$.

Examples.

1. If two different transversals cut the same pencil, their anharmonic ratios are equal.

2. If two equal anharmonic pencils have a common ray, the intersections of the remaining three homologous pairs of rays are collinear.

3. If F, G, H be three fixed points on a segment, A, B, C, D four other points on the same segment, such that the anharmonic ratios $(FGHA)$, $(FGHB)$, $(FGHC)$, $(FGHD)$ are respectively equal to $\alpha, \beta, \gamma, \delta$, prove that $(ABCD) = \dfrac{(\alpha - \beta)(\gamma - \delta)}{(\alpha - \gamma)(\beta - \delta)}$.

4. If the positions of four points A, B, C, X on a segment PQ be denoted by the ratios $a : 1$, $b : 1$, $c : 1$, $x : 1$; and the positions of four other points A', B', C', X' on another segment $P'Q'$ by the ratios $a' : 1$, $b' : 1$, &c.; then if $(ABCX) = (A'B'C'X')$, prove that x, x' are connected by an equation of the form

$$xx' - \mu' x - \mu x' + \lambda = 0. \qquad (118)$$

Dem.—From the hypothesis we have $\dfrac{(a-b)(c-x)}{(a-c)(b-x)} = \dfrac{(a'-b')(c'-x')}{(a'-c')(b'-x')}$, which, cleared of fractions, gives the required equation.

5. If two variable points on two different segments be connected by the equation (118), prove that the anharmonic ratio of any four positions of one of them is equal to the corresponding anharmonic ratio of their four homologous positions of the other.

6. If three sides of a variable triangle pass through three collinear points, and two of its vertices move on fixed lines, the locus of the third vertex is a right line.

34. *The equations of any four lines, no three of which are concurrent, are connected by an identical relation—that is, the equation of any one can be expressed in terms of the remaining three.*

Dem.—Let the four lines be

$$G \equiv g_1x + g_2y + g_3 = 0,$$
$$H \equiv h_1x + h_2y + h_3 = 0,$$
$$K \equiv k_1x + k_2y + k_3 = 0,$$
$$L \equiv l_1x + l_2y + l_3 = 0.$$

Now we can always determine four multiples a, b, c, d, such that the three following relations will be satisfied

$$ag_1 + bh_1 + ck_1 + dl_1 = 0,$$
$$ag_2 + bh_2 + ck_2 + dl_2 = 0,$$
$$ag_3 + bh_3 + ck_3 + dl_3 = 0.$$

Hence, for these multiples,

$$aG + bH + cK + dL \equiv 0,$$

which is the required identical relation.

This proposition may be stated and proved differently, as follows :—

If α, β, γ be any three lines forming a triangle A, B, C, the equation of any fourth line (DF) is of the form $l\alpha + m\beta + n\gamma = 0$.

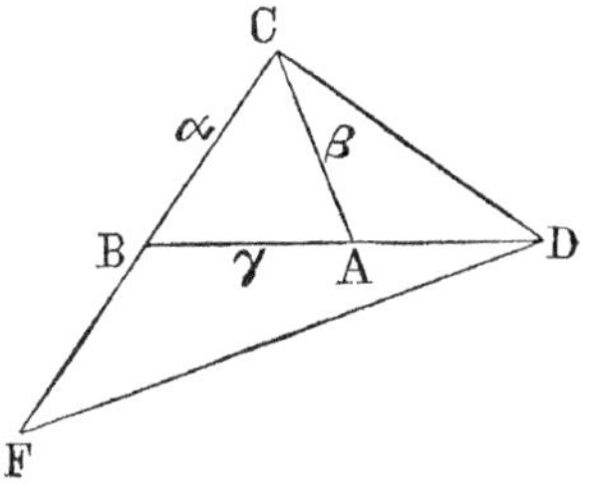

Dem.—Join CD. Now since CD passes through the intersection of α and β, its equation is of the form $l\alpha + m\beta = 0$ (Art. 25); and since DF passes through the intersection of $l\alpha + m\beta = 0$ and $\gamma = 0$, its equation is of the form $l\alpha + m\beta + n\gamma = 0$.

35. Def. I.—*The figure formed by four right lines produced indefinitely is called a complete quadrilateral.*

DEF. II.—*The triangle formed by the three diagonals of a complete quadrilateral is called its diagonal triangle.*—(CREMONA.)

DEF. III.—*The triangle whose vertices are the intersection of two diagonals and the extremities of the third diagonal is called the harmonic triangle of the quadrilateral.*—(SEQUEL TO EUCLID.)

DEF. IV.—*Two triangles which are such that the lines joining their vertices in pairs are concurrent are said to be in perspective; the point of concurrence is called their centre of perspective.*

36. *If thee quations of the sides of a complete quadrilateral be*

$$l\alpha + m\beta + n\gamma = 0, \quad (1)$$

$$m\beta + n\gamma - l\alpha = 0, \quad (2)$$

$$l\alpha - m\beta + n\gamma = 0, \quad (3)$$

$$l\alpha + m\beta - n\gamma = 0, \quad (4)$$

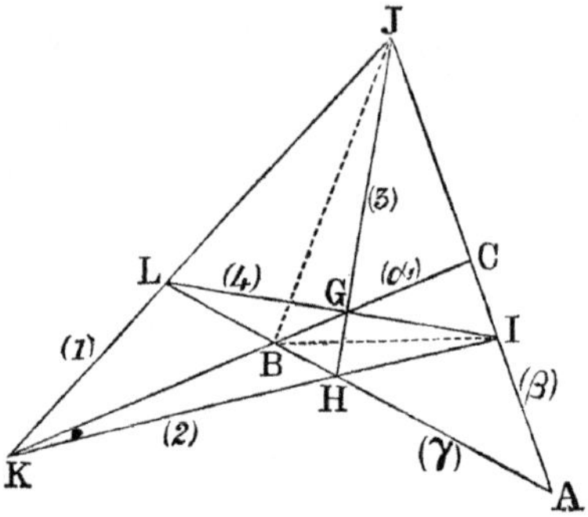

prove that the triangle of reference is its diagonal triangle.

Dem.—By subtraction of (1) and (2), and addition of (3) and (4), we see that α passes through the intersection of (1) and (2), and also through the intersection of (3) and (4). Hence $\alpha = 0$ is one of the diagonals. Similarly $\beta = 0$, $\gamma = 0$ are diagonals. Thus in the annexed diagram let $LKHG$ be the quadrilateral; then if the sides taken in order be (1), (2), (3), (4), the equations of the sides of the diagonal triangle BCA taken in order are, $\alpha = 0$, $\beta = 0$, $\gamma = 0$.

EXAMPLES.

1. The equations of the sides of the harmonic triangle IJB are

$$l\alpha + n\gamma = 0, \quad l\alpha - n\gamma = 0, \quad \text{and } \beta = 0.$$

For $l\alpha + n\gamma = 0$ evidently passes through B, which is the intersection of α and γ; and by adding the equations (1) and (3), we see that it passes through J. Hence $l\alpha + n\gamma = 0$ is the equation of BJ. Similarly $l\alpha - n\gamma = 0$ is the equation of BI.

2. The sides of the harmonic triangle taken in pairs form harmonic pencils with pairs of opposite sides, and also with the diagonals of the complete quadrilateral.

For the lines $l\alpha + n\gamma = 0$, $l\alpha - n\gamma = 0$ form a harmonic pencil with α and γ. (Art. 30, Def. II.)

3. The diagonal triangle is in perspective with the triangle formed by any three of the four sides of the complete quadrilateral. For the joins of the points A, B, C with the points G, I, H are

$$m\beta - n\gamma = 0, \quad n\gamma - l\alpha = 0, \quad l\alpha - m\beta = 0,$$

which are concurrent.

4. If the multiples l, m, n be variable, then the sides of the quadrilateral will vary in position; prove that if one of them passes through a fixed point, each of the others wiil pass through a fixed point.

5. If l, m, n be respectively equal to $\sin A$, $\sin B$, $\sin C$; prove that the lines (2), (3), (4) will each bisect two sides of the triangle of reference, and that (1) will represent the line at infinity.

6. If two triangles be such that the points of intersection of corresponding sides are collinear, the triangles are in *perspective*.

For if one be the triangle of reference, and the line of collinearity be $l\alpha + mb + n\gamma = 0$, the equations of the three sides of the other triangle will be of the forms

$$l'\alpha + m\beta + n\gamma = 0, \quad l\alpha + m'\beta + n\gamma = 0, \quad l\alpha + m\beta + n'\gamma = 0;$$

and taking the differences of these in pairs, we get the concurrent lines

$$(l - l')\,\alpha = (m - m')\,\beta = (n - n')\,\gamma,$$

which are evidently the joins of corresponding vertices.

DEF. I.—*The line of collinearity of the points of intersection of the corresponding sides of two triangles in perspective is called their axis of perspective.*

DEF. II.—*The centre of perspective of the diagonal triangle and the triangle formed by any three sides of the complete quadrilateral is called the pole of the fourth side with respect to the diagonal triangle.*

7. The co-ordinates of the poles of the four sides of the quadrilateral are—

$$\frac{1}{l},\ \frac{1}{m},\ \frac{1}{n}; \quad \frac{1}{l},\ \frac{1}{m},\ -\frac{1}{n}; \quad \frac{1}{l},\ -\frac{1}{m},\ \frac{1}{n}; \quad -\frac{1}{l},\ \frac{1}{m},\ \frac{1}{n}.$$

8. Being given the pole of a line with respect to a triangle, show how to construct the line; and, conversely, being given a line, show how to find its pole with respect to a given triangle.

37. A notation has been devised by Professor Cayley, which has the advantage of abridging long expressions. Thus an expression such as $ax^3 + 3bx^2y + 3cxy^2 + dy^3$ is denoted by $(a, b, c, d)(x, y)^3$, and $ax^2 + by^2 + cz^2 + 2fyz + 2gzx + 2hxy$ by $(a, b, c, f, g, h)(x, y, z)^2$.

38. *If the general equation* $(a, b, c, f, g, h)(\alpha, \beta, \gamma)^2 = 0$ *in trilinear co-ordinates represents two right lines, it is required to find the conditions of parallelism and perpendicularity, respectively.*

1°. Parallelism.—Let the given equation be transformed to Cartesian co-ordinates by the substitution of $x \cos \alpha + y \sin \alpha - p$ for α, &c. (*see* Art. 28); if the result be

$$(a', b', c', f', g', h')(x, y, 1)^2 = 0, \qquad (1)$$

by equating coefficients, we get

$$a' = (a, b, c, f, g, h)(\cos \alpha, \ \cos \beta, \ \cos \gamma)^2,$$

$$b' = (a, b, c, f, g, h)(\sin \alpha, \ \sin \beta, \ \sin \gamma)^2,$$

$$h' = a \sin \alpha \cos \alpha + b \sin \beta \cos \beta + c \sin \gamma \cos \gamma + f \sin (\beta + \gamma) + g \sin (\gamma + \alpha) + h \sin (\alpha + \beta).$$

Hence $a'b' - h'^2 = (A, B, C, F, G, H)(\sin A, \ \sin B, \ \sin C)^2$,

where $\sin A$, $\sin B$, $\sin C$ are the sines of the angles of the triangle of reference, and the coefficients A, B, &c., are the minors of the determinant (99), Art. 26, 3°.

If the equation (1) represents two parallel lines, $a'b' - h'^2 = 0$, Art. 27, *Cor.* 2; hence the condition that $(a, b, c, f, g, h)(\alpha, \beta, \gamma)^2$ represents two parallel lines is

$$A, B, C, F, G, H)(\sin A, \ \sin B, \ \sin C)^2, \text{ or say } \theta = 0. \qquad (119)$$

Cor.—The equation $\theta = 0$ may be written in determinant form thus:

$$\begin{vmatrix} a, & h, & g, & \sin A, \\ h, & b, & f, & \sin B, \\ g, & f, & c, & \sin C, \\ \sin A, & \sin B, & \sin C, & 0, \end{vmatrix} = 0. \qquad (120)$$

2°. PERPENDICULARITY.—By addition we get

$$a' + b' = a + b + c + 2f\cos(\beta - \gamma) + 2g\cos(\gamma - \alpha) + 2h\cos(\alpha - \beta),$$

or
$$a' + b' = a + b + c - 2f\cos A - 2g\cos B - 2h\cos C.$$

Hence, Art. 27, *Cor.* 2, the required condition is

$$a + b + c - 2f\cos A - 2g\cos B - 2h\cos C, \text{ or say } \theta' = 0. \quad (121)$$

39. *If the equation* $(a, b, c, f, g, h)(\alpha, \beta, \gamma)^2 = 0$ *represent two lines, it is required to find the angle between them.*

If we transform to Cartesian co-ordinates, so that the point of intersection may be the new origin, we have identically

$$(a, b, c, f, g, h)(\alpha, \beta, \gamma)^2 \equiv ky(y - x\tan\phi),$$

and, applying the results of Art. 38,

$$a'b' - h'^2 = \theta, \quad a' + b' = \theta';$$

but in the transformed equation

$$a' = 0, \quad b' = k, \quad h' = -\tfrac{1}{2}k\tan\phi.$$

Hence
$$k = \theta', \text{ and } -\frac{k^2\tan^2\phi}{4} = \theta;$$

and, eliminating k, we get

$$\tan^2\phi = -\frac{4\theta}{\theta'^2}. \quad (122)$$

Cor.
$$\cos^2\phi = \frac{\theta'^2}{\theta'^2 - 4\theta}. \quad (123)$$

EXAMPLES.

1. Find the condition that $l\alpha + m\beta + n\gamma = 0$, $l'\alpha + m'\beta + n'\gamma = 0$, shall be at right angles to each other.

Ans. $l(l' - m'\cos C - n'\cos B) + m(m' - n'\cos A - l'\cos C)$
$+ n(n' - l'\cos B - m'\cos A) = 0.$ (124)

2. Find the condition that $l\alpha + m\beta + n\gamma = 0$ shall be perpendicular to $\gamma = 0$.
Ans. $n = m\cos A + l\cos B$. (125)

3. Find the equation of the perpendiculars to the sides of the triangle of reference at their middle points.

Ans. $\alpha \sin A - \beta \sin B + \gamma \sin (A - B)$.

4. Find the angle between $l\alpha + m\beta + n\gamma = 0$, $l'\alpha + m'\beta + n'\gamma = 0$.

Here $\quad -\theta = \frac{1}{4}\{(mn' - m'n) \sin A + (nl' - n'l) \sin B + (lm' - l'm) \sin C\}^2,$

$\theta' = ll' + mm' + nn' - (mn' + m'n) \cos A - (nl' + n'l) \cos B - (lm' + l'm)\},$

Hence

$$\tan \phi = \frac{(mn' - m'n) \sin A + (nl' - n'l) \sin B + (lm' - l'm) \sin C}{ll' + mm' + nn' - (mn' + m'n) \cos A - (nl' + n'l) \cos B - (lm' + l'm) \cos C}. \tag{126}$$

Hence, if the lines are parallel, the numerator of this fraction vanishes; and if perpendicular, the denominator vanishes.

The condition of parallelism may be obtained more simply as follows:— If the given lines be parallel they will meet on the line at infinity, that is, on $\alpha \sin A + \beta \sin B + \gamma \sin C = 0$, for which the condition is expressed by the equation

$$\begin{vmatrix} l, & m, & n, \\ l', & m', & n', \\ \sin A, & \sin B, & \sin C, \end{vmatrix} = 0, \tag{127}$$

which is the same as the foregoing.

5. Find the equation of the perpendicular through $\alpha'\beta'\gamma'$ to the line

$$l\alpha + m\beta + n\gamma.$$

Let the required equation be $l'\alpha + m'\beta + n'\gamma = 0$. Then, since it passes through $\alpha'\beta'\gamma'$, we have $l'\alpha' + m'\beta' + n'\gamma' = 0$, and the condition of perpendicularity (Ex. 1) is

$$l'(l - m \cos C - n \cos B) + m'(m - l \cos C - n \cos A) + n'(n - m \cos A - l \cos B).$$

Hence, eliminating l', m', n', we get the determinant

$$\begin{vmatrix} \alpha, & \alpha', & l - m \cos C - n \cos B, \\ \beta, & \beta', & m - n \cos A - l \cos C, \\ \gamma, & \gamma', & n - l \cos B - m \cos A, \end{vmatrix} = 0. \tag{128}$$

6. Find the line through $\alpha'\beta'\gamma'$ parallel to $l\alpha + m\beta + n\gamma = 0$.

$$Ans. \quad \begin{vmatrix} \alpha, & \alpha', & m \sin C - n \sin B, \\ \beta, & \beta', & n \sin A - l \sin C, \\ \gamma, & \gamma', & l \sin B - m \sin A, \end{vmatrix} = 0. \tag{129}$$

*DEF.—*A line DE cutting the sides CA, CB of the triangle of reference in the points D, E, so that the triangle CDE is inversely similarly to CBA, is called an antiparallel to the base.*—LEMOINE.

7. Find the condition that $l\alpha + m\beta + n\gamma$ may be antiparallel to γ. If ϕ be the angle between $l\alpha + m\beta + n\gamma = 0$, and $\gamma = 0$,

$$\tan\phi = \frac{m\sin A - l\sin B}{n - m\cos A - l\cos B} \quad \text{(Ex. 4)};$$

but if $l\alpha + m\beta + n\gamma$ be antiparallel to γ, $-\phi = (A - B)$.

Hence
$$-\tan(A - B) = \frac{m\sin A - l\sin B}{n - m\cos A - l\cos B};$$

and reducing, we get

$$l\sin A - m\sin B - n\sin(A - B), \qquad (130)$$

which is the required condition.

8. Find the equation of the line through the symmedian point antiparallel to the base.

$$Ans. \quad \begin{vmatrix} \alpha, & \beta, & \gamma, \\ \sin A, & \sin B, & \sin C, \\ \sin A, & -\sin B, & -\sin(A - B), \end{vmatrix} = 0, \qquad (131)$$

or,
$$\alpha\sin B\cot A + \beta\sin A\cot B = \gamma. \qquad (132)$$

9. If through the symmedian point of the triangle of reference three antiparallels to the sides be drawn, they meet the sides in six points equally distant from the symmedian point.

10. Find the equations of the perpendiculars of the triangle whose sides are the lines

$$l_1\alpha + m_1\beta + n_1\gamma = 0, \quad l_2\alpha + m_2\beta + n_2\gamma = 0, \quad l_3\alpha + m_3\beta + n_3\gamma = 0.$$

The co-ordinates of the points of intersection of (2) and (3) are the three determinants

$$\begin{vmatrix} m_2 & n_2 \\ m_3 & n_3 \end{vmatrix}, \quad \begin{vmatrix} n_2 & l_2 \\ n_3 & l_3 \end{vmatrix}, \quad \begin{vmatrix} l_2 & m_2 \\ l_3 & m_3 \end{vmatrix}.$$

Denoting these by L, M, N, and substituting for α', β', γ' in Ex. 5, we have the equation of one of the perpendiculars; and, interchanging letters, we get the others.

11. Find the equation of the line through the middle point of BC, parallel to the external bisector of the vertical angle (see Ex. 6).

$$Ans. \quad \begin{vmatrix} \alpha, & 0, & \sin C - \sin B \\ \beta, & \sin C, & \sin A \\ \gamma, & \sin B, & -\sin A \end{vmatrix}.$$

12. Find the length of the perpendicuiar from $\alpha'\beta'\gamma'$ on $l\alpha + m\beta + n\gamma = 0$.

$$Ans. \quad \frac{l\alpha' + m\beta' + n\gamma'}{\sqrt{l^2 + m^2 + n^2 - 2mn\cos A - 2nl\cos B - 2lm\cos C}}. \qquad (133)$$

40. *To find the equation of the join of the points* $\alpha'\beta'\gamma'$, $\alpha''\beta''\gamma''$.

If $l\alpha + m\beta + n\gamma = 0$ be the required line, since it must pass through the given points, we have

$$l\alpha' + m\beta' + n\gamma' = 0, \quad l\alpha'' + m\beta'' + n\gamma'' = 0.$$

Hence, eliminating l, m, n, we get

$$\begin{vmatrix} \alpha, & \beta, & \gamma, \\ \alpha', & \beta', & \gamma', \\ \alpha'', & \beta'', & \gamma'', \end{vmatrix} = 0; \qquad (134)$$

or $L\alpha + M\beta + N\gamma = 0$, which is the required line.

It may be seen otherwise that (134) is a line through the given points; for it contains α, β, γ in the first degree, and is therefore a right line; secondly, if for α, β, γ be substituted the co-ordinates of either of the given points, the determinant will have two rows alike, and therefore vanishes identically. Hence the line passes through the given points.

41. It is easy to see that the coefficients L, M, N are equal to twice the areas of the triangles formed by the given points $\alpha'\beta'\gamma'$, $\alpha''\beta''\gamma''$, *and the vertices of the triangle of triangle of reference*, multiplied respectively by $\sin A$, $\sin B$, $\sin C$ (see Art. 3); but these triangles having a common base (the join of $\alpha'\beta'\gamma'$ and $\alpha''\beta''\gamma''$) are proportional to the

perpendiculars let fall on them from the vertices A, B, C. Hence we have the following theorem:—*If λ, μ, ν be the perpendiculars from the angular points of the triangle of reference on any line, the equation of the line may be written in the form*

$$(\lambda \sin A)\,\alpha + (\mu \sin B)\,\beta + (\nu \sin C)\,\gamma = 0,$$

or

$$\lambda a\alpha + \mu b\beta + \nu c\gamma = 0. \qquad (135)$$

42. If in equation (135) we write α for $a\alpha$, β for $b\beta$, γ for $c\gamma$, it is evident that the new co-ordinates of any point on the line will be proportional to the areas of the triangles formed by joining that point to the angles of the triangle of reference. These are called *areal co-ordinates*. Hence we infer the following theorem:—*If in the equation $\lambda\alpha + \mu\beta + \nu\gamma = 0$, α, β, γ denote areal co-ordinates, λ, μ, ν are proportional to the perpendiculars from the angular points of the triangle on the given line.*

Examples.

1. The point whose co-ordinates are $\alpha' + k\alpha''$, $\beta' + k\beta''$, $\gamma' + k\gamma''$, is collinear with the points $\alpha'\beta'\gamma'$, $\alpha''\beta''\gamma''$.

2. The determinant

$$\begin{vmatrix} b, & c, \\ M, & N, \end{vmatrix} = 2\Delta(\alpha' - \alpha''). \qquad (136)$$

For

$$\begin{vmatrix} b, & c, \\ M, & N, \end{vmatrix} = \begin{vmatrix} 0, & -c, & b, \\ \alpha', & \beta', & \gamma', \\ \alpha'', & \beta'', & \gamma'', \end{vmatrix} = \frac{1}{c}\begin{vmatrix} 0, & -c, & 0, \\ \alpha', & \beta', & 2\Delta, \\ \alpha'', & \beta'', & 2\Delta. \end{vmatrix}$$

3. Find the equations of the joins of the following pairs of points:—

1°. Orthocentre and centroid.

Ans. $\alpha \sin 2A \sin(B - C) + \beta \sin 2B \sin(C - A)$
$+ \gamma \sin 2C \sin(A - B) = 0. \qquad (137)$

2°. In-centre and circumcentre.

Ans. $\alpha(\cos B - \cos C) + \beta(\cos c - \cos A)$
$+ \gamma(\cos A - \cos B) = 0. \qquad (138)$

3°. Circumcentre and symmedian point.

Ans. $\alpha \sin(B - C) + \beta \sin(C - A) + \gamma \sin(A - B) = 0.$ (139)

4°. The *Brocard points* O, O'.

$$Ans.\ \frac{\alpha}{a}(a^4 - b^2c^2) + \frac{\beta}{b}(b^4 - c^2a^2) + \frac{\gamma}{c}(c^4 - a^2b^2) = 0. \quad (140).$$

4. Find the equation of the parallel through $\alpha'\beta'\gamma'$ to the join of $\alpha''\beta''\gamma''$, $\alpha'''\beta'''\gamma'''$.

$$Ans. \begin{vmatrix} \alpha, & \beta, & \gamma, \\ \alpha', & \beta', & \gamma', \\ \alpha'' - \alpha''', & \beta'' - \beta''', & \gamma'' - \gamma''', \end{vmatrix} = 0. \quad (141)$$

5. Prove that the join of the orthocentre and centroid is perpendicular to the line $\alpha \cos A + \beta \cos B + \gamma \cos C = 0.$

6. Prove that the join of the circumcentre and the symmedian points is perpendicular to the join of the *Brocard points*.

7. Prove that

$$2\Delta = \sqrt{\lambda^2a^2 + \mu^2b^2 + \nu^2c^2 - 2bc\,\mu\nu\cos A - 2ca\,\nu\lambda\cos B - 2ab\,\lambda\mu\cos C.} \quad (142)$$

For, denoting the perpendiculars of the triangle of reference by p, q, r, and the radical by Π, we have, letting fall from A—that is, from the point poo, a perpendicular on the line (135),

$$\lambda = \frac{\lambda a p}{\Pi}. \quad \text{Hence } ap = \Pi, \text{ that is, } 2\Delta = \Pi.$$

8. Prove that

$$2\Delta = \sqrt{a^2(\lambda - \mu)(\lambda - \nu) + b^2(\mu - \nu)(\mu - \lambda) + c^2(\nu - \lambda)(\nu - \mu)}. \quad (143)$$

Substitute, in Ex. 7, for $2bc \cos A$ its value, $b^2 + c^2 - a^2$, &c.

9. Prove

$$\frac{\lambda^2}{p^2} + \frac{\mu^2}{q^2} + \frac{\nu^2}{r^2} - 2\left(\frac{\mu\nu}{qr}\right)\cos A - 2\left(\frac{\nu\lambda}{rp}\right)\cos B - 2\frac{\lambda\mu}{pq}\cos C = 1. \quad (144)$$

10. Find the condition that the points $\alpha'\beta'\gamma'$, $\alpha''\beta''\gamma''$, may subtend a right angle at $\alpha\beta\gamma$.

Ans. $\alpha^2\{\beta'\beta'' + \gamma'\gamma'' + (\beta'\gamma'' + \beta''\gamma')\cos A\}$

$- \alpha\beta\{\alpha'\beta'' + \alpha''\beta' + (\gamma'\lambda'' + \gamma''\alpha')\cos A + (\beta'\gamma'' + \beta''\gamma')\cos B - 2\gamma'\gamma''\cos C\}$

$+$ two similar expressions got by interchange of letters $= 0.$ (145)

43. *To find the distance* δ *between two points* $\alpha_1\beta_1\gamma_1$, $\alpha_2\beta_2\gamma_2$. From the given points draw perpendiculars to the sides

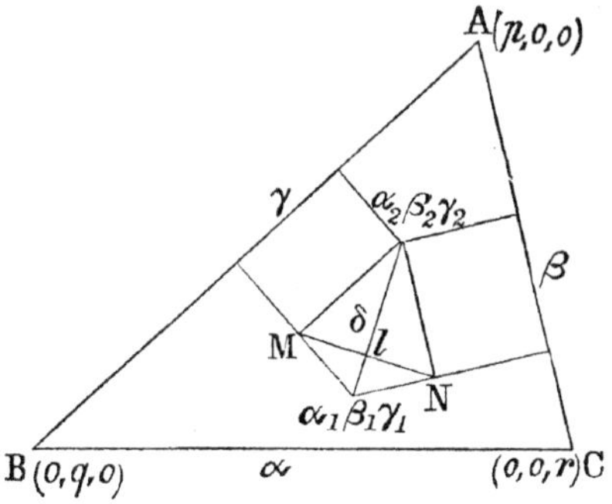

AB, AC, and from $\alpha_2\beta_2\gamma_2$ draw parallels to AB, AC; then, denoting the distance MN by l, we have

$$\delta^2 \sin^2 A = l^2 = (\beta_1 - \beta_2)^2 + (\gamma_1 - \gamma_2)^2 + 2(\beta_1 - \beta_2)(\gamma_1 - \gamma_2) \cos A. \qquad (1)$$

Again, evidently,

$$a(\alpha_1 - \alpha_2) + b(\beta_1 - \beta_2) + c(\gamma_1 - \gamma_2) = 0;$$

therefore

$$a^2(\alpha_1 - \alpha_2)^2 = b^2(\beta_1 - \beta_2)^2 + c^2(\gamma_1 - \gamma_2)^2 + 2(\beta_1 - \beta_2)(\gamma_1 - \gamma_2). \quad (2)$$

Hence, eliminating $(\beta_1 - \beta_2)(\gamma_1 - \gamma_2)$ between (1) and (2), we see that δ^2 is of the form $l(\alpha_1 - \alpha_2)^2 + m(\beta_1 - \beta_2)^2 + n(\gamma_1 - \gamma_2)^2$, where l, m, n are constants to be determined. For that purpose, suppose the given points to be in succession B, C; C, A; A, B, and we get the three equations

$$a^2 = mq^2 + nr^2; \quad b^2 = nr^2 + lp^2; \quad c^2 = lp^2 + mq^2;$$

therefore $$lp^2 = \frac{b^2 + c^2 - a^2}{2} = bc \cos A;$$

therefore $$l = \frac{bc \cos A}{p^2} = \frac{\sin 2A}{2 \sin A \sin B \sin C},$$

and similar values for m and n.

Therefore

$$\delta^2 = \frac{1}{2 \sin A \sin B \sin C} \{(\alpha_1 - \alpha_2)^2 \sin 2A + (\beta_1 - \beta_2)^2 \sin 2B + (\gamma_1 - \gamma_2)^2 \sin 2C\}. \qquad (146)$$

EXAMPLES.

1. Prove

$$\delta = \frac{\sqrt{L^2 + M^2 + N^2 - 2MN \cos A - 2NL \cos B - 2LM \cos C}}{\alpha \sin A + \beta \sin B + \gamma \sin C}. \qquad (147)$$

2. Prove

$$\delta^2 = -\frac{abc}{4\Delta^2} \{a(\beta_1 - \beta_2)(\gamma_1 - \gamma_2) + b(\gamma_1 - \gamma_2)(\alpha_1 - \alpha_2) + c(\alpha_1 - \alpha_2)(\beta_1 - \beta_2)\}. \quad (148)$$

44. *To find the area of the triangle whose vertices are the points*

$$\alpha'\beta'\gamma', \quad \alpha''\beta''\gamma'', \quad \alpha'''\beta'''\gamma'''.$$

If the axes be oblique, the area of the triangle formed by the points $x'y'$, $x''y''$, $x'''y'''$ (Art. 4), is—

$$\frac{\sin \omega}{2} \begin{vmatrix} x', & x'', & x''', \\ y', & y'', & y''', \\ 1, & 1, & 1. \end{vmatrix}$$

But taking as the oblique axes the lines $\alpha = 0$, $\beta = 0$, we evidently have

$$\sin \omega = \sin C, \quad x' \sin \omega = \alpha', \quad y' \sin \omega = \beta', \text{ \&c.,}$$

$$\Delta = \frac{\operatorname{cosec} C}{2} \begin{vmatrix} \alpha', & \alpha'', & \alpha''', \\ \beta', & \beta'', & \beta''', \\ 1, & 1, & 1, \end{vmatrix} = \frac{\operatorname{cosec} C}{2S} \begin{vmatrix} \alpha', & \alpha'', & \alpha''', \\ \beta', & \beta'', & \beta''', \\ S, & S, & S. \end{vmatrix}$$

Now taking $S \equiv \alpha \sin A + \beta \sin B + \gamma \sin C$, we get, diminishing the last row by the sum of the first multiplied by sin A, and second by sin B,

$$\Delta = \frac{1}{2S}\begin{vmatrix} \alpha', & \alpha'', & \alpha''', \\ \beta', & \beta'', & \beta''', \\ \gamma', & \gamma'', & \gamma'''. \end{vmatrix} \tag{149}$$

Observation.—For shortness of notation, determinants such as (149) are written in the form $(\alpha'\beta''\gamma''')$, and the area may be written thus:

$$\frac{(\alpha'\beta''\gamma''')}{2S}.$$

Or thus:—Writing the equations $\alpha = 0$ in Cartesian form,

$$x \cos \alpha + y \sin \alpha - p = 0.$$

By multiplication of determinants, we have—

$$\begin{vmatrix} \alpha', & \alpha'', & \alpha''', \\ \beta', & \beta'', & \beta''', \\ \gamma', & \gamma'', & \gamma''', \end{vmatrix} = \begin{vmatrix} x', & y', & 1, \\ x'', & y'', & 1, \\ x''', & y''', & 1, \end{vmatrix} \times \begin{vmatrix} \cos \alpha, & \sin \alpha, & -p, \\ \cos \beta, & \sin \beta, & -p', \\ \cos \gamma, & \sin \gamma', & -p'', \end{vmatrix} = 2\Delta S.$$

Therefore $$\text{area} = \frac{(\alpha'\beta''\gamma''')}{2S}. \tag{150}$$

Cor. 1.—If α', β', γ', &c., be not the actual lengths of the perpendiculars, let them be

$$(m'\alpha', m'\beta', m'\gamma'); \; (m''\alpha'', m''\beta'', m''\gamma''); \; (m'''\alpha''', m'''\beta''', m'''\gamma'''),$$

and we get $$\text{area} = \frac{m'm''m'''}{2S}(\alpha'\beta''\gamma'''). \tag{151}$$

Cor. 2.—To find m', m'', m'''. We have in this case

$$m'\alpha' \sin A + m'\beta' \sin B + m'\gamma' \sin C = S.$$

Hence $$m' = \frac{S}{\alpha' \sin A + \beta' \sin B + \gamma' \sin C} = \frac{S}{S'}. \tag{152}$$

Cor. 3.— $$2 \text{ area} = \frac{S^2(\alpha'\beta''\gamma''')}{S'S''S'''}. \tag{153}$$

EXAMPLES.

1. Find the value of m for the symmedian point $\sin A$, $\sin B$, $\sin C$.

$$Ans.\ m = \frac{S}{\sin^2 A + \sin^2 B + \sin^2 C}. \tag{154}$$

2. Find the value of m for the circumcentre.

$$Ans.\ m = \frac{2S}{\sin 2A + \sin 2B + \sin 2C}. \tag{155}$$

3. Find the value of m for the Brocard points.

$$Ans.\ m = \frac{\operatorname{cosec} A \operatorname{cosec} B \operatorname{cosec} C \,.\, S}{\operatorname{cosec}^2 A + \operatorname{cosec}^2 B + \operatorname{cosec}^2 C}. \tag{156}$$

4. Find the value for the orthocentre $\frac{1}{\cos A}$, $\frac{1}{\cos B}$, $\frac{1}{\cos C}$.

$$Ans.\ m = \frac{S}{\tan A + \tan B + \tan C}. \tag{157}$$

5. Find the area of the triangle formed by the lines

$$l_1\alpha + m_1\beta + n_1\gamma = 0, \quad l_2\alpha + m_2\beta + n_2\gamma = 0, \quad l_3\alpha + m_3\beta + n_3\gamma = 0.$$

Solving between the second and third, we get the co-ordinates of their point of intersection proportional to the minors L_1, M_1, N_1 of the determinant $(l_1 m_3 n_3)$. Hence, equation (153), the area

$$\frac{S^2 (L_1 M_2 N_3)}{(L_1 \sin A + M_1 \sin B + N_1 \sin C)(L_2 \sin A + M_2 \sin B + N_2 \sin C)(L_3 \sin A + M_3 \sin B + N_3 \sin C)} \tag{158}$$

6. The area of the triangle of reference is equal to

$$\frac{\{p \sin(\alpha' - \alpha'') + p' \sin(\alpha'' - \alpha) + p'' \sin(\alpha - \alpha')\}^2}{2 \sin(\alpha - \alpha') \sin(\alpha' - \alpha'') \sin(\alpha'' - \alpha)}. \tag{159}$$

7. Find the area of the triangle formed by $x \cos\alpha + y \sin\alpha - p = 0$, and the line pair $ax^2 + 2hxy + by^2 = 0$.

$$Ans.\ \text{Area} = \frac{p^2 \sqrt{h^2 - ab}}{a \sin^2\alpha - 2h \sin\alpha \cos\alpha + b \sin^2\alpha}. \tag{160}$$

SECTION III.—COMPARISON OF POINT AND LINE CO-ORDINATES.

45. DEF.—The coefficients in the equation of a line are called *line co-ordinates*. Because, if the coefficients be known the position of the line is fixed. Thus, let $\frac{x}{a} + \frac{y}{b} - 1 = 0$ be the equation of a line; then, putting $-\frac{1}{a} = u$, $-\frac{1}{b} = v$, we get

$$xu + yv + 1 = 0. \qquad (161)$$

In this equation u, v are called *line co-ordinates*, and x, y *point co-ordinates*. If x, y be fixed, and u, v variable, we shall have different lines, but each shall pass through the fixed point (xy). Thus, if xy be the point (ab); then, in Modern Geometry, the equation

$$au + bv + 1 = 0 \qquad (162)$$

is called the equation of the point (ab), and the variables u, v are the co-ordinates of any line passing through it. Hence we have the following general definition:—*The equation of a point is such a relation between the co-ordinates of a variable line which, if fulfilled, the line must pass through the point.*

46. The equation (161) expresses *the union of the positions of the point and the line*, in other words, it denotes that the point is found on the line, or what is the same thing, that the line passes through the point. And since it does not vary, if we interchange u, v with x, y, we have the following important result:—*In the equation which expresses the union of the positions of the point and the line, point and line co-ordinates enter symmetrically*. The point therefore enjoys in the geometry of the line the same *role* which the line does in the geometry of the point.

47. The following examples will illustrate the reciprocity between both systems of co-ordinates:—

EXAMPLES.

1°. Take the general equation.

Equation of the line,

$$Ax+By+C=0,$$

Equation of the point,

$$Au+Bv+C=0,$$

we shall have—

For the line co-ordinates,

$$u=\frac{A}{C},\quad v=\frac{B}{C}.$$

For the point co-ordinates,

$$x=\frac{A}{C},\quad y=\frac{B}{C}.$$

2°. Let there be given

Two points,

$$(x'y'),\quad (x''y''),$$

Two lines,

$$(u'v'),\quad (u''v''),$$

we shall have—

For the equation of their line connection, called *the join of the two points*,

$$\begin{vmatrix} x, & y, & 1, \\ x', & y', & 1, \\ x'', & y'', & 1, \end{vmatrix} = 0.$$

For the equation of their point of intersection, called the *join of the two lines*,

$$\begin{vmatrix} u, & v, & 1, \\ u', & v', & 1, \\ u'', & v'', & 1, \end{vmatrix} = 0.$$

The results and the operations which lead to them are the same in both cases. The significations of the variables only are different since the determinants will be satisfied if we put

$$x = lx' + mx'',\quad y = ly' + my'',\quad 1 = l + m.$$

$$u = lu' + mu'',\quad v = lv' + mv'',\quad 1 = l + m.$$

For, in fact, they are the results of eliminating l, m, 1. Between these two systems of equations, we shall have, putting $\lambda = \frac{m}{l}$,

$$x = \frac{x' + \lambda x''}{1 + \lambda},\quad y = \frac{y' + \lambda y''}{1 + \lambda}.$$

Supposing λ variable, these two equations represent the co-ordinates of any point of a row by means of two special ones. *It is the most general representation of a line as the base of a row of points.* Compare Art. 6, *Cor.* 1.

$$u = \frac{u' + \lambda u''}{1 + \lambda},\quad v = \frac{v' + \lambda v''}{1 + \lambda}.$$

Supposing λ variable, these two equations represent the co-ordinates of any ray of a pencil by means of two special rays. *It is the most general representation of a point as the vertex of a pencil of rays.* Compare Art. 25, *Cor.* 2.

Abridged from CLEBSCH 'Vorlesungen über Geometrie.'

Exercises on the Line.

1. Find the equation of the line joining the origin to the intersection of

$$\frac{x}{a} + \frac{y}{b} - 1 = 0, \quad \frac{x}{a'} + \frac{y}{b'} = 1.$$

2. Find the line through the intersection of $(x - a)$ and $(x + y + a) = 0$, and perpendicular to the latter.

3. Prove that $2x^2 + 3xy - 2y^2 - 8x + 4y = 0$ denotes two lines at right angles.

4. The opposite pairs of sides of a parallelogram are $x^2 - 5x + 6 = 0$ and $y^2 - 13y + 40 = 0$; find the equations of its diagonals.

5. Find the area of the figure included by the four lines

$$x \pm y = a, \quad x \pm y = b.$$

6. Find the area of the triangle whose angular points are the origin and the feet of perpendiculars from the origin on the lines

$$\frac{x}{a} + \frac{y}{b} - 1 = 0, \quad \frac{x}{a'} + \frac{y}{b'} - 1 = 0.$$

7. If $L = 0$, $L' = 0$ be two parallel lines, prove $L + L' = 0$ is midway between them.

8. If $\alpha = 0$, $\beta = 0$, $\gamma = 0$, $\delta = 0$ be the four sides of a quadrilateral, a, b, c, d their lengths, prove $a\alpha - b\beta + c\gamma - d\delta$ bisects the diagonals.

9. Find the locus of a point, the sum of whose distances from the sides of a given polygon is constant.

10. Find the locus of the intersection of the diagonals of the quadrilateral formed by the axes and the pairs of lines $\frac{x}{a} + \frac{y}{b} - 1$, $\frac{x}{\lambda a} + \frac{y}{\lambda b} - 1$, if λ be supposed to vary.

11. Find the equations of the line which is the join of the intersections of the transverse and direct joins of the pair of points where $x^2 + 2gx + c = 0$ meet the axis of x with the pair where $y^2 + 2fy + c = 0$ meet the axis of y.

12. Prove that the lines represented by $x^2 - xy - 6y^2 + 2x - y + 1 = 0$ include an angle of $45°$.

13. If A, B, C; A', B', C' be two triads of points on two lines intersecting in O; then if the anharmonic ratios $(OABC)$, $(OA'B'C')$ be equal, the three lines AA', BB', CC' joining homologous points are concurrent.

14. Find the equation of the parallel to α through the centre of any of the escribed circles.

15. Find the equation of the parallel to $\alpha \cos A - \beta \cos B = 0$ through $(\sin B, \sin A, 0)$.

$$Ans. \begin{vmatrix} \alpha, & \sin A, & \cos B, \\ \beta, & \sin A, & \cos A, \\ \gamma, & 0, & 1, \end{vmatrix}.$$

16. Prove that the join of $(1, 1, 1)$ and $(\cos(B - C), \cos(C - A), \cos(A - B))$, is perpendicular to

$$\frac{a\alpha}{b-c} + \frac{b\beta}{c-a} + \frac{c\gamma}{a-b} = 0.$$

17. Prove, by the properties of a harmonic pencil, that γ is parallel to

$$\alpha \sin A + \beta \sin B = 0.$$

18. Prove that the triangle whose sides are

$$\alpha + n\beta + \frac{\gamma}{m}, \quad \beta + l\gamma + \frac{\alpha}{n} = 0, \quad \gamma + m\alpha + \frac{\beta}{l} = 0$$

is inscribed in the triangle of reference.

19. If O be the circumcentre of the triangle of reference, and AG, AH be parallel to BO, CO respectively, prove that their equations are—

for AG, $\beta \cos C + \gamma \cos(C - A) = 0$; for AH, $\beta \cos(A - B) + \gamma \cos \beta = 0$.

20. Prove that the locus of the mean centre of the points, in which parallels to $l\alpha + m\beta + n\gamma = 0$ meet the sides of the triangle of reference, is

$$\frac{\alpha}{m \sin C - n \sin B} + \frac{\beta}{n \sin A - l \sin C} + \frac{\gamma}{l \sin B - m \sin A} = 0. \quad (163)$$

DEF.—*The line* (163) *is called the diameter of the triangle with respect to the line* $l\alpha + m\beta + n\gamma = 0$.

21. Find the equations of the parallel to the sides of the triangle of reference drawn—1°. through the incentre; 2°. the circumcentre; 3°. the symmedian point.

22. If on a variable line drawn through a fixed point O, meeting n fixed lines in the points R', R'', ... $R^{(n)}$, a point R be taken such that $\frac{n}{OR} = \frac{1}{OR'} + \frac{1}{OR''} \cdots \frac{1}{OR^{(n)}}$, the locus of R is a right line.

23. Find the length of the perpendicular from $(1, 1, 1)$ on

$$\frac{a\alpha}{b-c} + \frac{b\beta}{c-a} + \frac{c\gamma}{a-b} = 0.$$

24. Prove that the area of the parallelogram whose sides $\lambda x + \mu y \pm \delta = 0$ and $\lambda' x + \mu' y \pm \delta' = 0$ is $4\delta\delta' \div (\lambda\mu' - \lambda'\mu)^2$.

25. If $\alpha_1 \beta_1 \gamma_1$, $\alpha_2 \beta_2 \gamma_3$ be the areal co-ordinates of two points P', P'', and if $\alpha_1 \alpha_2 = \beta_1 \beta_2 = \gamma_1 \gamma_2$; prove, if we join these points to the three vertices, that the lines thus obtained cut the opposite sides in points that are symmetrical with respect to the middle point.

26. If three concurrent lines be drawn through the middle points of the sides of a triangle, three parallels to them through its vertices will be concurrent.

27. If $\lambda\alpha + \mu\beta + \nu\gamma = 0$ meet the sides AB, AC in the points D, E, and if O be the middle point of DE, the equation of OA is

$$(\mu \sin A - \lambda \sin B)\beta + (\lambda \sin C - \nu \sin A)\gamma = 0.$$

28. Prove that the sum of the tangents of the angles which $\lambda x + \mu y + \nu = 0$ makes with the lines $ax^2 + 2hxy + by^2 = 0$ is $\dfrac{2(a-b)\mu\mu - h(\lambda^2 - \mu^2)}{a\lambda^2 + 2h\lambda\mu + b\mu^2}$.

29. Find the value of m (see Art. 44, *Cor.* 2) for the point

$$\cos(B-C), \cos(C-A), \cos(A-B).$$

Ans. $\dfrac{S}{4 \sin A \sin B \sin C}$.

30. Prove that the ratio, in which the join of $x'y'$, $x''y''$ is divided by the line $Ax + By + C$, is

$$-(Ax'' + By'' + C) : (Ax' + By' + C). \qquad (164)$$

31. If a transversal cut the sides of a polygon of n sides, the ratio of the product of one set of alternate segments of the sides to the product of the remaining segments is $(-1)^n$.

32. The six anharmonic ratios of four collinear points A, B, C, D can be expressed in terms of the six trigonometric functions of an angle.

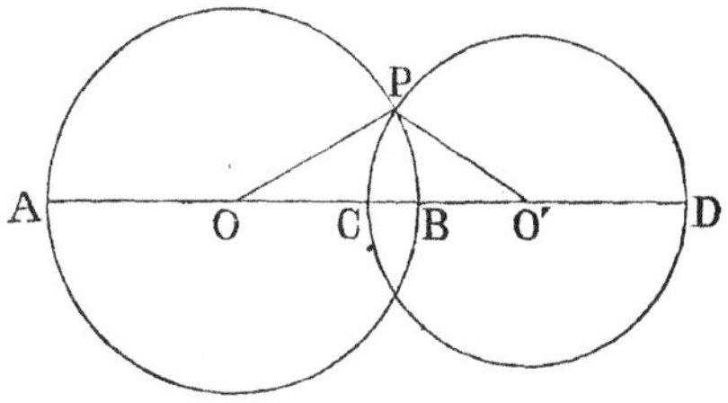

Dem.—On AB, CD describe semicircles. Let O, O' be the centres, P one of their points of intersection; then OPO is equal to one of the angles of intersection of the circles, denoting it by θ; then it is easy to see that

$$\frac{AB}{BC} : \frac{AD}{DC} = \sec^2 \tfrac{1}{2}\theta; \quad \frac{CA}{AB} : \frac{CB}{DB} = \sin^2 \tfrac{1}{2}\theta; \quad \frac{BC}{CA} : \frac{BD}{DA} = -\cot^2 \tfrac{1}{2}\theta,$$

$$\frac{BC}{AB} : \frac{DC}{AD} = \cos^2 \tfrac{1}{2}\theta; \quad \frac{AB}{CA} : \frac{DB}{CD} = \operatorname{cosec}^2 \tfrac{1}{2}\theta; \quad \frac{CA}{BC} : \frac{DA}{BD} = -\tan^2 \tfrac{1}{2}\theta,$$

and these are the six anharmonic ratios.

33. If α, β, γ, δ be the four sides of a quadrilateral, prove that $l\alpha+m\beta+n\gamma+p\delta=0$, $l(\alpha+m\beta)+n\gamma+p\delta=0$, $l(\alpha+m\beta+n\gamma)+p\delta=0$, $\alpha+m\beta+n\gamma+p\delta=0$ represent the sides of an inscribed quadrilateral.

34. If the joins of corresponding vertices of two triangles be concurrent, the points of intersection of corresponding sides are collinear.

For if the joins of corresponding vertices be the three lines $\alpha=\beta=\gamma$, the sides of the two triangles will be $\alpha+\beta+\delta=0$, $\beta+\gamma+\delta=0$, $\gamma+\alpha+\delta=0$; and $\alpha+\beta+\delta'=0$, $\beta+\gamma+\delta'=0$, $\gamma+\alpha+\delta'=0$, and each pair of corresponding vertices intersect on $\delta-\delta'=0$.

35. If the coefficients in the equation of a given line be connected by a given linear relation the line passes through a given point. *The given linear relation is the equation of the given point.*

36. If the vertical angle of a triangle be given in magnitude and position, and l times the reciprocal of one side plus m times the reciprocal of the other be given, the base passes through a given point.

37. Prove, by the method of complex variables, that if $ABCD$ be any plane quadrilateral, the rectangles $AB \,.\, CD$, $BC \,.\, AD$, $CA \,.\, BD$ are proportional to the sides of a triangle whose inclinations to the axis of x are inc. AB + inc. CD, inc. BC + inc. AD, inc. CA + inc. BD, respectively.

38. If a variable triangle ABC have its vertices on three concurrent lines OA, OB, OC, and if two of the sides pass through fixed points, the third side will pass through a fixed point.

For, if the reciprocals of OA, OB, OC be u, v, w, respectively, the conditions of the question give $au+bv-1=0$, $a'v+b'w-1=0$; hence, eliminating v, we get a linear relation between u and w, which is the equation of the point through which the third side passes.

Examples 39 to 42—Lemoine.

If through a point O we draw antiparallels—

1°. to BC, cutting BC in 1_1, AC in 1_2, AB in 1_3;

2°. to CA, " " 2_1, " 2_2, AB in 2_3;

3°. to AB, " " 3_1, " 3_2, AB in 3_3;

then, denoting the segments $2_1 3_1$, $3_2 1_2$, $1_3 2_3$ by ξ, η, ζ;

also, denoting the segments $1_2 1_3$, $2_3 2_1$, $3_1 3_2$ by ξ', η', ζ'.

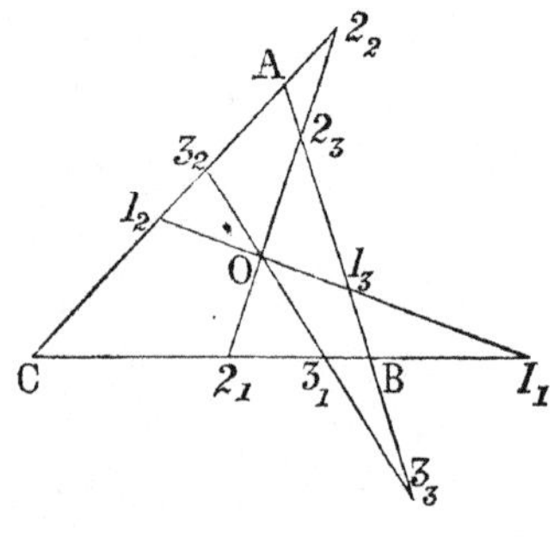

39. Prove $$\frac{\xi a^2}{\cos A} + \frac{\eta b^2}{\cos B} + \frac{\zeta c^2}{\cos C} = 2abc. \qquad (165)$$

40. Prove $$\frac{\xi' \cos A}{a} + \frac{\eta' \cos B}{b} + \frac{\zeta' \cos C}{c} = 1. \qquad (166)$$

41. Prove, if α, β, γ be the co-ordinates of O,

$$\xi = \frac{\alpha bc \cos A}{S}, \quad \eta = \frac{\beta ca \cos B}{S}, \quad \zeta = \frac{\gamma ab \cos C}{S}. \qquad (167)$$

42. $$\xi' = \frac{a(b\gamma + c\beta)}{S}, \quad \eta' = \frac{b(c\alpha + a\gamma)}{S}, \quad \zeta' = \frac{c(a\beta + b\alpha)}{S}. \qquad (168)$$

43. If Δ, Δ' be the areas of two triangles, the Cartesian co-ordinates of whose vertices are (ab), (bc), (ca); and $\{ac - b^2, ab - c^2\}$, $\{ab - c^2, cb - a^2\}$, $\{cb - a^2, ca - b^2\}$, respectively, prove $\Delta' = (a + b + c)^2 \Delta$.

44. Prove that $\alpha^2 + \beta^2 + \gamma^2 + \left(\frac{b}{c} + \frac{c}{b}\right)\beta\gamma + \left(\frac{c}{a} + \frac{a}{c}\right)\gamma\alpha + \left(\frac{a}{b} + \frac{b}{a}\right)\alpha\beta = 0$, represents only one finite line.

45. If α denote one of the six anharmonic ratios of four points, prove that the other five are $\frac{1}{\alpha}$, $1 - \alpha$, $\frac{1}{1-\alpha}$, $\frac{\alpha - 1}{\alpha}$, $\frac{\alpha}{\alpha - 1}$.

OBSERVATION.—*If* $\alpha = \frac{1}{1 - \alpha}$, *then* α *is one of the imaginary cube roots of* -1, *and the four points of the anharmonic ratio* $(ABCD)$ *cannot be all real. This special system has been named equianharmonic by* CREMONA.

46. If the equations of the three lines AO, BO, CO be $l\alpha = m\beta = n\gamma$, prove that

$$\frac{\sin AOB}{\sqrt{l^2 + m^2 - 2lm \cos C}} = \frac{\sin BOC}{\sqrt{m^2 + n^2 - 2mn \cos A}} = \frac{\sin COA}{\sqrt{n^2 + l^2 - 2nl \cos B}}. \qquad (169)$$

47. If λ, μ, ν be the sines of the angles which the line $l\alpha + m\beta + n\gamma = 0$ makes with the sides BC, CA, AB of the triangle of reference, prove

$$\lambda = \frac{m \sin C - n \sin B}{\Omega}, \quad \mu = \frac{n \sin A - l \sin C}{\Omega}, \quad \nu = \frac{l \sin B - m \sin A}{\Omega}, \qquad (170)$$

where $\Omega = \sqrt{l^2 + m^2 + n^2 - 2mn \cos A - 2nl \cos B - 2lm \cos C}$.

In the same case prove the following relations:—

48. $l\lambda + m\mu + n\nu = 0$. (171)

49. $a\lambda + b\mu + c\nu = 0$. (172)

50. $\mu^2 + \nu^2 + 2\mu\nu \cos A = \sin^2 A$, &c. (173)

51. $\lambda^2 \sin 2A + \mu^2 \sin 2B + \nu^2 \sin 2C = 2 \sin A \sin B \sin C$. (174)

52. $$\frac{\sin A}{\lambda} + \frac{\sin B}{\mu} + \frac{\sin C}{\nu} + \frac{\sin A}{\lambda} \cdot \frac{\sin B}{\mu} \cdot \frac{\sin C}{\nu} = 0. \qquad (175)$$

53. If λ, μ, ν denote the perpendiculars from any point on the lines $l\alpha = m\beta = n\gamma$, prove

$$\lambda\sqrt{m^2+n^2-2mn\cos A}+\mu\sqrt{n^2+l^2-2nl\cos B}+\nu\sqrt{l^2+m^2-2lm\cos C}=0. \qquad (176)$$

54. If λ, μ, ν, ρ be the perpendiculars from the point $\left(\frac{1}{l}, \frac{1}{m}, \frac{1}{n}\right)$ on the four lines $l\alpha \pm m\beta \pm n\gamma = 0$, prove that

$$\frac{1}{\lambda^2} + \frac{1}{\mu^2} + \frac{1}{\nu^2} + \frac{9}{\rho^2} = 4(l^2+m^2+n^2). \qquad (177)$$

55. If the side BC subtend a right angle at the point $(\alpha\beta\gamma)$, prove that

$$\beta\gamma = \alpha(\alpha\cos A - \beta\cos B - \gamma\cos C). \qquad (178)$$

56. In the figure, exercise 39, if O be the symmedian point, prove that the three extreme points 1_1, 2_2, 3_3, lie on the line

$$\frac{\alpha\cot A}{\sin A} + \frac{\beta\cot B}{\sin B} + \frac{\gamma\cos C}{\sin C} = 0. \qquad (179)$$

57. In the same case, if AD be perpendicular to BC, prove that the triangle DCA is inversely similar to the triangle whose angular points are 1_2, 2_1, 2_3.

58. If Θ, Θ_1, Θ_2, Θ_3 be four points, whose co-ordinates are—

For Θ,

$$\frac{\sin\frac{1}{2}A - \sin\frac{1}{2}B\sin\frac{1}{2}C}{\sin\frac{1}{2}A}, \quad \frac{\sin\frac{1}{2}B - \sin\frac{1}{2}C\sin\frac{1}{2}A}{\sin\frac{1}{2}B}, \quad \frac{\sin\frac{1}{2}C - \sin\frac{1}{2}A\sin\frac{1}{2}B}{\sin\frac{1}{2}C}.$$

For Θ_1,

$$-\left(\frac{\sin\frac{1}{2}A + \cos\frac{1}{2}B\cos\frac{1}{2}C}{\sin\frac{1}{2}A}\right), \quad \frac{\cos\frac{1}{2}B + \cos\frac{1}{2}C\sin\frac{1}{2}A}{\cos\frac{1}{2}B}, \quad \frac{\cos\frac{1}{2}C + \sin\frac{1}{2}A\cos\frac{1}{2}B}{\cos\frac{1}{2}C}.$$

For Θ_2,

$$\frac{\cos\frac{1}{2}A + \sin\frac{1}{2}B\cos\frac{1}{2}C}{\cos\frac{1}{2}A}, \quad -\left(\frac{\sin\frac{1}{2}B + \cos\frac{1}{2}C\cos\frac{1}{2}A}{\sin\frac{1}{2}B}\right), \quad -\frac{\cos\frac{1}{2}C + \cos\frac{1}{2}A\sin\frac{1}{2}B}{\cos\frac{1}{2}C}.$$

For Θ_3,

$$\frac{\cos\frac{1}{2}A + \cos\frac{1}{2}B\sin\frac{1}{2}C}{\cos\frac{1}{2}A}, \quad \frac{\cos\frac{1}{2}B + \sin\frac{1}{2}C\cos\frac{1}{2}A}{\cos\frac{1}{2}B}, \quad -\left(\frac{\sin\frac{1}{2}C + \cos\frac{1}{2}A\cos\frac{1}{2}B}{\sin\frac{1}{2}C}\right);$$

prove that their six joins are parallel respectively to the bisectors of the internal and external angles of the triangle of reference.—(LEMOINE.)

59. Prove a corresponding property for four points ω, ω_1, ω_2, ω_3, whose co-ordinates are—

For ω, $\operatorname{cosec}^2\frac{1}{2}A$, $\operatorname{cosec}^2\frac{1}{2}B$, $\operatorname{cosec}^2\frac{1}{2}C$.

For ω_1, $\operatorname{cosec}^2\frac{1}{2}A$, $-\sec^2\frac{1}{2}B$, $-\sec^2\frac{1}{2}C$.

For ω_2, $-\sec^2\frac{1}{2}A$, $\operatorname{cosec}^2\frac{1}{2}B$, $-\sec^2\frac{1}{2}C$.

For ω_3, $-\sec^2\frac{1}{2}A$, $-\sec^2\frac{1}{2}B$, $\operatorname{cosec}^2\frac{1}{2}C$.—(*Ibid.*).

CHAPTER III.

THE CIRCLE.

SECTION I.—CARTESIAN CO-ORDINATES.

48. *To find the general equation of a circle.*

Let (ab) be the centre, (xy) any point P in the circumference; then, if the radius OP be denoted by r, we have (Art. 1),

$$(x-a)^2+(y-b)^2=r^2; \quad (180)$$

or

$$x^2+y^2-2ax-2by+a^2+b^2-r^2=0,$$

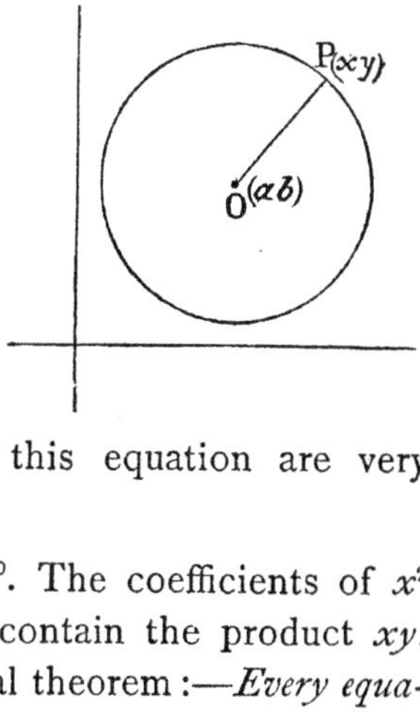

which is the required equation.

The following observations on this equation are very important:—

1°. It is of the second degree. 2°. The coefficients of x^2 and y^2 are equal. 3°. It does not contain the product xy. Hence we have the following general theorem:—*Every equation of the second degree which does not contain the product of the variables, and in which the coefficients of their second powers are equal, represents a circle.*

The following are special cases:—

1°. If the centre be origin, the equation is $x^2+y^2=r^2$, which is the standard form. (181)

2°. If the origin be on the circumference, $x^2+y^2-2ax-2by=0$. (182)

3°. If the axis of x pass through the centre, and the origin be on the circumference, $x^2 + y^2 = 2ax$. (183)

4°. If the axis of y pass through the centre, and the origin be on the circumference, $x^2 + y^2 = 2by$. (184)

Observation.—The criterion that the product xy must not be contained in the equation is true only when the axes are rectangular; for if they were oblique the equation would (Art. 1) be

$$(x - a)^2 + (y - b)^2 + 2(x - a)(y - b)\cos\omega = r^2. \quad (185)$$

49. *If the equation of a circle be given, we can construct it.* For let the equation be $ax^2 + ay^2 + 2gx + 2fy + c = 0$. Dividing by a, and completing squares, we get

$$\left(x + \frac{g}{a}\right)^2 + \left(y + \frac{f}{a}\right)^2 = \frac{g^2 + f^2 - ac}{a^2}. \quad (186)$$

Comparing this with the fundamental equation (180), we see that the co-ordinates of the centre are

$$-\frac{g}{a},\ -\frac{f}{a};\ \text{and that the radius is}\ \frac{\sqrt{g^2+f^2-ac}}{a}.$$

Hence the circle can be described. We have the following cases to consider: if $g^2 + f^2$ be greater than ac, the circle is real, and can be constructed; if $g^2 + f^2$ be equal to ac, the radius is zero, and the circle is indefinitely small, that is, it is a point; if $g^2 + f^2$ be less than ac, the radius is imaginary: there is no real circle corresponding to the equation; in other words, $ax^2 + ay^2 + 2gx + 2fy + c = 0$ represents in this case an imaginary circle.

Cor.—Since the co-ordinates of the centre of the circle $ax^2 + ay^2 + 2gx + 2fy + c = 0$ do not contain c, it follows *that two circles whose equations differ only in their absolute terms are concentric.*

50. Geometrical representation of the power of a point with respect to a circle. *The power of a point with respect to a circle* (Art. 20) *is positive, zero, or negative, according as the point is outside, on, or inside the circumference.*

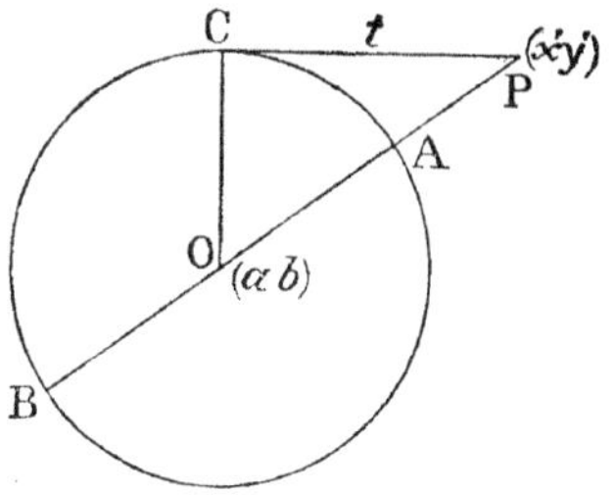

1°. Let $(x-a)^2+(y-b)^2-r^2=0$ be the circle $x'y'$ on external point; then the power of $x'y'$ with respect to the circle is

$$(x'-a)^2+(y'-b)^2-r^2;$$

that is (Art. 1) $OP^2 - r^2$, or t^2, since OCP is a right angle. *Hence the power of an external point with respect to a circle is equal to the square of the tangent drawn from that point to the circle.*

2°. When the point is on the circle its power is evidently zero.

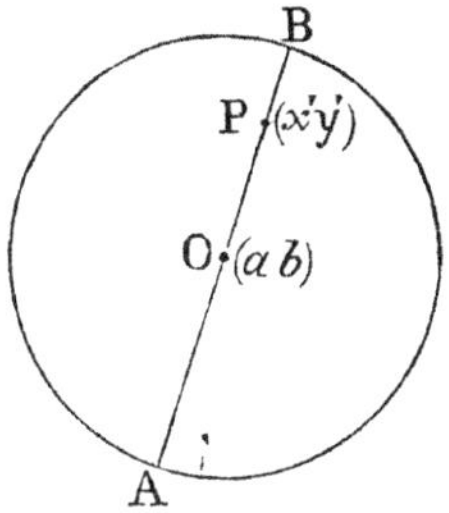

3°. Let $x'y'$ be an internal point; then denoting OP by δ, the power of OP with respect to the circle is

$$\delta^2 - r^2, \text{ or } -(r+\delta)(r-\delta);$$

that is $= - AP.PB$, a negative quantity.

Cor.—If for shortness the equation of a circle be denoted by $S=0$, the power of any point $x'y'$ with respect to S will be denoted by S', for this is the result of substituting the co-ordinates $x'y'$ in place of xy.

Examples.

1. If the equation of a line be added to the equation of a circle, the sum is the equation of a circle.

2. The sum of the equations of any number of circles is the equation cf a circle.

3. Construct the circles—

1°. $x^2+y^2-4x-8y=16$; 2°. $3x^2+3y^2+7x+9y+1=0$.

4. Find the equation of a circle, passing through the point (2, 4) through the origin, and having its centre on the axis of x.

5. Find the locus of the vertex of a triangle, being given the base and the sum of the squares of the sides.

6. Find the locus of the vertex of a triangle, being given the base and m squares of one side + n squares of the other.

7. If $S_1 = 0$, $S_2 = 0$, $S_3 = 0$, &c., be the equations of any number of circles; prove that the centre of $lS_1 + mS_2 + nS_3 +$ &c. $= 0$ is the mean centre of the centres of S_1, S_2, S_3, &c., for the system of multiples l, m, n, &c.

8. What is the locus of a point, the powers of which with respect to two given circles are equal?

9. Find the locus of a point, if the tangents from it to two given circles have a given ratio.

10. What does Ex. 9 become if the circles reduce to points?

11. Find the equation of the circle whose diameter is the join of the points $x'y'$, $x''y''$.

Ans. $(x - x')(x - x'') + (y - y')(y - y'') = 0.$ (187)

12. Given the base of a triangle and the vertical angle; prove that the locus of its vertex is a circle.

13. Given the base of a triangle and the vertical angle; prove that the locus of the intersection of perpendiculars is a circle.

14. Find the locus of a point at which two given circles subtend equal angles.

15. If a line of given length slide between two fixed lines, the locus of the centre of instantaneous rotation is a circle?

16. Given the base of a triangle and the ratio of the tangent of the vertical angle of the tangent of one of the base angles; prove that the locus of the vertex is a circle.

17. If the sum of the squares of the distances of a point from the sides of an equilateral triangle or of a square be given, the locus of the point is a circle.

18. If the sum of the squares of the distances from a variable point to any number of fixed points, each multiplied by a given constant, be given, the locus of the point is a circle.

19. If the base c of a triangle be given both in magnitude and position, and $ab \sin(C - \alpha)$, where α is a given angle, be given in magnitude, the locus of the vertex C is a circle.—(M'CAY).

51. *The equations of a line and a circle being given, it is required to find the equation of the circle whose diameter is the intercept which the latter makes on the former.*

Let the equations be—

$$x \cos \alpha + y \sin \alpha - p = 0. \quad (1) \qquad x^2 + y^2 - r^2 = 0. \quad (2)$$

Eliminating y and x in succession, we get

$$x^2 - 2px \cos \alpha + p^2 - r^2 \sin^2 \alpha = 0; \quad (3)$$

$$y^2 - 2py \sin \alpha + p^2 - r^2 \cos^2 \alpha = 0. \quad (4)$$

Equation (3), being a quadratic in x, denotes (Art. 26) two lines parallel to the axis of y through the points of intersection of (1) and (2). Similarly, equation (4) denotes two lines through the same points parallel to the axis of x. Hence, by addition, we get

$$x^2 + y^2 - 2p(x \cos \alpha + y \sin \alpha - p) - r^2 = 0, \quad (188)$$

which is evidently a circle passing through the four points in which the pair of lines (3) intersect the pair (4). Hence it has for diameter the intercept made by (2) on (1). See Art. 21, *Cor.* 4.

Examples.

1. Find the equation of the circle whose diameter is the intercept which the circle $x^2 + y^2 - 65 = 0$ makes on $3x + y - 25 = 0$.

Ans. $x^2 + y^2 - 15x - 5y + 60$.

2. Find the condition that the intercept which $x^2 + y^2 - r^2 = 0$ makes on $x \cos \alpha + y \sin \alpha - p = 0$ subtends a right angle at $x'y'$.

Ans. The circle (188) must pass through $x'y'$. Hence the required condition is $x'^2 + y'^2 - 2p(x' \cos \alpha + y' \sin \alpha - p) - r^2 = 0$. (189)

3. Find the condition that the intercept which $x \cos \alpha + y \sin \alpha - p = 0$ makes on $x^2 + y^2 + 2gx + 2fy + c = 0$ subtends a right angle at the origin. Eliminating x and y in succession between these equations, and adding, we get a circle whose diameter is the intercept; and by the given condition this must pass through the origin; therefore the absolute term must vanish. Hence

$$2p^2 + 2p(g \cos \alpha + f \sin \alpha) + c = 0. \quad (189)$$

4. If a variable chord of a circle subtend a right angle at a fixed point $x'y'$, find the locus of the middle point of the chord.

The middle point of the chord is evidently the centre of the circle (188), which has the chord for diameter. If, therefore, XY be the co-ordinates of the middle point, we have

$$X = p\cos\alpha, \quad Y = p\sin\alpha; \text{ therefore } X^2 + Y^2 = p^2;$$

and substituting in the equation (189), we get

$$(X - x')^2 + (Y - y')^2 + X^2 + Y^2 - r^2 = 0. \qquad (190)$$

52. *To find the equation of the tangent to a given circle* $(x-a)^2 + (y-b)^2 = r^2$ *at a given point* $(x'y')$.

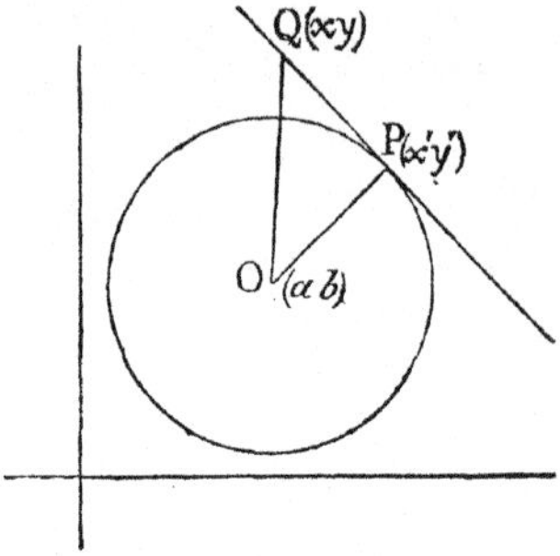

First method.—Let O be the centre, Q any point xy in the tangent. Join OQ; then, since the points (xy), (ab) subtend a right angle at $(x'y')$, we have (Art. 1, Ex. 5), $(x'-x)(x'-a) + (y'-y)(y'-b) = 0$; also, since the point $x'y'$ is on the circle, we have

$$(x'-a)^2 + (y'-b)^2 = r^2.$$

Hence, by subtraction,

$$(x-a)(x'-a) + (y-b)(y'-b) = r^2, \qquad (191)$$

which is the required equation.

Cor.—If the equation of the circle be given in the standard form $x^2 + y^2 = r^2$, the equation of the tangent is

$$xx' + yy' = r^2. \qquad (192)$$

Second method.—Taking the standard form of the equation of the circle, if $x'y'$, $x''y''$ be two points on its circumference, then the equations of the circle described on the join of $x'y'$, $x''y''$ as diameter is $(x-x')(x-x'') + (y-y')(y-y'') = 0$ (Art. 50, Ex. 11); and, subtracting this from the equation of the circle, we get

$$x^2 + y^2 - r^2 - \{(x-x') + (y-y')(y-y'')\} = 0,$$

or $$(x'+x'')x + (y'+y'')y - r^2 - x'x'' - y'y'' = 0, \qquad (193)$$

which (Art. 21, *Cor.* 4) is the equation of the secant through the two points $x'y'$, $x''y''$. Now suppose the points $x'y'$, $x''y''$ to become consecutive, the secant becomes a tangent, and this equation (193) reduces to

$$xx' + yy' - r^2 = 0.$$

Third method.—The polar co-ordinates of $x'y'$, $x''y''$ are $(r\cos\theta', r\sin\theta')$; $(r\cos\theta'', r\sin\theta'')$, and the equation of the join of these points is (Art 22, Ex. 3),

$$x\cos\tfrac{1}{2}(\theta' + \theta'') + y\sin\tfrac{1}{2}(\theta' + \theta'') = r\cos\tfrac{1}{2}(\theta' - \theta'');$$

and if the points be consecutive, this reduces to

$$x\cos\theta' + y\sin\theta' = r, \qquad (194)$$

which is another form of the equation of the tangent.

53. *From any point* (hk) *can be drawn to a circle two tangents, which are either real and distinct, coincident, or imaginary.*

For if $x'y'$ be the point of contact of a tangent from (hk), we get, substituting hk for xy in (192), $hx' + ky' = r^2$. Also, since $x'y'$ is on the circle, $x'^2 + y'^2 = r^2$. Eliminating y', we get

$$(h^2 + k^2)x'^2 - 2r^2hx' + r^4 - k^2r^2 = 0, \qquad \text{(I.)}$$

the discriminant of which is $r^2k^2(h^2 + k^2 - r^2)$; and according as this is positive, zero, or negative, the equation (I.) will be the product of two real and unequal, two equal, or two imaginary factors. Hence the proposition is proved.

54. If we omit the accents in equation (I.), we get

$$(h^2 + k^2)x^2 - 2r^2hx + r^4 - k^2r^2 = 0, \qquad \text{(II.)}$$

which represents two lines parallel to the axis of y, passing through the points of contact of tangents from hk to the circle. In like manner,

$$(h^2 + k^2)y^2 - 2r^2ky + r^4 - k^2r^2 = 4 \qquad \text{(III.)}$$

represents two parallels to the axis of x passing through the same points. Hence, by addition, we get

$$(h^2 + k^2)(x^2 + y^2 - r^2) - 2r^2(hx + ky - r^2) = 0, \qquad (195)$$

which is the equation of the circle whose diameter is the chord of contact of tangents from hk to $x^2+y^2-r^2=0$.

Cor.—If we multiply the equation $x^2+y^2-r^2=0$ by h^2+k^2, and subtract (195) from it, we get $hx+ky-r^2=0$, which is the common chord of the two circles (Art. 21, *Cor.* 4). Hence

$$hx+ky-r^2=0 \tag{196}$$

is the equation of the chord of contact of tangents from (hk). This can be shown otherwise. From the demonstration, Art. 53, we have $hx'+ky'-r^2=0$. In like manner, if $x''y''$ be the second point of contact, we have $hx''+ky''-r^2=0$. Hence the line $hx+ky-r^2=0$ is satisfied by the co-ordinates of each point of contact.

55. *To find the equation of the pair of tangents from* (hk) *to the circle.* On either of the tangents from (hk) to the circle take a point (xy); then twice the area of the triangle formed by the origin and the two points xy, hk, is $hx-ky$, and twice the same area is equal to the distance between the points multiplied by the radius of the circle. Hence

$$(hx-ky)^2=\{(x-h)^2+(y-h)^2\}r^2;$$

or, reducing,

$$(x^2+y^2-r^2)(h^2+k^2-r^2)=(hx+ky-r^2)^2. \tag{197}$$

56. *If* $(x-a)^2+(y-b)^2=r^2$, $(x-a')^2+(y-b')^2=r'^2$ *be the equations of two circles, it is required to find the equations of the chords of contact of common tangents.*

Let $x'y'$ be the point of contact on the first circle, then $(x-a)(x'-a)+(y-b)(y'-b)-r'=0$ is the tangent; and since this touches the second circle, the perpendicular on it from the centre of the second circle must be $=\pm r'$. Hence, remembering that $\sqrt{x'-a)^2+(y'-b)^2}=r$, we get

$$(x'-a)(a'-a)+(y'-b)(b'-b)-r^2\mp rr'=0,$$

the choice of sign depending on whether the common

tangent is direct or transverse. Hence the chords of contact are on—

1st circle,

$$(x-a)(a'-a)+(y-b)(b'-b)-r^2 \mp rr'=0; \quad (198)$$

2nd circle,

$$(x-a')(a-a')+(y-b')(b-b')-r'^2 \mp rr'=0. \quad (199)$$

EXAMPLES.

1. Find the equation, and the length of the common chord, of the two circles

$$(x-a)^2+(y-b)^2=r^2, \quad (x-b)^2+(y-a)^2=r^2.$$

2. Find the conditions that the lines $ax \pm by = 0$ may touch the circle $(x-a)^2+(y-b)^2=r^2$.

3. If tangents be drawn to $x^2+y^2-r^2=0$ from hk, the area of the triangle formed by the tangents and chord of contact is

$$\frac{r(h^2+k^2-r^2)^{\frac{3}{2}}}{h^2+k^2}. \quad (200)$$

4. Two circles whose radii are r, r' intersect at an angle θ; find the length of their common chord.

5. Find the equation of the diameter of $x^2+y^2-6x-2y+8=0$ passing through the origin.

6. Prove that the tangent to $x^2+y^2+2gx+2fy=0$ at the origin is $gx+fy=0$.

7. Prove that if tangents be drawn from the origin to $x^2+y^2+2gx+2fy+c=0$, the chord of contact is $gx+fy+c=0$.

8. If the chord of contact of tangents from a variable point hk subtend a right angle at a fixed point $x'y'$, the locus of hk is the circle

$$(x^2+y^2)(x'^2+y'^2-r^2)-2r^2(xx'+yy'-r^2)=0.$$

9. If R denote the radius of the circle in Ex. 8, δ the distance of its centre from the origin; prove

$$\frac{1}{(R+\delta)^2}+\frac{1}{(R-\delta)^2}=\frac{1}{r^2}.$$

10. PA, PB are two tangents to a circle, whose centre is O; Q any point in AP, QR a perpendicular on the chord of contact AB; prove $AP.AQ = QR.OP$, and thence infer the equation of the pair of tangents from R.

57. Def. i.—*If O be the centre of the circle $x^2 + y^2 - r^2 = 0$, P, Q two points collinear with O, such that the rectangle $OP \, . \, OQ = r^2$; P and Q are called inverse points with respect to the circle.*

Def. ii.—*A perpendicular at either of two inverse points to the line joining it to the centre is called the polar of the other.*

58. The co-ordinates $x'y'$ of a point P being given, it is required to find the co-ordinates of the point inverse to it with respect to the circle $x^2 + y^2 - r = 0$.

Using polar co-ordinates, we have $x' = \rho' \cos\theta'$, $y = \rho' \sin\theta'$, $x'' = \rho'' \cos\theta'$, $y'' = \rho'' \sin\theta'$; and by the condition of inversion, $\rho'\rho'' = r^2$. Hence $\dfrac{x''}{x'} = \dfrac{\rho''}{\rho'} = \dfrac{\rho'\rho''}{\rho'^2} = \dfrac{r^2}{x'^2 + y'^2}$.

Hence
$$x'' = \frac{r^2 x'}{x'^2 + y'^2}. \qquad (201)$$

In like manner
$$y'' = \frac{r^2 y'}{x'^2 + y'^2}. \qquad (202)$$

59. *The polar of the point $x'y'$* is $xx' + yy' - r^2 = 0$. For the equation of the perpendicular through $x''y''$ to the join of $x'y'$ to the centre is, Art. 24, *Cor.* 1,

$$x'(x - x'') + y'(y - y'') = 0;$$

and substituting the values (201), (202) for $x''y''$, we get

$$xx' + yy' - r^2 = 0. \qquad (203)$$

Cor. 1.—The polar of any point on the circumference of the circle is the tangent at that point.

Cor. 2.—The polar of any external point is the chord of contact of tangents drawn from that point.

Examples.

1. Find the equation of the inverse of the line $Ax + By + C = 0$ with respect to $x^2 + y^2 - r^2 = 0$. Substituting for x, y the co-ordinates (201), (202), and omitting accents, we get

$$C(x^2 + y^2) + Ar^2 x + Br^2 y = 0. \qquad (204)$$

2. Find the inverse of the circle $x^2 + y^2 + 2gx + 2fy + c = 0$, with respect to the circle $x^2 + y^2 - r^2 = 0$.

Ans. The circle $c(x^2 + y^2) + 2gr^2 x + 2fr^2 y + r^4 = 0$. (205)

3. Find the equation to the pair of tangents from the origin to

$$x^2 + y^2 + 2gx + 2fy + c = 0.$$

If the line $y = mx$ be a tangent to $x^2 + y^2 + 2gx + 2fy + c = 0$, substituting mx for y, the resulting equation, viz., $x^2(1 + m^2) + 2(g + mf)x + c = 0$ must have equal roots. Hence $(1 + m^2)c = (g + mf)^2$; but $m = \frac{y}{x}$; therefore

$$c(x^2 + y^2) = (gx + fy)^2, \qquad (206)$$

which is the pair of tangents required.

We get the same pair of tangents for the inverse circle $c(x^2 + y^2) + 2gr^2x + 2fr^2g + r^4 = 0$. Hence the pair of direct common tangents drawn to a circle, and to its inverse, passes through the centre of inversion.

4. Find the length of the direct common tangent drawn to the circles

$$x^2 + y^2 + 2gx + 2fy + c = 0, \quad x^2 + y^2 + 2g'x + 2f'y + c' = 0.$$

Ans. If R, R' denote the radii of the circles, the length of their direct common tangent

$$= \sqrt{c + c' - 2gg' - 2ff' + RR'}. \qquad (207)$$

5. The ratio of the square of the common tangent of two circles to the rectangle contained by their radii remains unaltered by inversion.

6. If A, B be any two points, A', B', their inverses with respect to $x^2 + y^2 - r^2 = 0$; prove that if p, p' be the perpendicular distances of the origin from AB, $A'B'$ respectively, $p : p' :: AB, A'B'$.

7. If two points A, B be so related that the polar of A passes through B, the polar of B passes through A. For if the co-ordinates of A be (aa'), and of B (bb'), the polar of A is $ax + a'y = r^2$, and the condition that this should pass through B is $ad' + bb' = r^2$, which, being symmetrical with respect to the co-ordinates of A and B, is also the condition that the polar of B should pass through A.

DEF.—*Two points so related that the polar of either passes through the other are called conjugate points, and their polars conjugate lines.*

8. If a variable point moves along a fixed line, its polar turns round a fixed point.

9. The join of any two points is the polar of the point of intersection of their polars.

10. Two triangles which are such that the angular points of one are the poles of the sides of the other are in perspective.

11. The anharmonic ratio of four collinear points is equal to the anharmonic ratio of the pencil formed by their four polars. For, let $x'y'$, $x''y''$ be two points, and P', P'' their polars; then if the join of $x'y'$, $x''y''$ be divided in two points in the ratios $k : 1$, $k' : 1$, the anharmonic ratio of the four points is $k \div k'$; and since the polars of the point of division are $kP'' + P' = 0$, $k'P'' + P' = 0$, the anharmonic ratio of their four polars is $k \div k'$.

60. *To find the angle of intersection of two given circles.*

DEF.—*The angle between the tangents to any two curves at a point of intersection is called the angle of intersection of the curves at that point.*

Let r, r' be the radii of the given circles, δ the distance between their centres, ϕ their angle of intersection; then, since radii drawn to the point of intersection are perpendicular to the tangents at that point, the angle between the radii is ϕ.

Hence $$\delta^2 = r^2 + r'^2 - 2rr' \cos\phi.$$

Now if the circles be

$$x^2 + y^2 + 2gx + 2fy + c = 0,$$

and $$x^2 + y^2 + 2g'x + 2f'y + c' = 0,$$

we have $\delta^2 = (g - g')^2 + (f - f')^2$, $r^2 = g^2 + f^2 - c^2$, $r'^2 = g'^2 + f'^2 - c'$.

Hence, by substitution, we get

$$c + c' + 2rr' \cos\phi - 2gg' - 2ff' = 0, \qquad (208)$$

which determines the angle ϕ.

Cor. 1.—If the circles cut orthogonally,

$$2gg' + 2ff' - c - c' = 0. \qquad (209)$$

Cor. 2.—If the circles touch,

$$c' \pm 2rr' - 2gg' - 2ff' + c = 0; \qquad (210)$$

the choice of sign being determined by the species of contact.

Cor. 3.—If a circle S cut three circles S', S'', S''' orthogonally, it cuts orthogonally any circle $\lambda S' + \mu S'' + \nu S'''$ expressed linearly in terms of S', S'', S'''. This is proved by writing the equations S', &c. in full, and applying the condition (209).

61. *To find the equation of a circle cutting three given circles* $x^2 + y^2 + 2g'x + 2fy' + c' = 0$, &c. *at given angles* ϕ', ϕ'', ϕ'''.

If we put $2r\cos\phi = k$, the equation (208) may be written $c' + kr' - 2gg' - 2ff' + c = 0$. Hence, if the circle $x^2 + y^2 + 2gx + 2fy + c$ intersect the three given circles at angles ϕ', ϕ'', ϕ''', we have three equations of the form $c' + k'r' - 2gg' - 2ff' + c_1 = 0$, &c., and, eliminating g, f, c between these and $x^2 + y^2 + 2gx + 2fy + c = 0$, we get

$$\begin{vmatrix} x^2+y^2, & -x, & -y, & 1, \\ c'+k'r', & g', & f', & 1, \\ c''+k''r'', & g'', & f'', & 1, \\ c'''+k'''r''', & g''', & f''', & 1 \end{vmatrix} = 0. \qquad (211)$$

If this determinant expanded be written in the form $A(x^2 + y^2) + 2Gx + 2Fy + C = 0$, and r denote the radius of the circle which it represents, we have

$$A^2r^2 = G^2 + F^2 - AC;$$

but the quantities G, F, C each contain r in the first degree. Hence we have a quadratic for determining r, either root of which substituted in the determinant (211) will give a circle cutting the given circles at the given angles.

Cor. 1.—If we suppose $\phi' = \phi'' = \phi''' = \dfrac{\pi}{2}$, we get the equation of the circle cutting the three given circles orthogonally, viz.,

$$\begin{vmatrix} x^2+y^2, & -x, & -y, & 1, \\ c', & g', & f', & 1, \\ c'', & g'', & f'', & 1, \\ c''', & g''', & f''', & 1 \end{vmatrix}. \qquad (212)$$

Cor. 2.—By first putting $\phi' = \phi'' = \phi'''$ = zero, and then $= \pi$, we get the equations of two circles touching the three given circles; or again, taking one or more of the angles ϕ', ϕ'', ϕ''' equal zero, and the remainder equal π, we get in this manner eight tangential circles, all whose equations are included in the form

$$\begin{vmatrix} x^2+y^2, & -x, & -y, & 1, \\ c' \pm rr', & g', & f', & 1, \\ c'' \pm rr'', & g'', & f'', & 1, \\ c''' \pm rr''', & g''', & f''', & 1 \end{vmatrix} = 0; \quad (213)$$

the choice of sign depending on the nature of the contact, the radius in each case being determined as above.

62. If four given circles be cut at given angles ϕ', ϕ'', ϕ''', ϕ'''' by a fifth circle; eliminating g, f, c from four equations of the form (208), we get the equation

$$\begin{vmatrix} c', & g', & f', & 1, \\ c'', & g'', & f'', & 1, \\ c''', & g''', & f''', & 1, \\ c'''', & g'''', & f''', & 1 \end{vmatrix} + r \begin{vmatrix} r' \cos\phi', & g', & f', & 1, \\ r'' \cos\phi'', & g'', & f'', & 1, \\ r''' \cos\phi''', & g''', & f''', & 1, \\ r''' \cos\phi''', & g''', & f'''', & 1 \end{vmatrix} = 0. \quad (214)$$

63. If the four angles ϕ', &c., be right, the second of these determinants vanishes, and the first equated to zero is the condition that one circle may be cut orthogonally by four given circles, viz.:—

$$\begin{vmatrix} c', & g', & f', & 1, \\ c'', & g'', & f'', & 1, \\ c''', & g''', & f''', & 1, \\ c'''', & g''', & f'''', & 1 \end{vmatrix} = 0. \quad (215)$$

Now since c' denotes the square of the tangent drawn from the origin to the first circle, Art. 50, and its minor in this determinant denotes twice the area of the triangle formed by the centres of the three remaining circles, we have the following theorem:—*If A, B, C, D be the centres of four co-orthogonal circles; t, t', t'', t''' tangents drawn to these circles* *from any arbitrary point,* (ABC) *the area of the triangle, whose angular points are* A, B, C, *&c., then*

$$t^2(BCD) - t'^2(CDA) + t''^2(DAB) - t'''^2(ABC) = 0. \qquad (216)$$

64. If xy, $x'y'$, $x''y''$, $x'''y'''$ be four concyclic points, they may be regarded as indefinitely small circles cutting a given circle orthogonally. Hence, substituting in the determinant $x^2 + y^2$ for c', and x, y for $-g'$, $-f'$, &c., we get

$$\begin{vmatrix} x^2 + y^2, & x, & y, & 1, \\ x'^2 + y'^2, & x', & y', & 1, \\ x''^2 + y''^2, & x'', & y'', & 1, \\ x'''^2 + y'''^2, & x''', & y''', & 1 \end{vmatrix} = 0. \qquad (217)$$

And the point xy *being supposed variable, we have the equation of a circle passing through three given points,* $x'y'$, $x''y''$, $x'''y'''$.

This may be shown otherwise as follows:—The determinant (217) evidently represents a circle, for the coefficients of x^2 and y^2 are equal, and the circle passes through the given points; for if in the determinant we substitute $x'y'$ for xy, it will vanish identically, having two rows alike.

65. If $S = 0$ *be the equation of any arbitrary circle,* S', S'', S''' *the powers of the points* $x'y'$, $x''y''$, $x'''y'''$ *with respect to it, then the determinant*

$$\begin{vmatrix} S, & x, & y, & 1, \\ S', & x', & y' & 1, \\ S'', & x'', & y'', & 1, \\ S''', & x''', & y''', & 1 \end{vmatrix} = 0, \qquad (218)$$

will represent a circle passing through the points $x'y'$, $x''y''$, $x'''y'''$. A form analogous to this is very important in Trilinear Co-ordinates.

EXAMPLES.

1. Find the condition that the radius of the circle $\lambda S' + \mu S'' + \nu S''' = 0$ may be zero.

If R denote the radius of $\lambda S' + \mu S'' + \nu S'''$, we have

$$R^2(\lambda+\mu+\nu)^2 = (\lambda g' + \mu g'' + \nu g''')^2 + (\lambda f' + \mu f'' + \nu f''')^2 - (\lambda+\mu+\nu)(\lambda c' + \mu c'' + \nu c''').$$

Hence, if $R = 0$,

$$(\lambda g' + \mu g'' + \nu g''')^2 + (\lambda f' + \mu f'' + \nu f''')^2 - (\lambda+\mu+\nu)(\lambda c' + \mu c'' + \nu c''') = 0.$$

If this be expanded, the coefficient of λ^2 is $g'^2 + f'^2 - c'$, that is r'^2, and the coefficient of $\lambda\mu$ is $2g'g'' + 2f'f'' - c' - c''$, which may be written $-2r'r'' \cos(S'S'')$, equation (208), where $(S'S'')$ denotes the angle of intersection of the circles. Hence, putting $\alpha = \lambda r'$, $\beta = \mu r''$, $\gamma = \nu r'''$, the required condition is

$$\alpha^2 + \beta^2 + \gamma^2 - 2\alpha\beta\cos(S'S'') - 2\beta\gamma\cos(S''S''') - 2\gamma\alpha\cos(S'''S') = 0. \quad (219)$$

2. If two circles be inverted into two others their angle of intersection remains unaltered by inversion.

3. Being given four points in a plane, the area of the triangle formed by any three of them, multiplied by the power of the fourth with respect to the circumcircle of the triangle, gives a constant product.—(STAUDT.)

4. If A_1, A_2, A_3, A_4; B_1, B_2, B_3, B_4 be two systems of four points, and if d_{rs} denote the distance A_rB_s, then

$$\begin{vmatrix} 0, & 1, & 1, & 1, & 1, \\ 1, & d_{11}^2, & d_{12}^2, & d_{13}^2, & d_{14}^2, \\ 1, & d_{21}^2, & d_{22}^2, & d_{23}^2, & d_{24}^2, \\ 1, & d_{31}^2, & d_{32}^2, & d_{33}^2, & d_{34}^2, \\ 1, & d_{41}^2, & d_{42}^2, & d_{43}^2, & d_{44}^2 \end{vmatrix} = 0.$$

(NEUBERG.)

5. In the same case, if the points A_1, A_2, A_3, A_4 be concyclic,

$$\begin{vmatrix} d_{11}^2, & d_{12}^2, & d_{13}^2, & d_{14}^2, \\ d_{21}^2, & d_{22}^2, & d_{23}^2, & d_{24}^2, \\ d_{31}^2, & d_{32}^2, & d_{33}^2, & d_{34}^2, \\ d_{41}^2, & d_{42}^2, & d_{43}^2, & d_{44}^2 \end{vmatrix} = 0.$$

66. DEF.—*If* $S = 0$, $S' = 0$ *denote two circles, the pencil**

* A system of curves of any order, passing through a number of points which is one less than the number required to determine a proper curve of that order, is called a pencil of curves.

$S - kS' = 0$, *where* k *receives all values from* $+\infty$ *to* $-\infty$ *is called a coaxal system.*

67. *One of the circles of a coaxal system is infinitely large, and two infinitely small.* For, let

$$S \equiv x^2 + y^2 + 2gx + 2fy + c = 0, \quad S' \equiv x^2 + y^2 + 2g'x + 2f'y + c';$$

then

$$S - kS' \equiv (1+k)(x^2+y^2) + 2(g-kg')x + 2(f-kf')y + c - kc' = 0 \quad (220)$$

is the general circle of the system. Now, in the special case where $k = 1$, this circle reduces to

$$S - S' \equiv 2(g-g')x + 2(f-f')y + c - c' = 0, \quad (221)$$

which represents a line that is an infinitely large circle. *This line is called the* RADICAL AXIS *of the coaxal system.*

Again, if R denote the radius of $S - kS'$, we have

$$R^2 = \frac{(g-kg')^2 + (f-kf')^2 - (1-k)(c-kc')}{(1-k)^2}.$$

Now, if $S - kS' = 0$ reduce to a point circle, $R = 0$;

hence $\quad (g-kg')^2 + (f-kf')^2 - (1-k)(c-kc') = 0,$

or $\quad (g^2+f^2-c) + k(c+c'-2gg'-2ff') + k^2(g'^2+f'^2-c'=0), \quad (222)$

which is a quadratic in k. If the roots be k_1, k_2, the circles $S - k_1S' = 0$, $S - k_2S' = 0$ reduce to points. *These are called the limiting points of the system.* Hence the proposition is proved.

68. *The limiting points of the coaxal system* $S - kS' = 0$ *are real when the circles* S, S' *do not intersect, and imaginary when they do.*

The roots of the equation (222) will be real if

$4(g^2+f^2-c)(g'^2+f'^2-c')$ be less than $(c+c'-2gg'-2ff')^2$,

or if $\quad 4r^2r'^2$ be less than $(c+c'-2gg'-2ff')^2$;

but $\quad r^2 + r'^2 = g^2 + f^2 - c + g'^2 + f'^2 - c'.$

Hence the roots will be real if

$(r+r')^2$ be greater than δ^2,

or $\quad (r-r')$ be less than δ^2,

where δ is the distance between the centres of S, S', that is, the roots are real when the circles do not intersect. Again, if ϕ be the angle of intersection of S, S', the equation (222) may be written

$$r^2 - 2krr' \cos\phi + k^2 r'^2 = 0;$$

therefore $$kr' = r(\cos\phi \pm \sin\phi\sqrt{-1}). \tag{223}$$

Hence the values of k are imaginary when ϕ is real, and the proposition is proved.

69. *A coaxal system may be expressed linearly in terms of any two circles of the system* $S - k'S = 0$.

For, let $S - lS' \equiv (1 - l)\sigma$, $S - mS' \equiv (1 - m)\sigma'$; then S, S' can be expressed in terms of σ and σ', and if l, m be given, σ, σ' are given. Hence $S - kS'$ can be expressed in terms of two given circles σ, σ': k will be the only variable parameter, and it will be in the first degree.

Cor. 1.—If σ, σ' be the limiting points, and k a variable parameter, then the equation $\sigma - k\sigma' = 0$ represents the coaxal system.

Cor. 2.—Similarly, if $L = 0$ denote the radical axis, any circle of the system may be expressed in the form $S - kL = 0$. Thus $x^2 + y^2 \pm d^2 - 2kx = 0$ denotes a coaxal system, having $x = 0$ for the radical axis, and real or imaginary limiting points, according as the sign of d^2 is *plus* or *minus*.

Examples.

1. The radical axes of any three circles are concurrent.

For if S, S', S'' be the circles, then (Art. 67) the radical axes are $S - S' = 0$, $S' - S'' = 0$, $S'' - S = 0$, which, added, vanish identically.

2. Tangents from any point on a fixed circle of a coaxal system to two other fixed circles of the system are in a given ratio.

For let tangents be drawn from any point P of the circle $S - k^2S' = 0$ to the circles S, S'; then denoting these tangents by t, t', we have, since the power of P with respect to $S - k^2S'$ is zero,

$$t^2 - k^2t'^2 = 0.$$

Hence $t : t' :: k : 1$, that is, in a given ratio. See also Ex. 9, p. 73.

The following are special cases :—

1°. *Tangents from any point in the radical axis to all the circles of the system are equal to one another. For in this case $k = 1$. Hence $t = t'$.*

2°. *The distances from any point of a fixed circle of the system to the two limiting points are in a given ratio.*

3. The limiting points are harmonic conjugates to the extremities collinear with them of the diameter of any circle of the system; because the ratio of the distances of the limiting points from one extremity is equal to the ratio of their distances from the other extremity of the diameter.

4. The limiting points are inverse points with respect to each circle.

5. The distance of any point in a given circle of a coaxal system from the radical axis is proportional to the square of the tangent from the same point to any other given circle of the system.

This follows from the equation $S - kL = 0$.

6. Any two circles and their circle of inversion are coaxal.

For the inverse of $x^2 + y^2 + 2gx + 2fy + c = 0$, with respect to $x^2 + y^2 - r^2 = 0$, is $c(x^2 + y^2) + 2gr^2x + 2fr^2y + r^4 = 0$; and the first, multiplied by r^2 and subtracted from the last, gives $(c - r^2)(x^2 + y^2 - r^2) = 0$.

7. The polars of any point with respect to the circles of a coaxal system are concurrent.

For if P, P' be the polars of the point with respect to S, S', its polar with respect to $S - kS'$ is $P - kP' = 0$, a line passing through the intersection of P, P'.

DEF.—*The* RADICAL CENTRE *of three given circles is the point of concurrence of their radical axes.*

8. The radical centre of three given circles is the centre of a circle, cutting them orthogonally.

9. The inverse of a coaxal system is a coaxal system.

For the inverse of $S - kS'$ is of the same form.

10. The inverse of a system of concurrent lines is a coaxal system of circles.

11. The inverse of a system of concentric circles is a coaxal system, of which the centre of inversion is one of the limiting points.

For the inverse of $(x - a)^2 + (y - b)^2 - R^2 = 0$ with respect to $x^2 + y^2 - r^2$ is $S - R^2S' = 0$, where $S \equiv (a^2 + b^2)(x^2 + y^2) - 2ar^2x - 2br^2y + r^4$, $S' \equiv x^2 + y^2$. Hence $S = 0$, $S' = 0$, are point circles.

12. A coaxal system having real limiting points is the inverse of a concentric system, and a system having imaginary limiting points the inverse of a pencil of lines.

13. If a variable circle cut two given circles of a coaxal system at given angles, it cuts every circle of the system at a constant angle. This may be seen at once by inversion : or without inversion, as follows :—If $S \equiv x^2 + y^2 + 2gx + 2fy + c = 0$ cuts $S' \equiv x^2 + y^2 + 2g'x + 2f'y + c' = 0$ and $S'' \equiv x^2 + y^2 + 2g''x + 2f''y + c'' = 0$ at angles ϕ', ϕ'', it cuts the circle $S' - kS'' = 0$ at the angle

$$\cos^{-1}\left\{\frac{r' \cos \phi' - r'' \cos \phi''}{R\,(1 - k)}\right\},$$

where R denotes the radius of $S' - kS'' = 0$.

14. The radical axes of the circles of a coaxal system and a circle which is not one of the system are concurrent.

15. The circles $x^2 + y^2 - 2hx + b^2 = 0$, $x^2 + y^2 - 2ky - b^2 = 0$, cut orthogonally.

DEF.—*The two points which divide the distance between the centres of two circles internally and externally in the ratio of their radii are called the centres of similitude of the circles.*

Thus if $x^2 + y^2 + 2gx + 2fy + c = 0$, $x^2 + y^2 + 2g'x + 2f'y + c' = 0$ be two circles, their centres of similitude are—

internal, the point, $$\left\{\frac{-gr' + g'r}{r + r'},\quad \frac{-fr' + f'r}{r + r'}\right\};$$

and external, $$\left\{\frac{-(g'r - gr')}{r - r'},\quad \frac{-(f'r - fr')}{r - r'}\right\}.$$

16. If S, S' be two circles whose radii are r, r'; prove that their internal centre of similitude is the centre of $\frac{S}{r} + \frac{S'}{r'} = 0$, and the external one, the centre of $\frac{S}{r} - \frac{S'}{r'} = 0$.

17. If S, S' be two circles, $\frac{S}{r} \pm \frac{S'}{r'} = 0$ will invert one into the other. In what respect do these inversions differ ?

18. If S, S' be two circles, the circle described on the distance between their centres of similitude as diameter is $\frac{S}{r^2} - \frac{S'}{r'^2} = 0$. This is called their circle of similitude.

19. Given any three circles, taking them two by two they have three circles of similitude ; prove that these circles are coaxal.

20. Given any three circles S', S'', S''', their six centres of similitude lie three by three on four right lines.

For if r', r'', r''' be the radii of the circles, the three external centres of similitude are the centres of the three circles,

$$\frac{S'}{r'} - \frac{S''}{r''} = 0, \quad \frac{S''}{r''} - \frac{S'''}{r'''} = 0, \quad \frac{S'''}{r'''} - \frac{S'}{r'} = 0;$$

that is, they are the centres of three coaxal circles. Hence they are collinear. In like manner, it may be proved that any two internal centres of similitude are collinear with one of the external centres of similitude.

21. If the three given circles be $x^2 + y^2 + 2g'x + 2f'y + c' = 0$, &c., the equations of the four axes of similitude are—

$$\begin{vmatrix} 0, & -x, & -y, & 1, \\ \pm r', & g', & f', & 1, \\ \pm r'', & g'', & f'', & 1, \\ \pm r''', & g''', & f''', & 1 \end{vmatrix} = 0. \quad (224)$$

Where the choice of signs in the first column is thus determined for the external axis of similitude the signs are all positive, and for each of the others, two are positive and one negative.

22. If a variable circle touch two fixed circles, the chord of contact passes through one of the centres of similitude of the two fixed circles.

23. In the same case the variable circle is cut orthogonally by one of the two circles of inversion of the fixed circles.

24. A system of circles cutting three given circles isogonally are coaxal, their radical axis being one of the axes of similitude of the three given circles.

*SECTION II.—A SYSTEM OF TANGENTIAL CIRCLES.

70. *To find the equations of the circles in pairs, touching three given circles.*

This depends on the following theorem, which is an extension of Ptolemy's theorem. (See *Sequel to Euclid*, p. 103):—

If S_1, S_2, S_3, S_4 *be four circles which have a common tangential circle* Ω, *and if the length of the common tangent to* S_1, S_2 *be denoted by* $\overline{12}$, *then*

$$\overline{23} \cdot \overline{14} + \overline{31} \cdot \overline{24} + \overline{12} \cdot \overline{34} = 0. \quad (225)$$

In this notation it is to be observed that the common tangent $\overline{31}$, in which the numerical order is transposed, is negative. In order to apply equation (225), suppose the circle S_4 to reduce to a point. In this case the common tangents $\overline{14}$, $\overline{24}$, $\overline{34}$ will be the square roots of the power of that point with respect to S_1, S_2, S_3; and may therefore be denoted by $\sqrt{S_1}$, $\sqrt{S_2}$, $\sqrt{S_3}$, respectively; and since the point to which S_4 reduces may be any point on the circumference of Ω, we have, for any point on that circle,

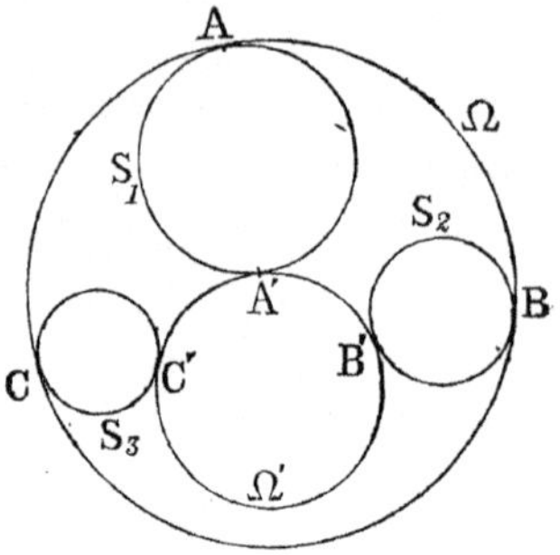

$$\overline{23}\sqrt{S_1} + \overline{31}\sqrt{S_2} + \overline{12}\sqrt{S_3} = 0;$$

or, denoting

$\overline{23}$, $\overline{31}$, $\overline{12}$ by $\sqrt{l}$, $\sqrt{m}$, $\sqrt{n}$, respectively,

$$\sqrt{lS_1} + \sqrt{mS_2} + \sqrt{nS_3} = 0; \qquad (226)$$

this, cleared of radicals, becomes

$$l^2S_1^2 + m^2S_2^2 + n^2S_3 - 2lmS_1S_2 - 2mnS_2S_3 - 2nlS_3S_1 = 0. \qquad (227)$$

Now if we substitute for S_1, S_2, S_3 their full expressions in x and y co-ordinates, the equation (227) will be of the fourth degree; it must, therefore, be the equation of a pair of circles tangential to S_1, S_2, S_3, as in the annexed diagram.

DEF.—*The equation* (226) *is called the* NORM *of* (227).

71. Since the points A, A' are common to $\Omega\Omega'$ and S_1, and since if in the equation (227) of $\Omega\Omega'$ we make $S_1 = 0$, we get $(mS_2 - nS_3)^2 = 0$, the circle $mS_2 - nS_3 = 0$ passes through the points A, A'; therefore the line AA' is the radical axis of S_1 and $mS_2 - nS_3$. Hence its equation is

$$(m - n) S_1 - (mS_2 - nS_3) = 0.$$

For this denotes a line, namely,

$$m(S_1 - S_2) - n(S_1 - S_3) = 0.$$

Now $S_1 - S_2 = 0$ is the radical axis of S_1, S_2; and $S_1 - S_3 = 0$ is the radical axis of S_1, S_3; denoting these by A_3, A_2, we have $mA_3 - nA_2 = 0$ as the equation of AA'. Therefore the equations of the three chords AA', BB', CC' may be written

$$\frac{A_1}{l} = \frac{A_2}{m} = \frac{A_3}{n}. \tag{228}$$

This theorem gives a new method of describing a circle touching three given circles. For drawing the three lines (228), the two triads of points A, B, C; A', B', C' are determined.

72. If the lengths of the transverse common tangents to S_1, S_2, S_3 be denoted by $\sqrt{l'}$, $\sqrt{m'}$, $\sqrt{n'}$, respectively, the norms of the other three pairs of tangentical circles will be—

$$\sqrt{lS_1} + \sqrt{m'S_2} + \sqrt{n'S_3} = 0. \tag{229}$$

$$\sqrt{l'S_1} + \sqrt{mS_2} + \sqrt{n'S_3} = 0. \tag{230}$$

$$\sqrt{l'S_1} + \sqrt{m'S_2} + \sqrt{nS_3} = 0. \tag{231}$$

73. If we denote the angles of intersection of the circles thus: $(\widehat{S_2S_3})$ by A, $(\widehat{S_3S_1})$ by B, and $(\widehat{S_1S_2})$ by C, we have

$$2\cos\tfrac{1}{2}A = \sqrt{\frac{l}{r_2r_3}}; \quad 2\sin\tfrac{1}{2}A = \sqrt{\frac{-l'}{r_2r_3}}, \text{ \&c.}$$

Hence the norm (226) may be written

$$\cos\tfrac{1}{2}A\sqrt{\frac{S_1}{r_1}} + \cos\tfrac{1}{2}B\sqrt{\frac{S_2}{r_2}} + \cos\tfrac{1}{2}C\sqrt{\frac{S_3}{r_3}} = 0; \tag{232}$$

and expanded, this may be written in determinant form:

$$\begin{vmatrix} 0, & \cos^2\frac{1}{2}C, & \cos^2\frac{1}{2}B, & \frac{S_1}{r_1}, \\ \cos\frac{1}{2}c, & 0, & \cos^2\frac{1}{2}A, & \frac{S_2}{r_2}, \\ \cos^2\frac{1}{2}B, & \cos^2\frac{1}{2}A, & 0, & \frac{S_3}{r_3}, \\ \frac{S'}{r_1}, & \frac{S_2}{r_2}, & \frac{S_3}{3}, & 0 \end{vmatrix} = 0, \tag{233}$$

and similarly for the others.

EXAMPLES.

1. The poles of the chords AA', BB', CC', with respect to the circles S_1, S_2, S_3, are collinear, their line of collinearity being the radical axis of Ω, Ω'.

2. The radical axis of Ω, Ω' is the external axis of similitude of S_1, S_2, S_3.

3. The circle which cuts S_1, S_2, S_3 orthogonally inverts Ω into Ω'.

4. If the join of the points A, B, fig. Art. 70, intersect the circles S_1, S_2 in the points D, E, respectively, prove that the rectangle $AE \cdot DB$ is = to the square of the common tangent of S_1, S_2, and thence prove the theorem of Art. 71.

5. Prove in the same manner the extension of Ptolemy's theorem, equation (225). See Euclid, page 262, second edition.

6. If Σ be the orthogonal circle of S_1, S_2, S_3, the radical axis of Σ and S_1 meets the radical axis of Ω and Ω' in the pole of AA' with respect to S_1.

7. The circles Ω, Ω' are tangential to the three circles

$$lS_1 - 2mS_2 - 2nS_3 = 0, \quad mS_2 - 2nS_3 - 2lS_1 = 0, \quad nS_3 - 2lS_1 - 2mS_2 = 0.$$

8. The three systems of points A, A', B, B'; B, B', C, C'; C, C', A, A' are concyclic, the circles through them being respectively

$$lS_1 + mS_2 - nS_3 = 0, \quad mS_2 + nS_3 - lS_1 = 0, \quad nS_3 + lS_1 - mS_2 = 0.$$

74. *To investigate the general condition that any number of circles may have one common tangential circle.*

LEMMAS.—If $f(x) = 0$ be an algebraic equation of the nth degree, whose roots, taken in order of magnitude, are $a, b, c \ldots l$, then

$$1°. \quad \frac{a-b}{(x-a)(x-b)} + \frac{b-c}{(x-b)(x-c)} + \ldots \frac{(l-a)}{(x-l)(x-a)} = 0. \quad (234)$$

$$2°. \quad \frac{a^{n-2}}{f'(a)} + \frac{b^{n-2}}{f'(b)} + \ldots \frac{l^{n-2}}{f'(l)} = 0. \quad (235)$$

Lemma 1° may be proved by dividing each fraction into the difference of two partial fractions. Lemma 2° is well known to those acquainted with the theory of equations.

When $n = 4$, which is the only case in which we shall use this lemma here, it may be stated thus:—If a, b, c, d be any four quantities, then

$$\frac{a^2}{(a-b)(a-c)(a-d)} + \frac{b^2}{(b-a)(b-c)(b-d)} + \frac{c^2}{(c-a)(c-b)(c-d)} + \frac{d^2}{(d-a)(d-b)(d-c)} = 0. \qquad (236)$$

75. If O be the origin, and $A, B, C \ldots, L$ any number of fixed points on a right line passing through O; X any variable point on the same line; then, if $OA, OB, OC, \ldots OL, OX$ be denoted by $a, b, c, \ldots l, x$, we have, from lemma 1°,

$$\frac{AB}{AX.BX} + \frac{BC}{BX.CX} + \ldots \frac{LA}{LX.AX} = 0. \qquad (237)$$

Now, if circles whose diameters are $\delta_a, \delta_b, \delta_c, \ldots \delta_l, \delta_x$ touch the line OX at the points $A, B, C, \ldots L, X$, then from (237) we get

$$\frac{AB}{\sqrt{\delta_a . \delta_b}} \div \frac{AX.BX}{\sqrt{\delta_a . \delta_x . \delta_b . \delta_x}} + \frac{BC}{\sqrt{\delta_b . \delta_c}} \div \frac{BX.CX}{\sqrt{\delta_b . \delta_x . \delta_c . \delta_x}} + \ldots \frac{LA}{\sqrt{\delta_l . \delta_a}} \div \frac{LX.AX}{\sqrt{\delta_l . \delta_x . \delta_a . \delta_x}} = 0.$$

Then, inverting from any arbitrary point, since, Art. 59, Ex. 5, the square of the common tangent of any two circles divided by the rectangle contained by their diameters remains unaltered by inversion, we have, after omitting common factors, the following general theorem:—*If a circle Ω touch any number of circles $S_1, S_2, \ldots S_l, S_x$, and if common tangents be denoted by $\overline{12}$, &c., then*

$$\frac{\overline{12}}{\overline{1x}.\overline{2x}} + \frac{\overline{23}}{\overline{2x}.\overline{3x}} + \ldots \frac{\overline{l1}}{\overline{lx}.\overline{1x}} = 0. \qquad (238)$$

76. If S_x reduce to a point, this will be a point on the circle Ω, and $\overline{1x}$, $\overline{2x}$, $\overline{3x}$, &c., may be replaced by $\sqrt{S_1}$, $\sqrt{S_2}$, $\sqrt{S_3}$, &c. Hence we have the following theorem:—*If a circle Ω be touched by any number of circles $S_1, S_2, S_3, \ldots$, the equation of Ω will be contained as a factor in the equation*

$$\frac{\overline{12}}{\sqrt{S_1 S_2}} + \frac{\overline{23}}{\sqrt{S_2 S_3}} + \frac{\overline{34}}{\sqrt{S_3 S_4}} + \text{\&c.} = 0. \tag{239}$$

Cor. 1.—If there be only three tangential circles this equation reduces to equation (226), Art. 70.

77. From lemma 2°, supposing $f(x)$ to be of the fourth degree, we get in the same manner the following theorem:—

If a circle Ω be tangential to five circles S_0, S_1, S_2, S_3, S_4, then

$$\frac{\overline{01}^2}{\overline{12} \cdot \overline{13} \cdot \overline{14}} + \frac{\overline{02}^2}{\overline{12} \cdot \overline{23} \cdot \overline{24}} + \frac{\overline{03}^2}{\overline{13} \cdot \overline{23} \cdot \overline{34}} + \frac{\overline{04}^2}{\overline{14} \cdot \overline{24} \cdot \overline{34}} = 0;$$

and supposing S_0 to reduce to a point, and denoting by $P(1)$ the product of all the common tangents from S_1 to all the other circles, then

$$\frac{S_1}{P(1)} + \frac{S_2}{P(2)} + \frac{S_3}{P(3)} + \frac{S_4}{P(4)} = 0. \tag{240}$$

Examples.

1. The circle through the middle points of the sides of a triangle touches both the inscribed and the escribed circles. For, let S_1, S_2, S_3 denote the middle points of the sides, S_x one of the circles touching the sides, say the inscribed circle; then $\overline{1x}$, $\overline{2x}$, $\overline{3x}$ are equal to $\frac{1}{2}(b - c)$, $\frac{1}{2}(c - a)$, $\frac{1}{2}(a - b)$ respectively, and $\overline{12}$, $\overline{23}$, $\overline{31}$, equal to $\frac{1}{2}c$, $\frac{1}{2}a$, $\frac{1}{2}b$, and these substituted in the equation

$$\frac{\overline{12}}{\overline{1x} \cdot \overline{2x}} + \frac{\overline{23}}{\overline{2x} \cdot \overline{3x}} + \frac{\overline{31}}{\overline{3x} \cdot \overline{1x}} = 0,$$

it vanishes identically.

2. The circle through the middle points of the sides passes through the feet of the perpendiculars. For, taking S_1, S_2, S_3, as in Ex. 1, and S_x the foot of the perpendicular on the side a, then

$$\overline{1x} = b \cos C - \tfrac{1}{2}a, \quad \overline{2x} = -\tfrac{1}{2}b, \quad \overline{3x} = \tfrac{1}{2}c,$$

and substituting as before.

3. If S_1, S_2, S_3, S_4 be the inscribed and escribed circles, then, Ex. 1, they have a common tangential circle Ω (called the 'Nine-points Circle'). Its equation in terms of these four circles is

$$\frac{S_1}{(a-b)(b-c)(c-a)} + \frac{S_2}{(a+b)(b-c)(c+a)} + \frac{S_3}{(a+b)(b+c)(c-a)} + \frac{S_4}{(a-b)(b+c)(c+a)} = 0. \qquad (241)$$

4. The equation (239) may be written thus:

$$\frac{\cos\frac{1}{2}(12)\sqrt{r_1 r_2}}{\sqrt{S_1 S_2}} + \frac{\cos\frac{1}{2}(23)\sqrt{r_2 r_3}}{\sqrt{S_2 S_3}} + \ldots \frac{\cos\frac{1}{2}(l1)\sqrt{r_l r_1}}{\sqrt{S_l S_1}} = 0. \qquad (242)$$

5. If a circle Ω touch four circles whose radii are $r_1 \ldots r_4$, then

$$\Omega = \frac{S_1}{r_1 \cos\frac{1}{2}(12)\cos\frac{1}{2}(13)\cos\frac{1}{2}(14)} + \frac{S_2}{r_2 \cos\frac{1}{2}(21)\cos\frac{1}{2}(23)\cos\frac{1}{2}(24)}$$
$$+ \frac{S_3}{r_3 \cos\frac{1}{2}(31)\cos\frac{1}{2}(32)\cos\frac{1}{2}(34)} + \frac{S_4}{r_4 \cos\frac{1}{2}(41)\cos\frac{1}{2}(42)\cos\frac{1}{2}(43)}$$

6. If S be a circle, O a point, and OPQ a line through O and the centre of S, meeting the circumference in P and Q, then we have $\dfrac{S}{2r} = \dfrac{OP \cdot OQ}{PQ}$.

Hence if S open out into a right line, $\dfrac{S}{2r}$ becomes equal to OQ; that is, equal to the perpendicular from O on the right line, into which S opens out. By means of this principle we can express the equations of the escribed and inscribed circles in terms of the sides of the triangle of reference and the 'Nine-points Circle.' Thus, in Ex. 5, let S_1, S_2, S_3 be the sides α, β, γ of the triangle of reference, S_4 the 'Nine-points Circle;' then, denoting the angles of intersection of the sides with S_4 by A, B, C, respectively, the equation of the inscribed circle is

$$\frac{2}{\cos\frac{1}{2}A\cos\frac{1}{2}B\cos\frac{1}{2}C}\left\{\frac{\alpha\cos\frac{1}{2}A}{\sin\frac{1}{2}A_1} + \frac{\beta\cos\frac{1}{2}B}{\sin\frac{1}{2}B_1} + \frac{\gamma\cos\frac{1}{2}C}{\sin\frac{1}{2}C_1}\right\}$$
$$+ \frac{S_4}{r_4 \sin\frac{1}{2}A' \sin\frac{1}{2}B' \sin\frac{1}{2}C} = 0. \qquad (243)$$

7. The tangent to the 'Nine-points Circle' at its point of contact with the inscribed circle is

$$\frac{a\alpha}{b-c} + \frac{b\beta}{c-a} + \frac{c\gamma}{a-b} = 0. \qquad (244)$$

For $$\frac{\cos\frac{1}{2}A}{\sin\frac{1}{2}A'} = \frac{\cos\frac{1}{2}A}{\sin\frac{1}{2}(B-C)} = \frac{a}{b-c}, \text{ \&c.}$$

*SECTION III.—TRILINEAR CO-ORDINATES.

78. *To find the equation of the circumcircle of the triangle of reference.*

Let A', B', C' be three collinear points, then we have

$$B'C' + C'A' + A'B' = 0.$$

Hence, if p denote the perpendicular from any point O on the line $A'C'$,

$$\frac{B'C'}{p} + \frac{C'A'}{p} + \frac{A'B'}{p} = 0.$$

Therefore, inverting from the point O, and denoting the inverses of A', B', C', by A, B, C, and the perpendiculars from O on the liues BC, CA, AB by α, β, γ, we have (Art. 59, Ex. 6),

$$\frac{B'C'}{p} = \frac{BC}{\alpha}, \text{ \&c.};$$

therefore $$\frac{BC}{\alpha} + \frac{CA}{\beta} + \frac{AB}{\gamma} = 0;$$

or, denoting the lengths of the sides of the triangle ABC by a, b, c, and calling it the triangle of reference, we have the equation of its circumcircle, viz..

$$\frac{a}{\alpha} + \frac{b}{\beta} + \frac{c}{\gamma} = 0, \qquad (245)$$

or
$$\frac{\sin A}{\alpha} + \frac{\sin B}{\beta} + \frac{\sin C}{\gamma} = 0. \qquad (246)$$

Or thus :—If in the general equation of the second degree,

$$a\alpha^2 + b\beta^2 + c\gamma^2 + 2h\alpha\beta + 2f\beta\gamma + 2g\gamma\alpha = 0,$$

the coefficient of α^2 vanish, the resulting equation will denote a curve passing through the point A, for it will be satisfied by the co-ordinates $\beta = 0$, $\gamma = 0$. Similarly, if the coefficients of β^2, γ^2 vanish, the curve will pass through the points B, C. Hence the equation $l\beta\gamma + m\gamma\alpha + n\alpha\beta = 0$ denotes a curve of the second degree circumscribing the triangle of reference. In order to find the conditions that it should be a circle, transforming it into Cartesian co-ordinates, equating the coefficients of x^2 and y^2, and putting the coefficients of $xy = 0$, we get,

$$l\cos(\beta+\gamma) + m\cos(\gamma+\alpha) + n\cos(\alpha+\beta) = 0,$$

$$l\sin(\beta+\gamma) + m\sin(\gamma+\alpha) + n\sin(\alpha+\beta) = 0.$$

Hence, eliminating l, m, n, we get

$$\begin{vmatrix} \beta\gamma, & \gamma\alpha, & \alpha\beta, \\ \cos(\beta+\gamma), & \cos(\gamma+\alpha), & \cos(\alpha+\beta), \\ \sin(\alpha+\beta), & \sin(\gamma+\alpha), & \sin(\alpha+\beta) \end{vmatrix} = 0;$$

or
$$\beta\gamma\sin A + \gamma\alpha\sin B + \alpha\beta\sin C = 0;$$

the same equation as before.

79. It may be proved exactly as in the first method, Art. 78, *that if a polygon of any number of sides be inscribed in a circle, and if the lengths of these sides be a, b, c, d, &c., and their standard equations* $\alpha = 0$, $\beta = 0$, $\gamma = 0$, *&c., then for any point in the circle*

$$\frac{a}{\alpha} + \frac{b}{\beta} + \frac{c}{\gamma} + \frac{d}{\delta} + \&c. = 0. \qquad (247)$$

80. *To find the equations of the tangents to the circumcircle at the angular points of the triangle of reference.*

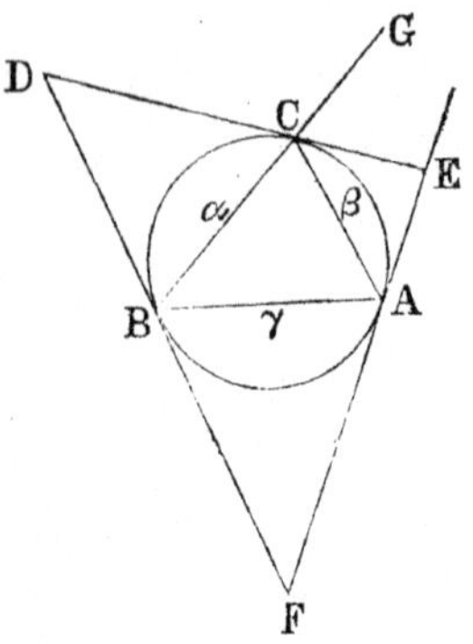

Let CE be the tangent. Produce BC to G; then, since CE passes through the intersection of α and β, its equation is of the form $l\alpha + m\beta = 0$, but the angles into which CE divides the angle GCA are (Euc. III., xxxii.) equal to A and B, respectively. Hence l, m are inversely proportional to $\sin A$, $\sin B$, or to a and b.

Hence, the equation of CE is

$$\frac{\alpha}{a} + \frac{\beta}{b} = 0. \tag{248}$$

Similarly the tangent at A is $\frac{\beta}{a} + \frac{\gamma}{c} = 0$; (249)

and the tangent at B is $\frac{\gamma}{c} + \frac{\alpha}{a} = 0.$ (250)

81. From these equations we have the following inferences:—

1°. *The triangles DEF, ABC are in perspective.* For, subtracting (248) from (249) we get $\frac{\gamma}{c} - \frac{\alpha}{a} = 0$, which is evidently the equation of BE. Similarly the equations CF, AD are $\frac{\alpha}{a} - \frac{\beta}{b} = 0$, $\frac{\beta}{b} - \frac{\gamma}{c} = 0$. Hence the three lines AD, BE, CE are concurrent; therefore the triangles are in perspective.

2°. *The axis of perspective is*

$$\frac{\alpha}{a} + \frac{\beta}{b} + \frac{\gamma}{c} = 0; \tag{251}$$

for this evidently passes through the intersections of (248) and γ, (249) and α, (250) and β.

3°. *The co-ordinates of the centre of perspective are a, b, c;* for these satisfy the equations of the lines AD, BE, CF. Hence the centre of perspective of the triangles ABC, DEF is the symmedian point of ABC.

82. *The chord joining two points* $\alpha'\beta'\gamma'$, $\alpha''\beta''\gamma''$, *is*

$$\frac{a\alpha}{\alpha'\alpha''} + \frac{b\beta}{\beta'\beta''} + \frac{c\gamma}{\gamma'\gamma''} = 0; \tag{252}$$

for, since the points are on the circle, we have

$$\frac{a}{\alpha'} + \frac{b}{\beta'} + \frac{c}{\gamma'} = 0,$$

$$\frac{a}{\alpha''} + \frac{b}{\beta''} + \frac{c}{\gamma''} = 0.$$

Hence

$$a = k\begin{vmatrix} \frac{1}{\beta'}, & \frac{1}{\gamma'} \\ \frac{1}{\beta''}, & \frac{1}{\gamma''} \end{vmatrix}, \quad b = k\begin{vmatrix} \frac{1}{\gamma'}, & \frac{1}{\alpha'} \\ \frac{1}{\gamma''}, & \frac{1}{\alpha''} \end{vmatrix}, \quad c = k\begin{vmatrix} \frac{1}{\alpha'}, & \frac{1}{\beta'} \\ \frac{1}{\alpha''}, & \frac{1}{\beta''} \end{vmatrix},$$

where k is any multiple. Therefore

$$\beta'\gamma'' - \beta''\gamma' = -\frac{a\beta'\beta''\gamma'\gamma''}{k}, \text{ \&c.};$$

and substituting in

$$\alpha(\beta'\gamma'' - \beta''\gamma') + \beta(\gamma'\alpha'' - \gamma''\alpha') + \gamma(\alpha'\beta'' - \alpha''\beta') = 0,$$

which is the equation of the join of the points $\alpha'\beta'\gamma'$, $\alpha''\beta''\gamma''$, we get

$$\frac{a\alpha}{\alpha'\alpha''} + \frac{b\beta}{\beta'\beta''} + \frac{c\gamma}{\gamma'\gamma''} = 0.$$

Cor.—Hence it follows that the tangent at the point $\alpha'\beta'\gamma'$ is

$$\frac{a\alpha}{\alpha'^2} + \frac{b\beta}{\beta'^2} + \frac{c\gamma}{\gamma'^2} = 0. \tag{253}$$

82. *To find the equation of the circle inscribed in the triangle of reference.*

The general equation of the second degree, viz., $a\alpha^2 + b\beta^2 + c\gamma^2 + 2h\alpha\beta + 2f\beta\gamma + 2g\gamma\alpha = 0$, represents a curve of the second degree cutting each side of the triangle of reference in two points; thus, if we make $\gamma = 0$, we get $a\alpha^2 + 2h\alpha\beta + b\beta^2 = 0$, which represents two lines passing through the vertex C of the triangle, and through the points where the curve meets γ. Hence, if it touches γ, these lines must coincide, and $a\alpha^2 + 2h\alpha\beta + b\beta^2 = 0$ must be a perfect square. *Hence it follows that the general equation of a curve of the second degree which touches the three sides of the triangle of reference must be such, that if any of the variables be made to vanish, the result will be a perfect square.* Therefore the equation $l^2\alpha^2 + m^2\beta^2 + n^2\gamma^2 - 2lm\alpha\beta - 2mn\beta\gamma - 2nl\gamma\alpha = 0$ represents a curve of the second degree inscribed in the triangle of reference, because making any of the variables to vanish, the result is a perfect square. The norm of this equation is $\sqrt{l\alpha} + \sqrt{m\beta} + \sqrt{n\gamma} = 0$ (Art. 70), and the problem to be solved is to find the values l, m, n, so that it may represent a circle. Now, making $\gamma = 0$, we get $(l\alpha - m\beta)^2 = 0$; hence the equation of CF is $l\alpha - m\beta = 0$, and this must be satisfied by the co-ordidates of F, which, from the figure, are evidently $2r\cos^2\frac{1}{2}B$, $2r\cos^2\frac{1}{2}A$, 0; r being the radius of the circle. Hence $l : m :: \cos^2\frac{1}{2}A : \cos^2\frac{1}{2}B$. Similarly $m : n :: \cos^2\frac{1}{2}B : \cos^2\frac{1}{2}C$. Therefore the equation of the circle is

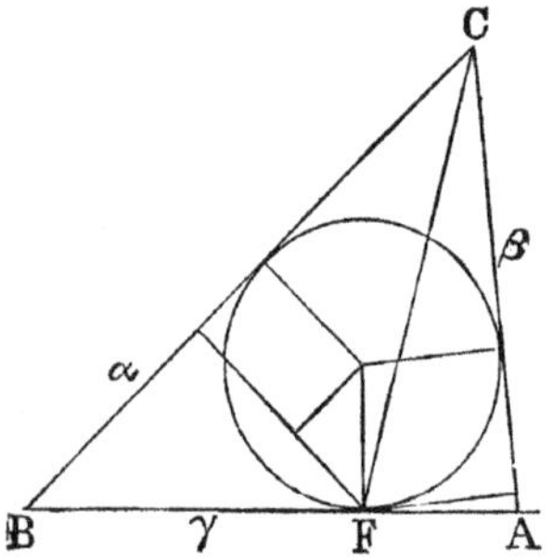

$$\cos\tfrac{1}{2}A\sqrt{\alpha} + \cos\tfrac{1}{2}B\sqrt{\beta} + \cos\tfrac{1}{2}C\sqrt{\gamma} = 0. \qquad (254)$$

This equation is a special case of equation (232), from which it may be inferred by the method of Ex. 6, Art. 77.

83. The equation of the incircle may be inferred from that of the circumcircle by the following method, which is due to Dr. Hart:—Let α', β', γ' be the standard equations of the sides of the triangle formed by joining the points of contact of the incircle on the sides of the triangle of reference; a', b', c', their lengths; then, since the incircle is described about this triangle, we have

$$\frac{a'}{\alpha'} + \frac{b'}{\beta'} + \frac{c'}{\gamma'} = 0;$$

but $$\alpha' = \sqrt{\beta\gamma}, \quad \beta' = \sqrt{\gamma\alpha}, \quad \gamma' = \sqrt{\alpha\beta},$$

since the perpendicular from any point on the circumference of a circle on the chord of contact of two tangents is a mean proportional between the perpendiculars from the same point on the tangents (Sequel, III., Prop. x.);

therefore $$\frac{a'}{\sqrt{\beta\gamma}} + \frac{b'}{\sqrt{\gamma\alpha}} + \frac{c'}{\sqrt{\alpha\beta}} = 0.$$

Again, if the angles between the lines $\alpha = 0$, $\beta = 0$ be denoted by $(\widehat{\alpha\beta})$, &c., it is evident that a', b', c' are proportional to $\cos\frac{1}{2}(\widehat{\alpha\beta})$, $\cos\frac{1}{2}(\widehat{\beta\gamma})$, $\cos\frac{1}{2}(\widehat{\gamma\alpha})$ respectively; hence the required equation is

$$\frac{\cos\frac{1}{2}(\widehat{\alpha\beta})}{\sqrt{\alpha\beta}} + \frac{\cos\frac{1}{2}(\widehat{\beta\gamma})}{\sqrt{\beta\gamma}} + \frac{\cos\frac{1}{2}(\widehat{\gamma\alpha})}{\sqrt{\gamma\alpha}} = 0.$$

Or, as it may be written,

$$\cos\tfrac{1}{2}A\sqrt{\alpha} + \cos\tfrac{1}{2}B\sqrt{\beta} + \cos\tfrac{1}{2}C\sqrt{\gamma} = 0.$$

In the same manner the equations of the escribed circles are

$$\cos\tfrac{1}{2}A\sqrt{-\alpha} + \sin\tfrac{1}{2}B\sqrt{\beta} + \sin\tfrac{1}{2}C\sqrt{\gamma} = 0, \quad (255)$$

$$\sin\tfrac{1}{2}A\sqrt{\alpha} + \sin\tfrac{1}{2}B\sqrt{-\beta} + \sin\tfrac{1}{2}C\sqrt{\gamma} = 0, \quad (256)$$

$$\sin\tfrac{1}{4}A\sqrt{\alpha} + \sin\tfrac{1}{2}B\sqrt{\beta} + \cos\tfrac{1}{2}C\sqrt{-\gamma} = 0. \quad (257)$$

84. *To find the equation of the chord joining the points $\alpha'\beta'\gamma'$, $\alpha''\beta''\gamma''$ on the incircle.*

Put for shortness $\cos\frac{1}{2}A = l^{\frac{1}{2}}$, $\cos\frac{1}{2}B = m^{\frac{1}{2}}$, $\cos\frac{1}{2}C = n^{\frac{1}{2}}$, and we have the two equations

$$l^{\frac{1}{2}}\sqrt{\alpha'} + m^{\frac{1}{2}}\sqrt{\beta'} + n^{\frac{1}{2}}\sqrt{\gamma'} = 0, \quad l^{\frac{1}{2}}\sqrt{\alpha''} + m^{\frac{1}{2}}\sqrt{\beta''} + n^{\frac{1}{2}}\sqrt{\gamma''} = 0.$$

Hence $l^{\frac{1}{2}} = k\{\sqrt{\beta'\gamma''} - \sqrt{\beta''\gamma'}\}$, where k denotes some constant, with similar values for $m^{\frac{1}{2}}$ and $n^{\frac{1}{2}}$; therefore

$$\beta'\gamma'' - \beta''\gamma' = l\{\sqrt{\beta'\gamma''} + \sqrt{\beta''\gamma'}\} \div k, \text{ \&c.}$$

But the join of the given points is

$$\alpha(\beta'\gamma'' - \beta''\gamma') + \beta(\gamma'\alpha'' - \gamma''\alpha') + \gamma(\alpha'\beta'' - \alpha''\beta') = 0$$

Hence, by substitution, we get

$$l^{\frac{1}{2}}\alpha\{\sqrt{\beta'\gamma''} + \sqrt{\beta''\gamma'}\} + m^{\frac{1}{2}}\beta\{\sqrt{\gamma'\alpha''} + \sqrt{\gamma''\alpha'}\}$$
$$+ n^{\frac{1}{2}}\gamma\{\sqrt{\alpha'\beta''} + \sqrt{\alpha''\beta'}\} = 0, \qquad (250)$$

which is the required equation. This result is due to Dr. Hart.

85. If the points $\alpha'\beta'\gamma'$, $\alpha''\beta''\gamma''$ become consecutive, the equation (258) reduces to

$$\frac{l^{\frac{1}{2}}\alpha}{\sqrt{\alpha'}} + \frac{m^{\frac{1}{2}}\beta}{\sqrt{\beta'}} + \frac{n^{\frac{1}{2}}\gamma}{\sqrt{\gamma'}} = 0, \qquad (259)$$

which is the equation of the tangent to the incircle at the point $\alpha'\beta'\gamma'$.

86. If the equation (247) be transformed by Dr. Hart's method (see Art. (83)), we get the following general theorem:—*If a polygon of any number of sides whose equations are $\alpha = 0$, $\beta = 0$, $\gamma = 0$, $\delta = 0$, &c., be circumscribed to a circle, the equation of the circle is a factor in the general equation*

$$\frac{\cos\frac{1}{2}(\alpha\beta)}{\sqrt{\alpha\beta}} + \frac{\cos\frac{1}{2}(\beta\gamma)}{\sqrt{\beta\gamma}} + \ldots + \frac{\cos\frac{1}{2}(\omega\alpha)}{\sqrt{\omega\alpha}} = 0. \qquad (260)$$

87. *If the equation* $(a, b, c, f, g, h)(\alpha, \beta, \gamma)^2$ *represent a circle, it is required to find the invariant relations between the coefficients.*

Transforming to Cartesian co-ordinates and equating the coefficients of x^2, y^2, and putting the coefficient of $xy = 0$, we get at once (see demonstration of Art. (38)), $\theta'^2 = 4\theta$. This condition however, though necessary, is not sufficient, since two relations must be fulfilled in order that an equation of the second degree may represent a circle. The following solution is from Salmon's *Conics*, page 121:—Since $\alpha \sin A + \beta \sin B + \gamma \sin C$ = twice the area of the triangle of reference, the equation $kS+(l\alpha+m\beta+n\gamma)(\alpha \sin A+\beta \sin B+\gamma \sin C)$, if S denote any circle, must represent a circle. Hence, taking S to denote the circumcircle, equating the coefficients $\alpha^2, \beta^2, \gamma^2$ in $kS + (l\alpha + m\beta + n\gamma)(\alpha \sin A + \beta \sin B + \gamma \sin C)$, and in the given equation, we get

$$l = \frac{a}{\sin A}, \quad m = \frac{b}{\sin B}, \quad n = \frac{c}{\sin C}.$$

Hence, substituting these values, and equating the remaining coefficients, we get, after eliminating k, the two following relations:—

$$b \sin^2 C + c \sin^2 B - 2f \sin B \sin C = c \sin^2 A + a \sin^2 C - 2g \sin C \sin A$$
$$= a \sin^2 B + b \sin^2 A - 2h \sin A \sin B. \qquad (261)$$

Examples.

1. If the area of the triangle formed by joining the feet of the perpendicular from a point P on the sides of the triangle of reference be given, prove that the locus of P is a circle concentric with the circumcircle.

2. If through P parallels EPF', FPD', DPE' to BC, CA, AB be drawn; prove that the locus of P is a circle, if the sum of the rectangles $EP.PF'$, $FP.PD$, $DP.PE'$ be given.

The three rectangles are, respectively, equal

$$\frac{\alpha\beta}{\sin A \sin B}, \quad \frac{\beta\gamma}{\sin B \sin C}, \quad \frac{\gamma\alpha}{\sin C \sin A}.$$

Hence the locus is $\alpha\beta \sin C + \beta\gamma \sin A + \gamma\alpha \sin B$ = constant.

3. If P be the symmedian point of the triangle, the six points D, D', E, E', F, F' are concyclic.

4. If the intercepts DD', EE', FF' (Ex. 2), be denoted by X, Y, Z, and the corresponding intercepts made by the antiparallels (Chap. II., Ex. 39) by ξ, η, ζ, prove that the locus of P is a circle, if $\xi X + \eta Y + \zeta Z$ = constant.

5. If in the same case the intercepts on the parallels made by the sides of the triangle be denoted by X_1, Y_1, Z_1; prove, if $\xi X_1 + \eta Y_1 + \zeta Z_1$ = constant, that the locus of P is a circle concentric with the circumcircle.

6. Find the equation of a circle through $\alpha'\beta'\gamma'$, $\alpha''\beta''\gamma''$, $\alpha'''\beta'''\gamma'''$. If $S = 0$, denote any circle, say, for instance, the circumcircle, then

$$\begin{vmatrix} S, & \alpha, & \beta, & \gamma, \\ S', & \alpha', & \beta', & \gamma', \\ S'', & \alpha'', & \beta'', & \gamma'', \\ S''', & \alpha''', & \beta''', & \gamma''' \end{vmatrix} = 0 \qquad (262)$$

is evidently the required equation.

7. Find the pedal circle of $\alpha'\beta'\gamma'$. The co-ordinates of the feet of perpendiculars are—0, $\beta' + \alpha' \cos C$, $\gamma' + \alpha' \cos B$; $\alpha' + \beta' \cos C$, 0 $\gamma' + \beta' \cos A$; $\alpha' + \gamma' \cos B$, $\beta' + \gamma' \cos A$, 0. These substituted in (262) give, by expansion,

$$(\beta\gamma \sin A + \gamma\alpha \sin B + \alpha\beta \sin C)(\beta'\gamma' \sin A + \gamma'\alpha' \sin B + \alpha'\beta' \sin C)(\alpha' \sin A + \beta' \sin B + \gamma' \sin C)$$

$$= \sin A \sin B \sin C (\alpha \sin A + \beta \sin B + \gamma \sin C)\left\{\frac{\alpha\alpha'(\beta' + \gamma' \cos A)(\gamma' + B' \cos A)}{\sin A}\right.$$

$$\left. + \frac{\beta\beta'(\gamma' + \alpha' \cos \beta)(\alpha' + \gamma' \cos B)}{\sin B} + \frac{\gamma\gamma'(\alpha' + \beta' \cos C)(B' + \alpha' \cos C)}{\sin C}\right\}. \quad (263)$$

This equation remains unaltered if we substitute for α', β', γ' their reciprocals $\frac{1}{\alpha'}$, $\frac{1}{\beta'}$, $\frac{1}{\gamma'}$. Hence the pedal circle of a point and its reciprocal are the same.

8. The Simson's line of any point $\alpha'\beta'\gamma'$ on the circumcircle is

$$\frac{\alpha\alpha'(\beta' + \gamma' \cos A)(\gamma' + \beta' \cos A)}{\sin A} + \frac{\beta\beta'(\gamma' + \alpha' \cos B)(\alpha' + \gamma' \cos B)}{\sin B}$$

$$+ \frac{\gamma\gamma'(\alpha' + \beta' \cos C)(\beta' + \alpha' \cos C)}{\sin C} = 0. \qquad (264)$$

9. Prove that $\beta^2 + \gamma^2 - 2\beta\gamma \cos A =$ constant represents a circle.

10. If $S = 0$, $S' = 0$ represent two circles whose radii are r, r'; prove that the circles

$$\frac{S}{r} + \frac{S'}{r'} = k(r + r'), \quad \frac{S}{r} - \frac{S'}{r'} = k(r - r')$$

cut orthogonally.—(CROFTON.)

11. If $(a, b, c, f, g, h)(\alpha, \beta, \gamma)^2$ represent a circle, and if the same, when transformed to Cartesian co-ordinates, becomes

$$\equiv m\{x - f)^2 + (y - g)^2 - r^2\};$$

find the value of m in terms of the invariants.

Ans. $m = \frac{1}{2}\theta'$.

DEF.—*We shall call m the modulus of the equation.*

12. Find the modulus for $\beta\gamma \sin A + \gamma\alpha \sin B + \alpha\beta \sin C$.

Ans. $-\sin A \sin B \sin C$.

13. Find the modulus for the incircle.

Ans. $4 \cos^2 \frac{A}{2} \cos^2 \frac{B}{2} \cos^2 \frac{C}{2}$.

14. If a, b, c denote the lengths of the sides of the triangle of reference, prove that $a\alpha^2 + b\beta^2 + c\gamma^2 + [(a + b + c)(\alpha\beta + \beta\gamma + \gamma\alpha) = 0$ denotes a circle through the centres of the three escribed circles.

15. If $R =$ radius of circumcircle; prove that the modulus of the circle in Ex. 14 is $2R \sin A \sin B \sin C$.

16. The equation $b\beta^2 + c\gamma^2 - a\alpha^2 + 2(S - a)\{\beta\gamma - \gamma\alpha - \alpha\beta\} = 0$ denotes the circle through the incentre and two excentres, and its modulus is $-2R \sin A \sin B \sin C$.

17. If $\delta =$ distance of incentre from circumcentre; prove, by aid of the modulus of the equation of the circumcircle, that

$$\frac{1}{R + \delta} + \frac{1}{R - \delta} = \frac{1}{r}.$$

18. If on the sides AB, BC, CA of the triangle of reference portions BF, CD, AE be cut off equal to

$$\lambda\left(\frac{a}{b}\right), \lambda\left(\frac{b}{c}\right), \lambda\left(\frac{c}{a}\right),$$

respectively, where λ denotes a line of any given length, the triangle EDF is similar to ABC. For, by an easy calculation,

$$DF^2 = \frac{\lambda^2(a^2b^2 + b^2c^2 + c^2a^2) - \lambda abc(a^2 + b^2 + c^2) + a^2bc^2}{a^2b^2},$$

with similar values for FE^2, ED^2.

19. The equation of the circle Σ described about triangle EDF (see Ex. 6) is—

$$\begin{vmatrix} S, & \alpha, & \beta, & \gamma, \\ 1, & 0, & (\lambda c)^{-1}, & (ac-\lambda b)^{-1}, \\ 1, & (ab-\lambda c)^{-1}, & 0, & (\lambda a)^{-1}, \\ 1, & (\lambda b)^{-1}, & (bc-\lambda a)^{-1}, & 0 \end{vmatrix}. \qquad (265)$$

This is the trilinear equation of what NEUBERG has called 'Tucker's Circles,' and includes several important cases. For example, if $\lambda = 0$, we get the circumcircle; if $\lambda = \dfrac{abc}{a^2+b^2+c^2}$, the Lemoine Circle of Ex. 3; if $\lambda = \dfrac{2abc}{a^2+b^2+c^2}$, the cosine circle, &c.

20. If the other points in which the circle of Ex. 19 cuts the same sides be denoted by F', D', E, prove that the triangles EDF, $E'D'F'$ are equal and similar.

21. Find the equation of the circle (called the 'Brocard Circle') through the circumcentre and the Brocard points

$$\frac{c}{b}, \ \frac{a}{c}, \ \frac{b}{a}; \ \frac{b}{c}, \ \frac{c}{a}, \ \frac{a}{b}.$$

Ans. $abc(\alpha^2+\beta^2+\gamma^2) = a^3\beta\gamma + b^3\gamma\alpha + c^3\alpha\beta.$

22. Find the radical axis of the incircle and the circle through the middle point of the sides.

In Ex. 6 let S denote the incircle, and S', S'', S''' the powers of the middle points of the sides with respect to the incircle; then if $\alpha'\beta'\gamma'$, $\alpha''\beta''\gamma''$, $\alpha'''\beta'''\gamma'''$ be the middle points, the required equation is,

$$\begin{vmatrix} 0, & \alpha, & \beta, & \gamma, \\ (b-c)^2, & 0, & q, & r, \\ (c-a)^2, & p, & 0, & r, \\ (a-b)^2, & p, & q, & 0 \end{vmatrix}, \qquad (266)$$

where p, q, r denote the perpendiculars of the triangle of reference. Expanding and putting for p, q, r their values in terms of the sides, we get, after an easy reduction, the same result as in equation (244), which, by using areal co-ordinates, may be written

$$\frac{\alpha}{b-c} + \frac{\beta}{c-a} + \frac{\gamma}{a-b} = 0. \qquad (267)$$

Section IV.—Tangential Equations.

88. *To find the tangential equation of the circumcircle of the triangle of reference.*

First method.—If we eliminate γ between the equation of the circumcircle $\frac{a}{\alpha} + \frac{b}{\beta} + \frac{c}{\gamma} = 0$ and the line $\lambda\alpha + \mu\beta + \nu\gamma = 0$, we get $(b\lambda)\alpha^2 + (a\lambda + b\mu - c\nu)\alpha\beta + (a\mu)\beta^2 = 0$.

Now this denotes two lines passing through the point $(\alpha\beta)$ and the points where the line $\lambda\alpha + \mu\beta + \nu\gamma = 0$ meets the circle. Hence, if it be a perfect square, the line touches the circle; that is, if

$$a^2\lambda^2 + b^2\mu^2 + c^2\nu^2 - 2ab\lambda\mu - 2bc\mu\nu - 2ca\nu\lambda = 0.$$

But the norm of this is

$$\sqrt{a\lambda} + \sqrt{b\mu} + \sqrt{c\nu} = 0.$$

Hence $$\sqrt{a\lambda} + \sqrt{b\mu} + \sqrt{c\nu} = 0 \qquad (268)$$

is the condition that the line $\lambda\alpha + \mu\beta + \nu\gamma = 0$ should touch the circle, and is on that account called its tangential equation.

Second method.—The same equation can be obtained otherwise as follows. Since $\lambda\alpha + \mu\beta + \nu\gamma = 0$ is a tangent to the circle, if the point of contact be $\alpha'\beta'\gamma'$, comparing it with equation (253), we have

$$\lambda = \frac{a}{\alpha'^2}, \quad \mu = \frac{b}{\beta'^2}, \quad \nu = \frac{c}{\gamma'^2}.$$

Hence $$\frac{a}{\alpha'} + \frac{b}{\beta'} + \frac{c}{\gamma'} = \sqrt{a\lambda} + \sqrt{b\mu} + \sqrt{c\nu}.$$

But, since $\alpha'\beta'\gamma'$ is a point on the circumcircle, we have

$$\frac{a}{\alpha'} + \frac{b}{\beta'} + \frac{c}{\gamma'} = 0.$$

Hence $$\sqrt{a\lambda} + \sqrt{b\mu} + \sqrt{c\nu} = 0.$$

89. *To find the tangential equations of a circle circumcribed to a polygon of any number of sides.*

This problem requires the following lemma:—*If AB be a chord of a circle APB, and* λ, μ *denote the perpendiculars from A, B on the tangent at P; α the perpendicular from P on AB; then* $\alpha^2 = \lambda\mu$. [Euclid, VI. xvii., Ex. 11.]

Now, if a polygon *ABCD*, &c., of *n* sides be inscribed in the circle, and if the standard equations of the sides be $\alpha = 0$, $\beta = 0$, &c., we have by equation (247)

$$\frac{AB}{\alpha} + \frac{BC}{\beta} + \frac{CD}{\gamma} + \frac{DE}{\delta} + \&c. = 0.$$

Hence, if the perpendiculars from *A*, *B*, *C*, &c., on any tangent to the circle be denoted by λ, μ, ν, ρ, &c., we have

$$\frac{AB}{\sqrt{\lambda\mu}} + \frac{BC}{\sqrt{\mu\nu}} + \frac{CD}{\sqrt{\nu\rho}} + \&c. \ldots + \frac{LA}{\sqrt{\omega\lambda}} = 0, \qquad (269)$$

which is the required equation.

Cor.—If the polygon reduce to a triangle, the equation (269) becomes

$$\frac{c}{\sqrt{\lambda\mu}} + \frac{a}{\sqrt{\mu\nu}} + \frac{b}{\sqrt{\nu\lambda}} = 0;$$

or

$$a\sqrt{\lambda} + b\sqrt{\mu} + c\sqrt{\nu} = 0. \qquad (270)$$

It will be observed that λ, μ, ν have different significations from those in equation (268). In fact the λ, μ, ν in (268) are equivalent to $a\lambda$, $b\mu$, $c\nu$ in (270); and this difference can be explained; for in (268) the three ratios $\lambda : \mu$, $\mu : \nu$, $\nu : \lambda$ are those of the sines of the angles into which the angles of the triangle of reference are divided by lines from its vertices to the intersections of $\lambda\alpha + \mu\beta + \nu\gamma$ with the opposite sides; and in (270) they denote the ratios of the segments into which the tangent divides the sides of the triangle of reference. Compare Art. 29.

90. *To find the tangential equation of the incircle of the triangle of reference.*

If $\lambda\alpha + \mu\beta + \nu\gamma = 0$ be a tangent to the circle, comparing it with equation (259), viz.

$$\frac{l^{\frac{1}{2}}\alpha}{\sqrt{\alpha'}} + \frac{m^{\frac{1}{2}}\beta}{\sqrt{\beta'}} + \frac{n^{\frac{1}{2}}\gamma}{\sqrt{\gamma'}} = 0,$$

we have $\dfrac{l^{\frac{1}{2}}}{\sqrt{\alpha'}} = \lambda$, &c. Hence $l^{\frac{1}{2}}\sqrt{\alpha'} = \dfrac{l}{\lambda}$, &c.

But, since $\alpha'\beta'\gamma'$ is a point on the circle,

$$l^{\frac{1}{2}}\sqrt{\alpha'} + m^{\frac{1}{2}}\sqrt{\beta'} + n^{\frac{1}{2}}\sqrt{\gamma'} = 0;$$

therefore $$\frac{l}{\lambda} + \frac{m}{\mu} + \frac{n}{\nu} = 0;$$

and restoring the values of l, m, n (see Art. 84), we get

$$\frac{\cos^2\frac{1}{2}A}{\lambda} + \frac{\cos^2\frac{1}{2}B}{\mu} + \frac{\cos^2\frac{1}{2}C}{\nu} = 0, \qquad (271)$$

which is the required equation.

91. *To find the tangential equation of the incircle of an n-sided polygon.*

If AB be any chord of a circle, P any point in its circumference, Q the pole of AB; then if α, λ be the perpendiculars from P on AB, and from Q on the tangent at P respectively, it may be easily proved that $\alpha \div \lambda = \sin\frac{1}{2}AQB$; but if R be the radius of the circle, $AB = 2R\cos\frac{1}{2}AQB$. Hence

$$\frac{AB}{\alpha} = \frac{2R\cot\frac{1}{2}AQB}{\lambda}.$$

Now, for any inscribed polygon we have, by equation (247),

$$\frac{AB}{\alpha} + \frac{BC}{\beta} + \frac{CD}{\gamma} + \&c. = 0.$$

Hence, for a circumscribing polygon whose angles are A, B, C, &c., we have

$$\frac{\cot\frac{1}{2}A}{\lambda} + \frac{\cot\frac{1}{2}B}{\mu} + \frac{\cot\frac{1}{2}C}{\nu} + \&c. = 0. \qquad (272)$$

where λ, μ, ν, &c., are the perpendiculars from the angles on any tangent to the circle.

Cor.—In the case of a triangle we get

$$\frac{\cot\frac{1}{2}A}{\lambda} + \frac{\cot\frac{1}{2}B}{\mu} + \frac{\cot\frac{1}{2}C}{\nu} = 0,$$

which may be written

$$\frac{\cos^2\frac{1}{2}A}{a\lambda} + \frac{\cos^2\frac{1}{2}B}{b\mu} + \frac{\cos^2\frac{1}{2}C}{c\nu} = 0;$$

and putting (see Art. 89) λ, μ, ν for $a\lambda$, $b\mu$, $c\nu$, we get the same result as in equation (271).

Miscellaneous Exercises on the Circle.

1. Find the centre and radius of $x^2 + y^2 - 6x + 8y - 11 = 0$.

2. Find the value of m if $y = mx$ be a tangent to $x^2 + y^2 - 6x - 2y + 8 = 0$.

3. Find the points where $x^2 + y^2 - 7x - 8y + 12 = 0$ cuts the axes.

4. Find the circle through the origin, and making intercepts h, k on the axes.

5. If the axes be oblique, find the equation of a circle touching each at a distance a from the origin.

6. Find the circle through the points $(7, 5)$, $(-2, 4)$, $(3, -3)$.

7. Find the circle whose diameter is the intercept made by

$$x^2 + y^2 = r^2 \quad \text{on} \quad \frac{x}{a} + \frac{y}{b} - 1 = 0.$$

8. Find in the same case the pair of lines from the origin to the points of intersection.

9. Find the length of the common chord of $(x-a)^2 + (y-b)^2 = r^2$, $(x-b)^2 + (y-a)^2 = r^2$.

10. Find the equation of the circle whose centre is $(2, 3)$, and which touches $3x + 4y + 12 = 0$.

11. Find the condition that the line $\lambda x + \mu y + \nu = 0$ may touch the circle $(x-a)^2 + (y-b)^2 = r^2$.

12. Find the radical centre of the circles $x^2 + y^2 + 6x + 4y + 12 = 0$, $x^2 + y^2 - 6x + 4y + 12 = 0$, $x^2 + y^2 + 6x - 4y + 12 = 0$.

13. Through O, the origin, a line OPQ cuts $x^2 + y^2 + 2gx + 2fy + c = 0$ in the points P, Q; find the locus of R in each of the following cases:—

1°. When OR is an arithmetic mean between OP, OQ. 2°. A geometric mean. 3°. A harmonic mean.

14. If two tangents be drawn to $x^2 + y^2 - r^2 = 0$ from the point $(a, 0)$, find the equation of the incircle of the triangle formed by the tangents and the chord of contact.

15. If O be the centre of a circle whose radius is r, prove that the area of the triangle which is the polar reciprocal of a given triangle ABC is

$$r^4 (ABC)^2 \div 4(AOB) . (BOC) . (COA). \qquad (273)$$

16. Prove that a triangle and its polar reciprocal with respect to any given circle are in perspective.

17. If a chord of a given circle of a coaxal system pass through either limiting point, the rectangle contained by the perpendiculars from its extremities on the radical axis is constant.

18. The three circles whose diameters are the three diagonals of a complete quadrilateral are coaxal.

19. If from a given S in the axis of x a perpendicular SY be drawn to the tangent at any point P of the circle $x^2 + y^2 = r^2$, and the ordinate PM at P of the circle be produced to Q until $MQ = SY$, the locus of Q is a right line.

20. Find the polar equation of the circle whose diameter is the join of the points $(\rho' \theta')$, $(\rho'' \theta'')$.

21. The equations of any two circles can be written in the forms $x^2 + y^2 + 2kx + \delta = 0$, $x^2 + y^2 + 2k'x + \delta = 0$, and one is within the other if kk' and δ are both positive.

22. If three given circles be cut by a fourth circle Ω which is variable, the radical axis of Ω and the given circles form systems of triangles in perspective.

23. If R be the circumradius of the triangle ABC, prove that the distance between its orthocentre and circumcentre is

$$R\sqrt{1 - 8 \cos A \cos B \cos C}. \qquad (274)$$

24. The locus of the radical centre of the circles $(x - a)^2 + (y - b)^2 = (r + \rho)^2$, $(x - a')^2 + (y - b')^2 = (r + \rho')^2$, $(x - a'')^2 + (y - b'')^2 = (r + \rho'')^2$, where r is a variable quantity, is a right line.

25. If $\alpha\gamma = k\beta\delta$ represent a circle; prove that $k = 1$, and give the geometrical interpretation.

26. If $\alpha\gamma = k\beta^2$ represent a circle; prove $k = 1$, and give the interpretation.

27. If $l\alpha^2 + m\beta^2 + n\gamma^2 = 0$ represent a circle; prove

$$l = \sin 2A, \quad m = \sin 2B, \quad n = \sin 2C.$$

28. Prove that the tangential equation of the circle whose radius is r, and centre $\alpha'\beta'\gamma'$, is

$$r^2(\lambda^2 + \mu^2 + \nu^2 - 2\mu\nu\cos A - 2\nu\lambda\cos B - 2\lambda\mu\cos C) = (\lambda\alpha' + \mu\beta' + \nu\gamma')^2. \quad (275)$$

29. If the four lines $\alpha = 0$, $\beta = 0$, $\gamma = 0$, $\delta = 0$, have a common tangential circle; prove

$$\frac{\alpha}{\cos\frac{1}{2}(\alpha\beta)\cos\frac{1}{2}(\alpha\gamma)\cos\frac{1}{2}(\alpha\delta)} + \frac{\beta}{\cos\frac{1}{2}(\beta\alpha)\cos\frac{1}{2}(\beta\gamma)\cos\frac{1}{2}(\beta\delta)}$$

$$+ \frac{\gamma}{\cos\frac{1}{2}(\gamma\alpha)\cos\frac{1}{2}(\gamma\beta)\cos\frac{1}{2}(\gamma\delta)} + \frac{\delta}{\cos\frac{1}{2}(\delta\alpha)\cos\frac{1}{2}(\delta\beta)\cos\frac{1}{2}(\delta\gamma)} = 0. \quad (276)$$

30. Show that the equation, Ex. 28, is of the form $r^2\omega\omega' = L^2$, and give the interpretation of $\omega\omega'$.

31. If the sum of the perpendiculars on a variable line from any number of given points, each multiplied by a constant, be given, the envelope of the line is a circle.

32. Find the condition that the points are concyclic in which the circles $x^2 + y^2 + gx + fy + c = 0$, $x^2 + y^2 + g'x + f'y + c' = 0$ meet respectively the lines $\lambda x + \mu y + \nu = 0$, $\lambda' x + \mu' y + \nu' = 0$.

33. Find the equations of the tangents to the 'Nine-points Circle' at its points of contact with the escribed circles.

34. The circle which passes through the symmedian point P and the points B, C of the triangle of reference is $S - 3\alpha \sin B \sin C = 0$, (277)

$$\text{where } S \equiv \alpha\beta \sin C + \beta\gamma \sin A + \alpha\gamma \sin B.$$

35. If P be the symmedian point of the triangle ABC; prove that the diameters of the circles APB, BPC, CPA are inversely proportional to the medians of the sides AB, BC, CA.

36. If G be the centroid of the triangle ABC, the diameters of the circles AGB, BGC, CGA, are inversely proportional to the symmedians of the triangle.

37. The circle, whose diameter is the side a of the triangle of reference, is

$$a^2 \cos A = \beta\gamma + \alpha(\beta \cos B + \gamma \cos C). \quad (278)$$

This may be inferred from Ex. 55, but we indicate an independent proof here. The equation will evidently be of the form

$$k\alpha(\alpha \sin A + \beta \sin B + \gamma \sin C) + (\alpha\beta \sin C + \beta\gamma \sin A + \gamma\alpha \sin B) = 0.$$

Now, put $\beta = 0$ in this, and equate the result to $\alpha \cos A - \gamma \cos C$, and we get $k = -\cos A$: this gives the required equation.

38. To find the equation of the circle which passes through the feet of the perpendiculars. The line $\beta \cos B + \gamma \cos C - \alpha \cos A = 0$ will evidently be the radical axis of this circle and the last. Hence the equation will be of the form

$$(\beta \cos B + \gamma \cos C - \alpha \cos A)(\beta \sin B + \gamma \sin C + \alpha \sin A)$$
$$= k\{\alpha^2 \cos A - \beta\gamma - \alpha(\beta \cos B + \gamma \cos C)\};$$

and this must pass through the point whose co-ordinates are 0, $\cos C$, $\cos B$. Hence $k = -2\sin A$; and by substitution and reduction we get

$$\alpha^2 \sin 2A + \beta^2 \sin 2B + \gamma^2 \sin 2C - 2(\beta\gamma \sin A + \gamma\alpha \sin B + \alpha\beta \sin C) = 0. \quad (279)$$

39. Deduce the 'Nine-points Circle' equation from Ptolemy's theorem. Let A', B', C' be the middle points of the sides of the triangle of reference; P any point in the circle. Let fall the perpendiculars PD, PE, PF on $B'C'$, $C'A'$, $A'B'$, respectively; then we have, by equation (245),

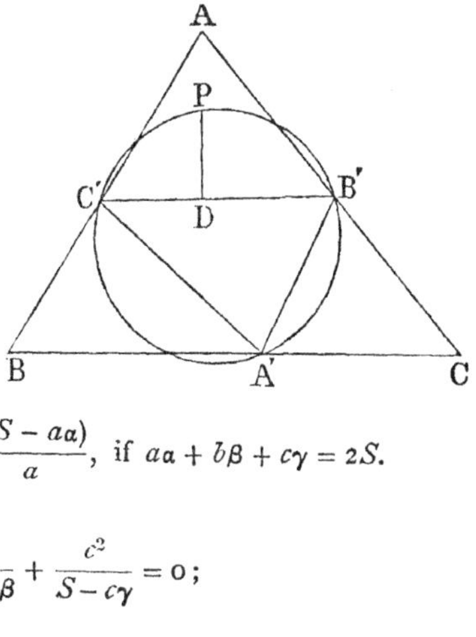

$$\frac{B'C'}{PD} + \frac{C'A'}{PE} + \frac{A'B'}{PF} = 0;$$

but PD is evidently

$$= \alpha - \tfrac{1}{2}\frac{(a\alpha + b\beta + c\gamma)}{a} = -\frac{(S - a\alpha)}{a}, \text{ if } a\alpha + b\beta + c\gamma = 2S.$$

Hence we get

$$\frac{a^2}{S - a\alpha} + \frac{b^2}{S - b\beta} + \frac{c^2}{S - c\gamma} = 0;$$

or, in areal co-ordinates,

$$\frac{a^2}{S - \alpha} + \frac{b^2}{S - \beta} + \frac{c^2}{S - \gamma} = 0: \quad (280)$$

this is a new form of the equation.

40. If α, β, γ denote the tangents drawn from any point to three coaxal circles whose centres are A, B, C; prove that

$$BC\alpha^2 + CA\beta^2 + AB\gamma^2 = 0. \quad (281)$$

41. Prove that a common tangent to any two circles of a coaxal system subtends a right angle at either limiting point.

42. If through the symmedian point an antiparallel be drawn to one of the sides of the triangle of reference; find the equation of the circle described on the intercept made by the other sides on it as diameter. This will pass through the three points $\tan A$, $\sin C$, o; o, $\tan B$, $\sin A$; $\sin B$, o, $\tan C$.

43. *Pascal's Theorem.*—The intersections of opposite sides of a hexagon inscribed in a circle are collinear.

Let the equations of BC be $\alpha = 0$; BE, $\gamma = 0$; EF, $\beta = 0$; CF, $\delta = 0$; then the equation of the circle will be $\alpha\beta - \gamma\delta = 0$.

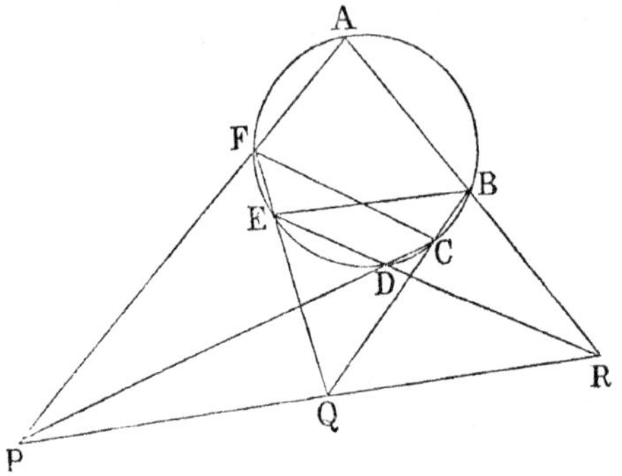

The equation of AB will be of the form $l\alpha - \gamma = 0$; of AF, $\beta - l\delta = 0$; of DE, $\beta - m\gamma = 0$; of CD, $m\alpha - \delta = 0$; and the equation of the line PQR is $lm\alpha - \beta = 0$; for it will be seen that this passes through each pair of opposite sides.

44. If t', t'', t''' be the tangents drawn to a circle from the vertices of a self-conjugate triangle; R the radius of the circle, and Δ the area of the triangle; then

$$-4\Delta^2R^2 = t'^2\,t''^2\,t'''^2. \tag{281}$$

(Prof. Curtis, S.J.)

For if $(x'y')$, $(x''y'')$, $(x'''y''')$ be the vertices of the triangle, multiplying the determinants

$$\begin{vmatrix} x', & y', & R, \\ x'', & y''', & R, \\ x''', & y''', & R, \end{vmatrix} \begin{vmatrix} x', & y', & -R, \\ x'', & y'', & -R, \\ x''', & y''', & -R, \end{vmatrix} \text{ we get } \begin{vmatrix} t'^2, & 0, & 0, \\ 0, & t''^2, & 0, \\ 0, & 0, & t'''^2, \end{vmatrix}$$

which proves the proposition.

45. Find the equation of the circle whose diameter is any of the perpendiculars of the triangle of reference.

46. If $\alpha = 0$, $\beta = 0$, $\gamma = 0$, $\delta = 0$ be the standard equations of the sides of a cyclic quadrilateral, and their lengths a, b, c, d, the equation of the third diagonal is

$$\frac{\alpha}{a} + \frac{\beta}{b} + \frac{\gamma}{c} + \frac{\delta}{d} = 0. \qquad (282)$$

47. In the same case, if $\epsilon = 0$, $\phi = 0$ denote the other diagonals of the quadrilateral, and e, f their lengths, the equations of the remaining sides of the harmonic triangle of the quadrilateral are

$$\frac{\alpha}{a} + \frac{\epsilon}{e} + \frac{\gamma}{c} + \frac{\phi}{f} = 0, \quad \frac{\beta}{b} + \frac{\epsilon}{e} + \frac{\delta}{d} + \frac{\phi}{f} = 0. \qquad (283)$$

48. The circumcircle of the triangle formed by the side $\alpha = 0$, and the internal and external bisectors of the opposite angle, is

$$\sin(B - C)(\alpha\beta \sin C + \beta\gamma \sin A + \gamma\alpha \sin B)$$
$$+ (\beta \sin C - \gamma \sin \beta)(\alpha \sin A + \beta \sin B + \gamma \sin C) = 0. \qquad (284)$$

This circle and its two analogues are coaxal, the radical axis being

$$\sin(B - C)\,\alpha + \sin(C - A)B + \sin(A - B)\,\gamma = 0. \qquad (285)$$

(Prof. J. Purser.)

49. Find the equation of the pair of lines, from the origin to the intersection of the circles

$$x^2 + y^2 + 2gx + 2fy + c = 0, \quad x^2 + y^2 + 2g'x + 2f'y + c' = 0.$$

50. With the same hypothesis as in Ex. 44, prove

$$\frac{1}{t'^2} + \frac{1}{t''^2} + \frac{1}{t'''^2} = \frac{1}{R^2}. \quad \text{(Prof. Curtis, S.J.)} \qquad (286)$$

Equate to zero the product of the two matrices

$$\begin{vmatrix} x', & y', & -R, \\ x'', & y'', & -R, \\ x''', & y''', & -R, \\ 0, & 0, & -R, \end{vmatrix} \quad \begin{vmatrix} x', & y', & R, \\ x'', & y'', & R, \\ x''', & y''', & R, \\ 0, & 0, & R \end{vmatrix}.$$

51. If Ω be the equation of the 'Nine-points Circle,' prove that the circle whose diameter is the median that bisects α is

$$\Omega - 2\alpha \cos A(\alpha \sin A + \beta \sin B + \gamma \sin C) = 0. \qquad (287)$$

52. The radical axis of the circumcircle and the circle whose diameter is the median that bisects α is

$$\beta \cos B + \gamma \cos C = 0.$$

53. Find the equations of the circles whose diameters are the joins of the feet of the perpendiculars of the triangle of reference.

54. If the three sides of a plane triangle be replaced by three circles, then the circles tangential to those corresponding to the inscribed and escribed circles of a plane triangle are all touched by a fourth circle (Dr. Hart's), which corresponds to the 'Nine-points Circle' of the plane triangle. Its equation is

$$\frac{S_1}{\overline{12'}\,\overline{13'}\,\overline{14}} + \frac{S_2}{\overline{21'}\,\overline{23}\,\overline{24}} + \frac{S_3}{\overline{31'}\,\overline{32}\cdot\overline{34}} + \frac{S_4}{\overline{41'}\cdot\overline{42}\cdot\overline{43}} = 0, \qquad (288)$$

where S_1, S_2, &c., correspond to the inscribed and escribed circles of the plane triangle, and $\overline{12'}$, &c., denote a transverse common tangent.

55. Find the equations of the circles whose diameters are the joins of the middle points of the triangle of reference.

56. Find the equation of the circle which passes through the points of intersection of bisectors of angles with opposite sides.

57. If $ABCD$ be a cyclic quadrilateral, AC the diameter of its circumcircle; prove the difference of the triangles BAD, $BCD = \frac{1}{4}AC^2 \sin^2 BAD$. —STEINER.

58. If a point in the plane of a polygon be such, that the area of the figure formed by joining the feet of perpendiculars from it on the sides of the polygon be given, its locus is a circle.

59. If any hexagon be described about a circle, the joins of the three pairs of opposite angles are concurrent.

Let the equation of the circle be $\sqrt{l\alpha} + \sqrt{m\beta} + \sqrt{n\gamma} = 0$; ABC the tri-

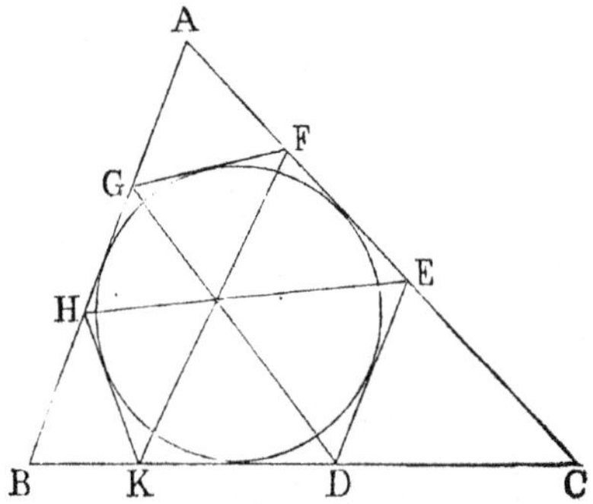

angle of reference; and let the equations of the alternate sides DE, FG, HK of the hexagon be respectively

$$\lambda\alpha + \mu\beta + \nu\gamma = 0, \quad \lambda'\alpha + \mu'\beta + \nu'\gamma = 0. \quad \lambda''\alpha + \mu''\beta + \nu''\gamma = 0.$$

Hence, equation (271),

$$\frac{l}{\lambda}+\frac{m}{\mu}+\frac{n}{\nu}=0,\quad \frac{l}{\lambda'}+\frac{m}{\mu'}+\frac{n}{\nu'}=0,\quad \frac{l}{\lambda''}+\frac{m}{\mu''}+\frac{n}{\nu''}=0. \qquad (1.)$$

Again, the equations of the three diagonals are easily seen to be—

for $\quad GD,\quad \frac{\alpha}{\mu'\nu}+\frac{\beta}{\lambda'\nu}+\frac{\gamma}{\lambda'\mu}=0;$

" $\quad HE,\quad \frac{\alpha}{\mu''\nu}+\frac{\beta}{\lambda''\nu}+\frac{\gamma}{\mu''\lambda}=0;$

" $\quad KF,\quad \frac{\alpha}{\mu''\nu'}+\frac{\beta}{\lambda'\nu''}+\frac{\gamma}{\lambda'\mu''}=0.$

And the condition of concurrence is the vanishing of the determinant,

$$\begin{vmatrix} \frac{1}{\mu'\nu}, & \frac{1}{\lambda'\nu}, & \frac{1}{\lambda'\mu}, \\ \frac{1}{\mu''\nu}, & \frac{1}{\lambda''\nu}, & \frac{1}{\mu''\lambda}, \\ \frac{1}{\mu''\nu'}, & \frac{1}{\lambda'\nu''}, & \frac{1}{\lambda'\mu''} \end{vmatrix}:$$

this differs only by the factor $\frac{1}{\lambda'\mu''\nu}$ from the determinant got by eliminating l, m, n from the equations (1.). Hence the proposition is proved.— See WRIGHT'S *Trilinear Co-ordinates.*

60. One circle lies entirely within another; a tangent at any point P to the inner meets the outer in M, N and the radical axis in Q; prove, if S be the internal limiting point, that the angle MSN is bisected.

61. The ratio of $\sin PSN : \cos \frac{1}{2} SQN$ is constant.

62. The envelope of the circle about the triangle MSN is a circle.

63. The diameter of the circle which cuts the three escribed circles orthogonally is $\frac{a}{\sin A}(1+\cos A\cos B+\cos B\cos C+\cos C\cos A)^{\frac{1}{2}}$.

64. The diameters of the circles cutting the inscribed circle and two escribed circles orthogonally are

$$\frac{a}{\sin A}(1+\cos A\cos B-\cos B\cos C+\cos C\cos A)^{\frac{1}{2}}, \text{ \&c.}$$

65. If δ be the distance between the incentre and the circumcentre of

the triangle of reference, prove by the modulus of the equation of the circumcircle that

$$\frac{1}{R+\delta}+\frac{1}{R-d}=\frac{1}{r}.$$

66. For any five circles prove—

$$\begin{vmatrix} 0, & 1, & 1, & 1, & 1, & 1, \\ 1, & 0, & \overline{12}^2, & \overline{13}^2, & \overline{14}^2, & \overline{15}^2, \\ 1, & \overline{21}^2, & 0, & \overline{23}^2, & \overline{24}^2, & \overline{25}^2, \\ 1, & \overline{31}^2, & \overline{32}^2, & 0, & \overline{34}^2, & \overline{35}^2, \\ 1, & \overline{41}^2, & \overline{42}^2, & \overline{43}^2, & 0, & \overline{45}^2, \\ 1, & \overline{51}^2, & \overline{52}^2, & \overline{53}^2, & \overline{54}^2, & 0 \end{vmatrix} = 0. \qquad (289)$$

(SALMON.)

Multiply together the two matrices, each of six rows and five columns—

$$\begin{Vmatrix} 1, & 0, & 0, & 0, & 0, \\ x'^2+y'^2-r'^2, & -2x', & -2y', & -2r', & 1, \\ x''^2+y''^2-r''^2, & -2x'', & -2y'', & -2r'', & 1, \\ & & \&c., & & \end{Vmatrix} \begin{Vmatrix} 0, & 0, & 0, & 0, & 1, \\ 1, & x', & y', & r', & x'^2+y'^2-r'^2, \\ 1, & x'', & y'', & r'', & x''^2+y''-r''^2, \\ & & \&c. & & \end{Vmatrix}$$

By supposing the circle S to touch all the others, $\overline{15}$, $\overline{25}$, $\overline{35}$, $\overline{45}$, all vanish, and we get a new proof of my extension of Ptolemy's theorem.

67. Prove the following relation between the angles of intersection of four circles:—

$$\begin{vmatrix} 0, & \frac{1}{r'}, & \frac{1}{r''}, & \frac{1}{r'''}, & \frac{1}{r''''}, \\ \frac{1}{r'}, & 1, & \cos\overline{12}, & \cos\overline{13}, & \cos\overline{14}, \\ \frac{1}{r''}, & \cos\overline{21}, & 1, & \cos\overline{23}, & \cos\overline{24}, \\ \frac{1}{r'''}, & \cos\overline{31}, & \cos\overline{32}, & 1, & \cos\overline{34}, \\ \frac{1}{r''''}, & \cos\overline{41}, & \cos\overline{42}, & \cos\overline{43}, & 1 \end{vmatrix} = 0. \qquad (290)$$

68. Prove by the modulus of the equation of the 'Nine-points Circle' that it touches the inscribed and escribed circles.

69. Prove that the determinant

$$\begin{vmatrix} x+g', & y+f', & g'x+f'y+c', \\ x+g'', & y+f'', & g''x+f''y+c'', \\ x+g''', & y+f''', & g'''x+f'''y+c''' \end{vmatrix} = 0 \qquad (291)$$

is the circle orthogonal to the three circles $x^2 + y^2 + 2gx' + 2fy' + c' = 0$, &c.

70. The circumcircle of the triangle, found by drawing through the vertices of the triangle of reference parallels to the sides, in areal co-ordinates, is

$$\frac{a^2}{\beta+\gamma} + \frac{b^2}{\gamma+\alpha} + \frac{c^2}{\alpha+\beta} = 0.$$

71. Find the equation of the circle through the points

$$(a \cos \alpha,\ b \sin \alpha); \quad (a \cos \beta,\ b \sin \beta); \quad (a \cos \gamma,\ b \sin \gamma).$$

(R. A. ROBERTS.)

72. Find the equation of the 'Nine-points Circle' of the triangle formed by the same points. (*Ibid.*)

73. Prove that the sides of a triangle *DEF*, homothetic to the triangle of reference with respect to its symmedian point, determine upon it six concyclic points; and that the locus of the centre of the circle passing through these points is a right line.—(NEUBERG.)

74. If the base *BC* of a triangle be given in magnitude and position, and the Brocard angle ω in magnitude, find the locus of the vertex.

Ans. If the middle point of *BC* be taken as origin, and the base and a perpendicular to it as axes, the locus is $x^2 + y^2 - (a \cot \omega)\, y + \frac{3a^2}{4} = 0$.

(*Ibid.*)

75. Find the equation of the circle whose diameter is the join of the orthocentre and centroid of the triangle of reference.

Ans. $\alpha^2 \sin 2A + \beta^2 \sin 2B + \gamma^2 \sin 2C - (\alpha\beta \sin C + \beta\gamma \sin A + \gamma\alpha \sin B) = 0$.

(BROCARD.)

CHAPTER IV.

THE GENERAL EQUATION OF THE SECOND DEGREE.

CARTESIAN CO-ORDINATES.

92. The equation $S \equiv ax^2 + 2hxy + by^2 + 2gx + 2fy + c = 0$, or, as it may be written, $u_2 + u_1 + u_0 = 0$, where u_2 denotes the terms of the second degree, &c., is the most general equation of the second degree. The object of this Chapter is to classify the curves represented by this equation, to reduce their equations to the normal forms, and to prove some of the properties common to all these curves. It will be shown in Chapter VIII. that every curve of the second degree can be obtained as the intersection of a cone standing on a circular base by a plane. In fact, it was from this point of view that these curves were first studied, and for this reason have been called "Conic Sections."

If we suppose the terms of the first degree removed, the equation will be of the form $ax^2 + 2hxy + by^2 + c = 0$, and this transformed into polar co-ordinates, gives

$$(a\cos^2\theta + b\sin\theta\cos\theta + b\sin^2\theta)\rho^2 + c = 0.$$

Now, since this quadratic in ρ wants its second term, its two roots will be equal in magnitude, but of opposite signs. Therefore to each value of θ there will be two equal values of ρ, of opposite signs; or, in other words, every line drawn through the origin is bisected at the origin. Hence, *When the equation of a curve of the second degree is of the form* $u_2 + u_0 = 0$, *the origin is the centre of the curve.*

93. *Terms of the first degree can be removed from the general equation* $S = 0$ *by transformation to parallel axes, unless* $ab - h^2 = 0$.

Dem.—Writing $x + \bar{x}$ for x, and $y + \bar{y}$ for y in $S = 0$, it becomes $\quad ax^2 + 2hxy + by^2 + 2g'x + 2f'y + c' = 0,$

where
$$g' \equiv a\bar{x} + h\bar{y} + g, \quad f' \equiv h\bar{x} + b\bar{y} + f,$$
$$c' \equiv a\bar{x}^2 + 2h\bar{x}\bar{y} + b\bar{y}^2 + 2g\bar{x} + 2f\bar{y} + c.$$

Now, if the new origin be the centre, we must (Art. 92) have g', f' each equal to zero. Hence, solving for $\bar{x}$, $\bar{y}$ from the equation
$$a\bar{x} + h\bar{y} + g = 0, \quad h\bar{x} + b\bar{y} + f = 0,$$
we get
$$\bar{x} = \frac{hf - bg}{ab - h^2}, \quad \bar{y} = \frac{gh - af}{ab - h^2}, \qquad (292)$$
or
$$\bar{x} = \frac{G}{C}, \quad \bar{y} = \frac{F}{C}. \quad \text{(See Art. 26.)}$$

Hence, except when $ab - h^2 =$ zero, the values of $\bar{x}$, $\bar{y}$ are finite, but these are the co-ordinates of the new origin; therefore, &c.

Cor.—The general equation $S = 0$ represents a central curve when the value of $ab - h^2$ differs from zero, and a non-central curve when it is equal to zero. In other words, *When the terms* u_2 *of* S *form a perfect square it represents a non-central curve; and when they do not form a perfect square it represents a central curve.*

94. *The lines* $ax + hy + g = 0$, $hx + by + f = 0$ *are diameters of* S.

For, solving from these equations, we get the co-ordinates of the centre. Hence each passes through the centre, and is therefore a diameter.

Or thus: the equation $S = 0$ may be written
$$(ax + hy + g)^2 - \{(h^2 - ab)y^2 + 2(gh - af)y + (g^2 - ac)\} = 0.$$
Or in the form
$$X^2 + Cy^2 - 2Fy + B = 0, \qquad \text{(see Art. 26)}$$
putting X for $(ax + hy + g)$.

It is evident that for each value of y there will be two values of X equal in magnitude, but of opposite signs: hence the line $X = 0$, or $ax + hy + g = 0$ is a diameter.

Cor. 1.—In non-central curves the lines $ax + hy + g = 0$, $hx + by + f = 0$ are parallel; for the condition of parallelism gives $ab - h^2 = 0$.

Cor. 2.—When the general equation $S = 0$ is referred to the centre as origin, and written in the form $ax^2 + 2hxy + by^2 + c' = 0$, then

$$c' = \frac{abc + 2fgh - af^2 - bg^2 - ch^2}{ab - h^2} \text{ or } \frac{\Delta}{C}; \qquad (293)$$

for the discriminant of $ax^2 + 2hxy + by^2 + 2gx + 2fy + c = 0$ is Δ, and the discriminant of $ax^2 + 2hxy + by^2 + c'$ is $abc' - c'h^2$;

and equating these we get $c' = \dfrac{\Delta}{C}$.

95. *In every curve of the second degree two real and distinct lines, two coincident lines, or two imaginary lines, can be drawn through the origin, each of which will meet the curve once at infinity.*

Dem.—Transforming S to polar co-ordinates, we get $(a \cos^2\theta + 2h \sin\theta \cos\theta + b \sin^2\theta)\rho^2 + 2(g\cos\theta + f\sin\theta)\rho + c = 0$;

or for shortness, $a'\rho^2 + 2b'\rho + c = 0$; and, putting $\rho = \dfrac{1}{\rho'}$, this becomes $c\rho'^2 + 2b'\rho' + a' = 0$. Now if $a' = 0$, one of the values of ρ' in this equation is zero, and the other value is finite. Again, if not only $a' = 0$, but $b' = 0$ also, then the second value of ρ' will be zero. Now when ρ' is zero, ρ is infinite. Hence, if in the equation $a'\rho^2 + 2b'\rho + c = 0$, $a' = 0$, one of the values of ρ will be infinite and the other finite; and if not only $a' = 0$ but also $b' = 0$, the two values of ρ will be infinite.

Now when $a' = 0$, we have $a \cos^2\theta + 2h \sin\theta \cos\theta + b \sin^2\theta = 0$; hence $a + 2h \tan\theta + b \tan^2\theta = 0$, an equation which gives two values for $\tan\theta$. Hence the proposition is proved.

96. If the two roots of the equation $a + 2h\tan\theta + b\tan^2\theta = 0$ (Art. 95) be real and unequal, the lines from the origin to meet the curve at infinity are real, as in the annexed diagram, the angles corresponding to the two values of θ being XOA, XOB. In order to find the equation of the lines OA, OB, let the co-ordinates of any point P in OA be xy; then we have $\dfrac{y}{x} = \dfrac{PM}{OM} = \tan\theta$.

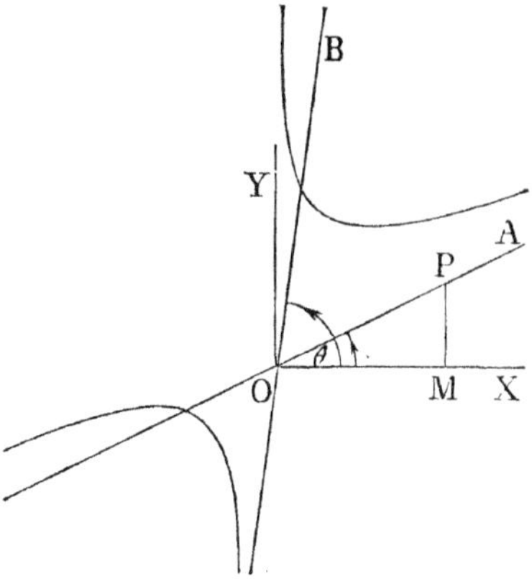

Hence, substituting $\dfrac{y}{x}$ for $\tan\theta$ in $a + 2h\tan\theta + b\tan^2\theta = 0$, we get

$$ax^2 + 2hxy + by^2 = 0. \qquad (294)$$

This form of the curve is called a hyperbola, and we see that $S = 0$ represents a hyperbola when $u_2 = 0$ represents two distinct lines. Now the condition that $u_2 \equiv ax^2 + 2hxy + by^2 = 0$ should denote two distinct lines is, that its discriminant $h^2 - ab$ should be positive. *Therefore if* $S = 0$ *represents a hyperbola,* $h^2 - ab$ *is positive.*

Secondly.—If the roots of $a + 2h\tan\theta + b\tan^2\theta = 0$ be equal, the two lines from the origin to meet the curve at infinity are coincident. *This variety of the curve is called a parabola.*

As before, to get the equation of these two coincident lines, put $\dfrac{y}{x} = \tan\theta$, and we get

$$ax^2 + 2hxy + by^2 = 0.$$

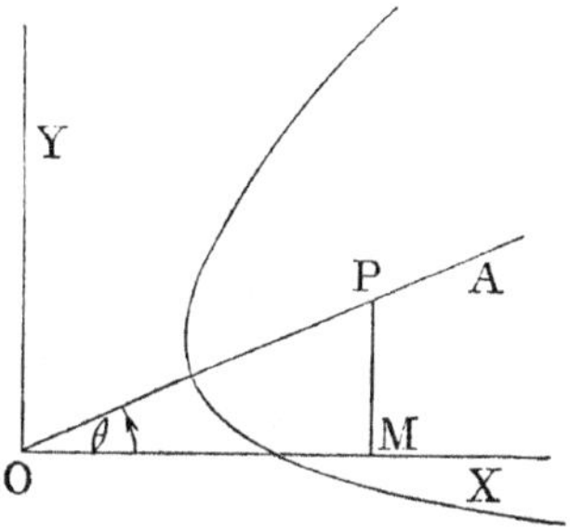

Hence, *When* $ax^2 + 2hxy + by^2$ *is a perfect square;* that is, *when* $h^2 - ab = 0$, *the curve is a parabola.*

Lastly.—Suppose the roots of $a + 2h \tan\theta + b \tan^2\theta = 0$ to be imaginary, then no real line can be drawn from the origin to meet the curve at infinity. *This species is closed in every direction, and is called an ellipse.* The equation of the imaginary lines from the origin to meet the curve at infinity is $ax^2 + 2hxy + by^2 = 0$, as before.

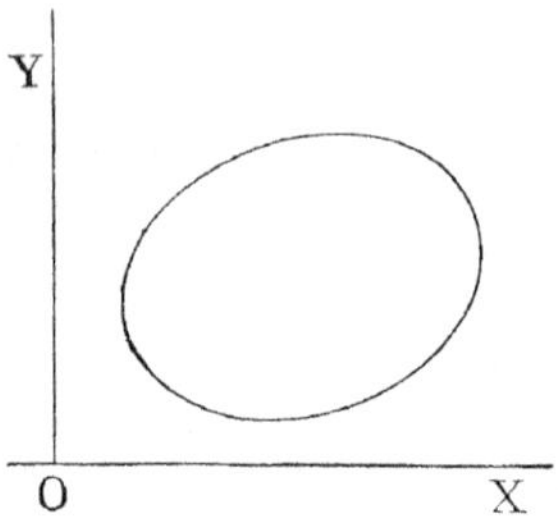

Now if this represents two imaginary lines, we must have $h^2 - ab$ negative. Hence the conditions for the three curves are—

For hyberbola, $h^2 - ab$ positive.

For parabola, $h^2 - ab$ equal zero.

For ellipse, $h^2 - ab$ negative.

Cor. 1.—The hyperbola meets the line at infinity in two real and distinct points—the parabola in two coincident points, and therefore touches it; and the ellipse in two imaginary points.

Cor. 2.—In the equation $S = 0$, if either a or b vanish, but not h, the curve is a hyperbola, for in either case $h^2 - ab$ is positive.

Cor. 3.—If a and b have contrary signs, the curve is a hyperbola.

Cor. 4.—If $a + b = 0$, the lines $ax^2 + 2hxy + by^2 = 0$ are at right angles to each other. *The curve in this case is called an equilateral hyperbola, and sometimes a rectangular hyperbola.*

Cor. 5.—The circle is a species of ellipse; for in the equation of the circle $h = 0$ and $a = b$. Hence $h^2 - ab$ is negative.

Cor. 6.—The ellipse and hyperbola are central curves, and the parabola non-central.

97. *The locus of the middle points of a series of parallel chords of a curve of the second degree is a right line.*

Dem.—Let S represent the curve given by its general equation; ADE one of the chords of the system: bisect the intercept DE in C; the locus of C is required. Let the equation of AE be $y = mx + n$; and supposing m constant, and n variable, we have a system of parallel lines. Now, substituting $mx + n$ for y in the general equation, we get

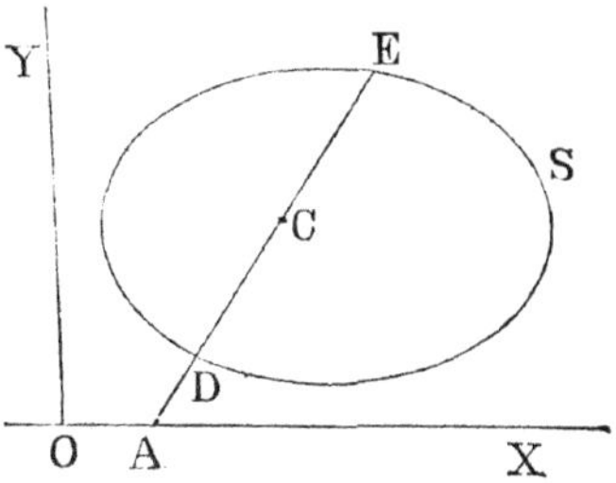

$$(a + 2mh + m^2b)\,x^2 + 2\,(hn + mbn + g + mf)\,x + n^2 + 2nf + c = 0.$$

Half the sum of the roots of this equation will be the ordinate of the middle point C. Hence, for that point, we have

$$x = -\frac{hn + bmn + g + mf}{a + 2mh + m^2b};$$

and, eliminating n between this and the equation $y = mx + n$, we get

$$ax + hy + g + m(hx + by + f) = 0, \qquad (295)$$

which is the locus required.

Cor. 1.—Since the lines $ax + hy + g = 0$, $hx + by + f = 0$ are diameters, the line $(ax + hx + g) + m\,(hx + by + f) = 0$ is a diameter, as is otherwise evident. By putting $m = \tan\theta$, this may be written $(ax + hy + g)\cos\theta + (hx + by + f)\sin\theta$. Hence, putting first $\theta = 0$, and then $\theta = \dfrac{\ }{2}$, we see that $ax + hy + g = 0$ is the equation of the diameter which bisects chords parallel to the axis of x, and $hx + by + f = 0$ of the diameter which bisects chords parallel to the axis of y. Employing the notation of the Differential Calculus, these propositions may be more simply stated, thus:—*If* $S = 0$ *be the general equation of the second degree,* $\dfrac{dS}{dx} = 0$ *is the equation of the diameter which bisects chords parallel to the axis of* x*, and* $\dfrac{dS}{dy} = 0$ *that which bisects chords parallel to the axis of* y.

Cor. 2.—If $S = 0$ be a parabola, the lines $ax + hy + g = 0$, $hx + by + f = 0$, are each parallel to the line which can be drawn from the origin to meet the curve at infinity; for in that case $h = \sqrt{ab}$; and, substituting in $ax^2 + 2hxy + by^2 = 0$, we get $(\sqrt{a} \,.\, x + \sqrt{b} \,.\, y)^2 = 0$. Hence, in the parabola, the line through the origin to meet the curve at infinity is $\sqrt{a} \,.\, x + \sqrt{b} \,.\, y = 0$; or, multiplying successively by $\sqrt{a}$, $\sqrt{b}$, the line may be written either $ax + by = 0$ or $hx + by = 0$, and the foregoing lines differ only by a constant from these.

98. *If two diameters be such that the first bisects chords parallel to the second, the second bisects chords parallel to the first.*

For if m be the tangent of the angle which the second diameter makes with the axis of x, the equation of the first diameter is

$$(ax + hy + g) + m(hx + by + f) = 0;$$

and if m' be the tangent of the angle which this makes with the axis of x, we get

$$m' = -\frac{a + mh}{h + mb},$$

or

$$a + (m + m')h + mm'b = 0; \qquad (296)$$

since this remains unaltered by the interchange of m and m', the proposition is proved.

Def.—*A pair of diameters, so related that each bisects chords parallel to the other, are called conjugate diameters.*

Cor. 1.—If in the general equation $h = 0$, the axes of x and y are parallel to a pair of conjugate diameters; for if $h = 0$, $ax + hy + g = 0$ reduces to $ax + g = 0$, which is parallel to the axis of y; that is, the diameter which bisects chords parallel to the axis of x is parallel to the axis of y.

99. *To find the ratio in which the join of the points $x'y'$, $x''y''$ is cut by S.*—(JOACHIMSTHAL.)

Let the ratio be $k : 1$; then the co-ordinates of the point of intersection are

$$\frac{x' + kx''}{1 + k}, \qquad \frac{y' + ky''}{1 + k};$$

and these substituted in S, give the quadratic

$$S' + 2kP'' + k^2S'' = 0, \qquad (297)$$

where S', S'' denote the powers of the given points with respect to S, and P'' the power of $x''y''$ with respect to the line

$$P \equiv (ax' + hy' + g)x + (hx' + by' + f)y + gx' + fy' + c = 0. \qquad (298)$$

The equation (297) is a fundamental one in the theory of conics. Several important theorems can be inferred from it by supposing its roots to have special relations to each other.

1°. *Suppose the sum of the roots to be zero.* Then $P'' = 0$, and the point $x''y''$ must be on the line P.

Let, in the annexed diagram, Q, R be the points where

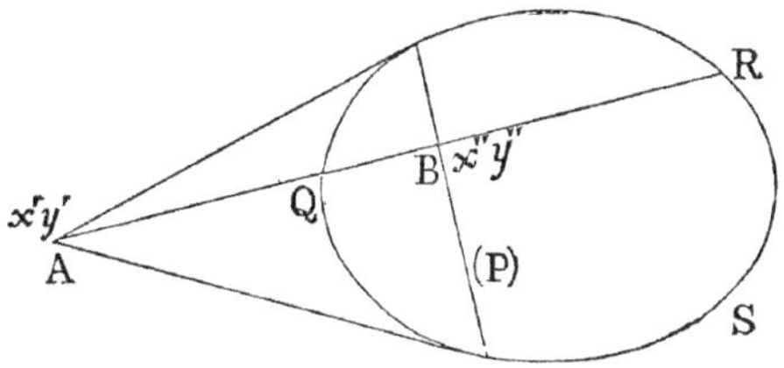

the join of the points A and B meets the curve, the values of k are $AQ : QB$, $AR : RB$, and these are equal, but with contrary signs. Hence AB is divided harmonically in Q and R.

Cor. 1.—Any line through A is divided harmonically by P and S.

Cor. 2.—P is the chord of contact of tangents from A. For if the line QR turn round A until the points Q, R coincide; then, since B is the harmonic conjugate of A with respect to Q, R, when Q, R come together, B coincides with them and AB will be a tangent.

DEF.—*The line P is called the polar of the point $x'y'$.*

Cor. 3.—If a point be external to a conic its polar cuts the conic. If the point be internal its polar is external. For the harmonic conjugate to an internal point on any line passing through it is external to the conic. Lastly, if a point be on the conic, its polar, being the secant through two consecutive points of contact, is a tangent.

Cor. 4.—If $x'y'$ be a point on S, the tangent at $x'y'$ is the equation (298).

2°. *Let the roots of* (297) *be equal.* Since the roots are the ratios $AQ : QB$, $AR : RB$, they will be equal only when the points Q, R coincide, that is, when the line AB is a tangent to the curve. The condition for equal roots in (297) is $S'S'' - P''^2 = 0$, which must be fulfilled when the point $x''y''$ is on either of the tangents from $x'y'$ to S. Hence, supposing the latter fixed and the former variable, we get the equation of tangents to S from $x'y'$, by removing the double accents, to be

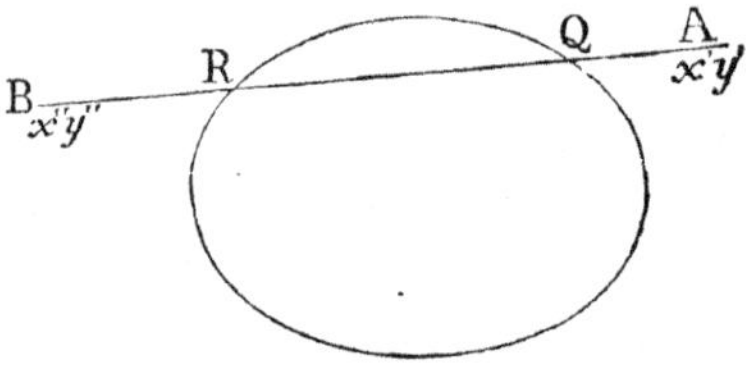

$$SS' - P^2 = 0. \qquad (299)$$

3°. *Let the anharmonic ratio of the four points A, B, Q, R be given*, then the roots of (297) have a given ratio. Let this ratio be λ, and changing k into $k\lambda$ in (297), we get

$$S' + 2\lambda kP'' + \lambda^2 k^2 S'' = 0.$$

Eliminating k between this and (297), and omitting double accents, we get the locus of a point B, which divides a secant of S passing through a given point in a given anharmonic ratio, viz.,

$$(1+\lambda)^2 SS' - 4\lambda P^2 = 0. \qquad (300$$

100. *If through any point P two chords, whose direction angles are θ, θ', be drawn cutting the conic*

$$ax^2 + 2hxy + by^2 + 2gx + 2fy + c = 0$$

in the points A, B; C, D respectively, then

$$\frac{PA \cdot PB}{PC \cdot PD} = \frac{a\cos^2\theta' + 2h\sin\theta'\cos\theta' + b\sin^2\theta'}{a\cos^2\theta + 2h\sin\theta\cos\theta + b\sin^2\theta}. \qquad (301$$

Dem.—Transforming the given equation to P as origin, we get

$$S \equiv ax^2 + 2hxy + by^2 + 2g'x + 2f'y + c' = 0; \quad \text{(see Art. 94)}$$

and transforming this to polar co-ordinates, we get

$$(a\cos^2\theta + 2h\sin\theta\cos\theta + b\sin^2\theta)\rho^2 + 2(g'\cos\theta + f'\sin\theta) + c' = 0,$$

and the roots of this quadratic are PA, PB. Hence

$$PA \cdot PB = \frac{c'}{a\cos^2\theta + 2h\sin\theta\cos\theta + b\cos^2\theta}.$$

Similarly,

$$PC \cdot PD = \frac{c'}{a\cos^2\theta' + 2h\sin\theta'\cos\theta' + b\cos^2\theta'};$$

and dividing one of these equalities by the other, the proposition is proved.

Cor.—If through any other point P' two lines $P'A'B'$, $P'C'D'$ be drawn parallel respectively to the former, and cutting the conic in the points A', B'; C', D', then

$$PA \cdot PB : PC \cdot PD :: P'A' \cdot P'B' : P'C' \cdot P'D'. \qquad (301)$$

101. The theorem of the last Article corresponds to Euclid III., XXXV., XXXVI. The following are special cases:—

1°. If P be the centre, then $PA = PB$, $PC = PD$, and we have the following theorem from (301):—*The rectangles con-*

tained by the segments of any two chords of a conic are proportional to the squares of the parallel semidiameters.

2°. If the lines PA, PC turn round the point P until they become tangents, $PA \, . \, PB$ becomes PA^2, and $PC \, . \, PD$ becomes PD^2, and we have the following theorem :—*The squares of two tangents drawn from any point to a conic are proportional to the rectangles contained by the segments of any two parallel chords. Also two tangents from any point to the conic are proportional to the parallel semidiameters.*

3°. Let the join of PP' produced be a diameter, and let the lines through P be this diameter and its conjugate CD, then the chords through P' will be AB and $C'D'$, of which the latter is bisected in P'. Denoting AP by a, PC by b, PP' by x, and $P'C'$ by y, we have, from (301),

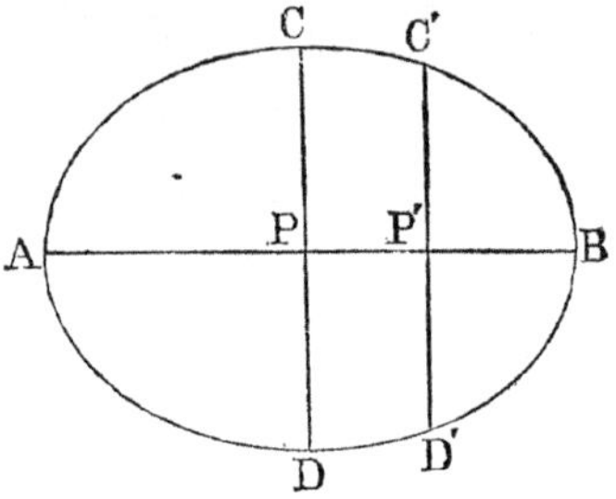

$$a^2 : b^2 :: (a + x)(a - x) : y^2;$$

or,

$$\frac{x^2}{a^2} + \frac{y^2}{b^2} = 1, \qquad (302)$$

which is the normal form of the equation for central conics.

4°. Let PB, $P'B'$ meet the curve at infinity, then in the proportion (301) it is evident that PB, $P'B'$ meeting the curve in the same point at infinity have to each other a ratio of equality. Hence in this case we have

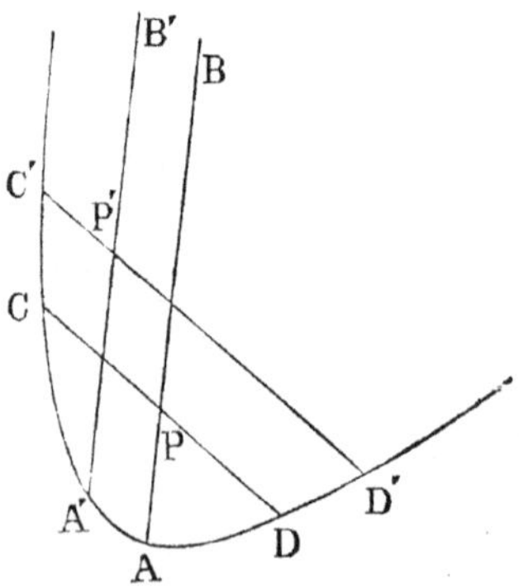

$$AP : A'P' :: CP \, . \, PD : C'P' \, . \, P'D'. \qquad (303)$$

5°. Let the curve be the parabola, and let the line joining the points P, P' have the direction which meets the curve at infinity; then if CD, $C'D'$ belong to the system of parallel chords (Art. 97) which this line bisects, we have, from (303),

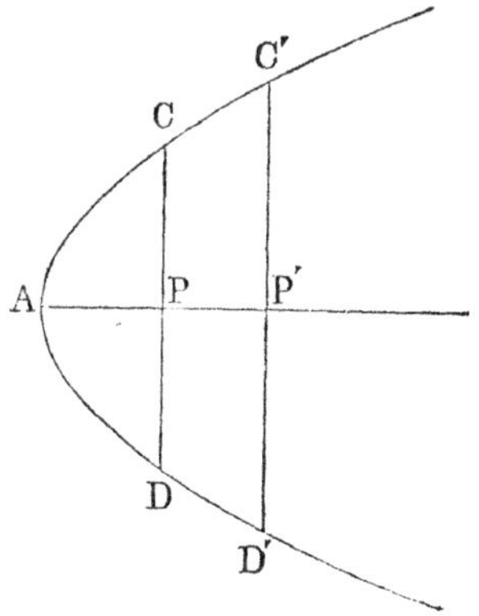

$$AP : AP' :: CP^2 : C'P'^2.$$

Hence, supposing P fixed, and P' variable, and denoting AP', $P'C'$ by x, y respectively, we have $y^2 : CP^2 :: x : AP$; hence, putting $CP^2 = 4a\,AP$, we have

$$y^2 = 4ax, \qquad (304)$$

which is the normal form of the equation of the parabola.

102. It has been proved in Art. 94 that when the centre is taken as origin the equation of the curve can be written in the form $ax^2 + 2hxy + by^2 + c' - 0$. We shall now show that, retaining the same origin (viz. the centre), this equation can be further simplified. Thus, transforming by the substitutions of Art. 10 to new rectangular axes inclined at an angle θ to the old, that is, putting $x = x\cos\theta - y\sin\theta$, $y = y\cos\theta + x\sin\theta$, we get $a'x^2 + 2h'xy + b'y^2 + c' = 0$,

where
$$a' = a\cos^2\theta + b\sin^2\theta + h\sin 2\theta, \qquad (305)$$

$$b' = a\sin^2\theta + b\cos^2\theta - h\sin 2\theta, \qquad (306)$$

$$2h' = 2h\cos 2\theta - (a - b\sin 2\theta). \qquad (307)$$

From these equations we get, after an easy calculation,

$$a' + b' = a + b, \text{ and } a'b' - h'^2 = ab - h^2. \qquad (308)$$

Hence $a + b$ and $ab - h^2$ are invariants. In other words, they are functions of the coefficients which are unaltered by transformation.

103. If $h' = 0$, we have, from equation (307),

$$\tan 2\theta = \left(\frac{2h}{a-b}\right), \tag{309}$$

and the equation of the curve is reduced to the normal form, $a'x^2 + b'y^2 + c' = 0$. The value of θ obtained from (309) substituted in this gives the values of the new coefficients a', b' in terms of the old; but we get them more simply from equation (308); for if $h' = 0$, we have $a' + b' = a + b$, and

$$a'b' = ab - h^2:$$

solving from these we get, putting

$$R^2 = 4h^2 + (a-b)^2,$$

$$a' = \tfrac{1}{2}(a + b - R), \qquad b' = \tfrac{1}{2}(a + b + R).$$

Hence the equation of the curve referred to rectangular conjugate diameters is

$$(a + b - R)x^2 + (a + b + R)y^2 + 2c' = 0. \tag{310}$$

Cor. 1.—The equation of the new axes, when referred to the old, are $x - y \cot\theta = 0$, $x + y \tan\theta = 0$. Hence, multiplying and making use of (309), the equation of the axes is

$$hx^2 - (a - b)xy - hy^2 = 0. \tag{311}$$

Cor. 2.—If the equation (310) be written in the form

$$\frac{x^2}{\alpha^2} + \frac{y^2}{\beta^2} = 1,$$

α^2, β^2 will be the roots of the quadratic

$$\xi^2 + \frac{\Delta}{C^2}\xi + \frac{\Delta}{C^3} = 0, \text{ where } C = ab - h^2. \tag{312}$$

104. Def.—*Any line, except the line at infinity, which touches a curve at infinity is called an asymptote to the curve.*

105. *Each of the lines represented by the equation* $ax^2 + 2hxy + by^2 = 0$ *is an asymptote to the conic* $ax^2 + 2hxy + by^2 + c = 0$.

Dem.—The equation of the tangent to $ax^2 + 2hxy + by^2 + c = 0$ at $x'y'$ is $(ax' + hy')x + (hx' + by')y + c = 0$ (see Art. 99, Cor. 4). Now put $\frac{y'}{x'} = m$, and we get

$$(a + mh)x + (h + mb)y + \frac{c}{x'} = 0.$$

If the point of contact be at infinity, x' is infinite, and $\frac{c}{x'} = 0$, and we get $(a + mh)x + (h + mb)y = 0$, which represents a line passing through the origin; and since it also passes through $x'y'$, we must have $\frac{y}{x} = \frac{y'}{x'} = m$. Hence, substituting $\frac{y}{x}$ for m,

we get $$ax^2 + 2hxy + by^2 = 0, \qquad (313)$$
which is the equation of the two asymptotes.

Hence every central conic has two asymptotes, which for the hyperbola are real, because $ax^2 + 2hxy + by^2 = 0$ is for that curve the product of two real factors, and imaginary for the ellipse.

Cor. 1.—When the equation of a conic is in the form $u_2 + u_0 = 0$, $u_2 = 0$ denotes the asymptotes.

Cor. 2.—If when a constant is subtracted from the equation of a conic the remainder is the product of two lines, these lines will be the asymptotes.

Cor. 3.—The line at infinity is the polar of the centre; for it is the chord of contact of tangents drawn from the centre (the asymptotes).

Cor. 4.—In order to find the asymptotes of the conic given by the general equation $S = 0$, equate the discriminant of

$S - \lambda$ to zero, and we get $\lambda = \dfrac{\Delta}{ab - h^2}$.

Hence

$$ax^2 + 2hxy + by^2 + 2gx + 2fy + c - \frac{\Delta}{ab - h^2} = 0,$$

or

$$(ab - h^2)(ax^2 + 2hxy + by^2 + 2gx + 2fy) - 2fgh + af^2 + bg^2 = 0, \qquad (314)$$

denotes the asymptotes.

Exercises on the General Equation.

1. Prove that five conditions are sufficient to determine a conic.

2. Transform the following curves to their centres :—

$$1^\circ. \quad 4x^2 - 6xy + 6y^2 + 10x - 12y + 13 = 0.$$
$$2^\circ. \quad xy + 4ax - 2by = 0.$$
$$3^\circ. \quad 3x^2 - 2xy - 3y^2 + 6x - 9y = 0.$$

3. What curves are represented by the equations

$$1^\circ. \quad \sqrt{x+a} - \sqrt{y+b} = \sqrt{a+b};$$
$$2^\circ. \quad (x+1)^{-1} + (y+2)^{-1} = 2;$$
$$3^\circ. \quad \cos^{-1} x + \cos^{-1} y = \frac{\pi}{3}?$$

4. Find the equation of the asymptotes of the hyperbola

$$3x^2 - 4xy - 5y^2 + 2x - 4y + 6 = 0.$$

5. Prove that the equation of the chord of the conic

$$ax^2 + 2hxy + by^2 + 2gx + 2fy + c = 0,$$

which passes through the origin, and is bisected at that point is $gx + fy = 0$.

6. The polar of the origin with respect to $S = 0$ is $gx + fy + c = 0$.

7. The maximum and minimum semi-diameters of a central conic are conjugate semi-diameters, and perpendicular to each other.

For, transforming $ax^2 + 2hxy + by^2 + c' = 0$ to polar co-ordinates, we get

$$(a \cos^2\theta + 2h \sin\theta \cos\theta + b \sin^2\theta)\rho^2 + c' = 0;$$

and ρ will be a minimum when $a \cos^2\theta + 2h \sin\theta \cos\theta + b \sin^2\theta$ is a maximum. Now this last will be a maximum when $(a - b) \cos 2\theta + 2h \sin 2\theta$ is a maximum ; but

$$\{(a-b)\cos 2\theta + 2h \sin 2\theta\}^2 + \{(a-b)\sin 2\theta - 2h\cos 2\theta\}^2 = R^2.$$
(Art. 103)

Hence the required maximum is when $\tan 2\theta = \frac{2h}{a-b}$. (See equation 309.)

From this equation we get two values of $\tan\theta$, the product of which is negative unity, showing that they belong to perpendicular semi-diameters. The values of $\tan\theta$ are $\frac{b - a \pm R}{2h}$. Hence the two semi-diameters are $2hy + (a - b \pm R)x = 0$; or, multiplying and reducing, $h(x^2 - y^2) - (a-b)xy = 0$, which is the equation of the pair of lines bisecting the angles between the asymptotes. Compare Cor. 1, Arts. 103, and 104.

8. If the line joining any fixed point O to a variable point P of a conic S meet a fixed line in the point Q; prove, if R be the harmonic conjugate of P with respect to O and Q, that the locus of R is a conic.

9. Find the locus of the centre of a conic passing through four given points. If S, S be two fixed conics passing through the given points, then $S + kS'$ is the most general equation of a conic passing through them, and the centre of this is the intersection of the diameters

$$\frac{dS}{dx} + k\frac{dS'}{dx} = 0; \quad \frac{dS}{dy} + k\frac{dS'}{dy} \ 0. \quad \text{(See Art. 97, Cor. 1.)}$$

Hence, eliminating k, the required locus is the determinant

$$\begin{vmatrix} \dfrac{dS}{dx}, & \dfrac{dS'}{dx}, \\ \dfrac{dS}{dy}, & \dfrac{dS'}{dy} \end{vmatrix} = 0. \qquad (315)$$

Thus, if one of the three pairs of lines passing through the four points be taken as axes, another pair may be written

$$\left(\frac{x}{\lambda} + \frac{y}{\mu} - 1\right)\left(\frac{x}{\lambda'} + \frac{y}{\mu'} - 1\right) = 0.$$

These being taken for S, S' respectively, the required locus will be

$$\left\{\frac{2x^2}{\lambda\lambda'} - x\left(\frac{1}{\lambda} + \frac{1}{\lambda'}\right)\right\} - \left\{\frac{2y^2}{\mu\mu'} - y\left(\frac{1}{\mu} + \frac{1}{\mu'}\right)\right\} = 0, \qquad (316)$$

the discriminant of which is

$$\frac{(\lambda + \lambda')^2}{\lambda\lambda'} - \frac{(\mu + \mu')^2}{\mu\mu'} = 0. \qquad (317)$$

10. With the same notation, find the value of k, in order that $S + kS'$ may be an equilateral hyperbola.

$$\textit{Ans.}\ k = \frac{1}{\lambda}\left\{\frac{1}{\lambda' \cos\omega} - \frac{1}{\mu'}\right\} + \frac{1}{\mu}\left\{\frac{1}{\mu' \cos\omega} - \frac{1}{\lambda'}\right\}. \qquad (318)$$

11. If the harmonic mean between the rectangles contained by the segments of two perpendicular chords of a conic be given, the locus of their point of intersection is a conic.

12. Prove that through four points can be drawn two parabolas, and that the directions of their diameters are at right angles to each other.

13. Find the equation of the pair of tangents from the origin to the conic, $ax^2 + 2hxy + by^2 + 2gx + 2fy + c = 0$.

The line $y = mx$ passes through the origin, and, eliminating y, we get

$$(a + 2mh + m^2b)\,x^2 + 2\,(g + mf)\,x + c = 0,$$

the discriminant of which is

$c(a + 2mh + m^2b) - (g + mf)^2 = 0$; and, substituting $\dfrac{y}{x}$ for m, we get

$$(ac - g^2)\,x^2 + 2\,(ch - fg)\,xy + (bc - f^2)\,y^2 = 0. \qquad (319)$$

14. Find the equation of the chord joining the points $x'y'$, $x''y''$ on the conic $S \equiv ax^2 + 2hxy + by^2 + 2gx + 2fy + c = 0$.

The conic

$$S' \equiv a\,(x - x')\,(x - x'')(+ h\,\{\,(x - x')(y - y'') + (x - x'')\,(y - y')\}$$
$$+ b\,(y - y')\,(y - y'')\} = 0$$

evidently passes through $x'y'$, $x''y''$. Hence $S - S' = 0$ is the required chord.

15. Find the condition that $\lambda x + \mu y + \nu = 0$ may be a tangent to $S = 0$.

Eliminating y between $\lambda x + \mu y + \nu = 0$ and $S = 0$, and forming the discriminant of the resulting equation in x, we get $A\lambda^2 + B\mu^2 + C\nu^2 + 2F\mu\nu + 2G\nu\lambda + 2H\lambda\mu = 0$, where A, B, &c., have the same meaning, as in Art. 26. (320)

16. If α, β denote the co-ordinates of the middle point of the chord in Ex. 14, we get

$$S - S' \equiv 2(a\alpha + h\beta + g)x + 2\,(h\alpha + b\beta + f\,)\,y - \{ax'x'' + h\,(x'y'' + x''y') + by'y''\} = 0.$$

If this chord make an angle θ with the axis of x, we have

$$\tan\theta = -\frac{a\alpha + h\beta + g}{h\alpha + b\beta + f}.$$

Hence, putting xy for $\alpha\beta$, the locus of the middle points of chords making an angle θ with the axis of x, is

$$(ax + hy + g)\cos\theta + (hx + by + f\,)\sin\theta = 0.$$

Compare Art. 97, equation (295).

17. If two points A, B be such that the polar of A passes through B, the polar of B passes through A.

18. To describe a conic section (x.) through five given points A, B, C, D, E.

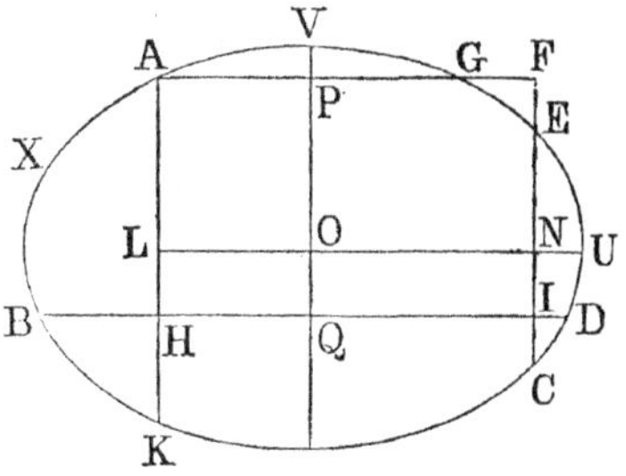

Join B, D, C, E. Through A draw AG parallel to BD, cutting the conic in G, and AK parallel to CE, cutting BD in H. Then $BI \cdot ID : CI \cdot IE :: BH \cdot HD : AH \cdot HK$; therefore K is a given point. In like manner, G is a given point. Hence, bisecting AK in L, CE in N, AG in P, and BD in Q, O, the point of intersection of LN and PQ, is given. Again (Art. 101), $PG^2 : QD^2 :: OV^2 - OP^2 : OV^2 - OQ^2$; hence V is a given point. In like manner U is a given point, and OV, OQ are semiconjugate axes. Hence, &c.

CHAPTER V.

THE PARABOLA.

106. Def. I.—*Being given in position a point S and a line NN'. The locus of a variable point P, whose distance SP from S is equal to its perpendicular distance PN from NN', is called* A PARABOLA.

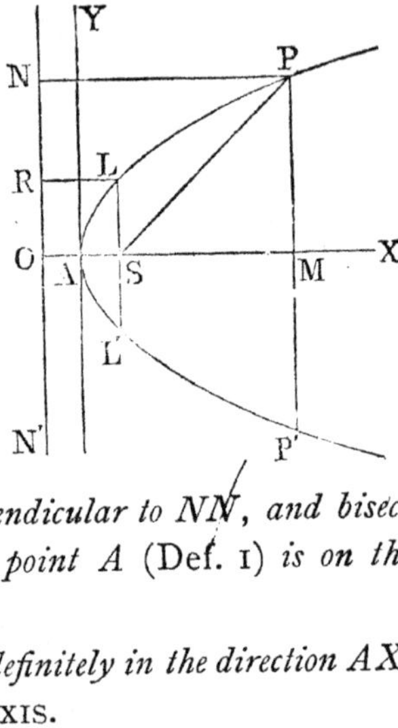

It will be seen subsequently that this definition agrees with that already given in p. 124.

II.—*The point S is called the* FOCUS, *and the line NN' the* DIRECTRIX.

III.—*If from S we draw SO perpendicular to NN', and bisect it in A; then, since $OA = AS$, the point A* (Def. I) *is on the parabola, and is called the* VERTEX.

IV.—*If the line AS be produced indefinitely in the direction AX, the whole line produced is called the* AXIS.

107. *To find the equation of the parabola.*

Let the vertex A be taken as origin, and AX and AY perpendicular to it as axes. Then denoting $OA = AS$ by a, and the co-ordinates of any point P in the curve by x, y, we have (Def. I.) $SP = PN$; but $PN = OM = OA + AM = a + x$; therefore $SP = a + x$.

Again, $\quad SM = AM - AS = x - a$, and $PM = y$.

Hence, from the right-angled triangle SMP, we have

$$(x - a)^2 + y^2 = (a + x^2); \text{ therefore } y^2 = 4ax, \qquad (321)$$

which is the standard form of the equation of the parabola. Compare Art. 101, *Cor.* 5, equation (304). From the equation of the parabola, we see that two values of y correspond to each value of x; and that these are equal in magnitude, but contrary signs. Hence, if PM be produced, it will meet the curve on the other side of the axis in a point P', such that $PM = MP'$. *Hence the axis of the parabola is an axis of symmetry of the figure.*

v.—*The double ordinate* LL' *through the focus is called the* LATUS RECTUM *of the parabola.*

Cor.—The latus rectum $= 4a$; for $SL = LR = OS = 2a$; therefore $LL' = 4a$.

108. *The co-ordinates of a point on the parabola can be expreseed in terms of a single variable.*

For, writing the equation in the form $2x \cdot 2a = y^2$, it is a special case of $LM = R^2$, a form in which each of the three conics may be written; and we may put $2x = y \tan \phi$, $2a = y \cot \phi$, or which is the same thing, $y = 2a \tan \phi$, $x = a \tan^2 \phi$. Hence the co-ordinates of a point on the parabola may be denoted by $a \tan^2 \phi$, $2a \tan \phi$. *We shall for shortness call it the point* ϕ, *and* ϕ *the* INTRINSIC ANGLE *of the point.*

Cor. 1.—Since $PS = a + x = a + a \tan^2 \phi = a \sec^2 \phi$, the distance of the point ϕ from the focus is $a \sec^2 \phi$.

Cor. 2.—The angle ASP is equal to twice the intrinsic angle of P.

For $$\cos MSP = \frac{MS}{SP} = \frac{a \tan^2 \phi - a}{a \sec^2 \phi} = -\cos 2\phi;$$

therefore $$ASP = 2\phi.$$

109. *To find the equation of the chord passing two points $x'y'$, $x''y''$ on the parabola.*

Let the intrinsic angles of the points be ϕ', ϕ''; then the required equation is (Art. 20, Ex. 3, 4°).

$$2x - (\tan\phi' + \tan\phi'')y + 2a\tan\phi'\tan\phi'' = 0; \qquad (322)$$

or, putting for $\tan\phi'$, $\tan\phi''$ their values in terms y', y'',

$$4ax = (y' + y'')y - y'y''. \qquad (323)$$

Examples.

1. If a chord of a parabola cut the axis in a fixed point, the rectangle contained by the tangents of the intrinsic angles of its extremities is constant.

Because if we put $x = AO$, $y = 0$, in equation (322), we get

$$\tan\phi' \,.\, \tan\phi'' = -\frac{OA}{a} \qquad (324)$$

2. If PM, $P'M'$ be the ordinates of the points P, P', and OQ the ordinate of O, $PM \,.\, P'M' = -OQ^2$.

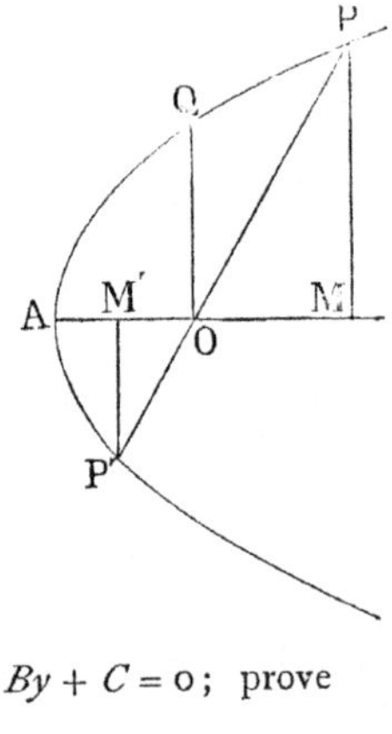

For, from equation (324), we get

$$(2a\tan\phi')(2a\tan\phi'') = -4a \,.\, OA = -OQ^2.$$

3. In the same case, $AM \,.\, AM' = AO^2$.

4. The direction tangent of PP' is

$$\frac{2}{\tan\phi' + \tan\phi''}. \qquad \text{(See equation (322).)}$$

Hence, if a chord of a parabola be parallel to a fixed line, the sum of the tangents of the intrinsic angles of its extremities is constant.

5. If PN, $P'N'$ be perpendiculars from the extremities of a focal chord on the line $Ax + By + C = 0$; prove

$$\frac{PN}{PS} + \frac{P'N'}{P'S} = \frac{Aa + C}{a\sqrt{A^2 + B^2}}.$$

6. If PP' cut the axis of y in a fixed point Q, from equation (323) we get $\cot\phi' + \cot\phi'' = \dfrac{2a}{AQ}$. Hence, *If through a fixed point on the tangent at the vertex of a parabola any secant be drawn, the sum of the cotangents of the intrinsic angles of its points of intersecting the parabola is constant.*

110. *To find the equation of the tangent to the parabola of the point $x'y'$.*

In equation (322), suppose the points ϕ', ϕ'' become consecutive, then their joining chord becomes a tangent, viz.

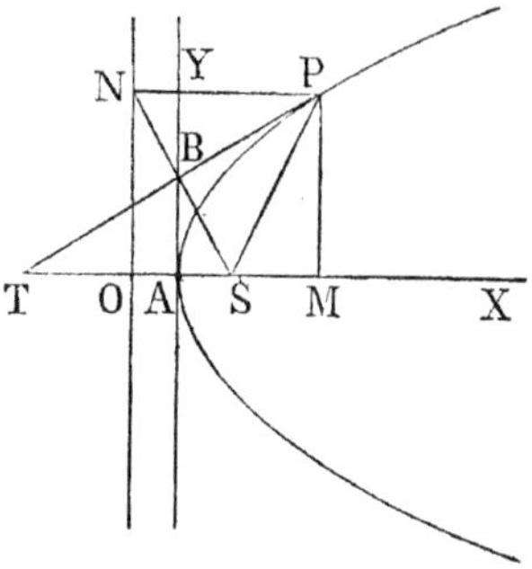

$$x - y\tan\phi' + a\tan^2\phi' = 0, \quad (324)$$

or putting $x' = a\tan^2\phi'$, $y' = 2a\tan\phi'$,

$$yy' = 2a(x + x'). \quad (325)$$

Cor. 1—If PT be the tangent, putting $y = 0$, we get from (325),

$$x = -x';$$

but when $y = 0$, $x = AT$. Hence, since $x' = AM$, we have $AT = -AM$; therefore $TA = AM$. Hence TM is bisected in A.

Def.—*The line TM, intercepted on the axis between the ordinate and the tangent, is called the sub-tangent.* Hence in the parabola the subtangent is bisected at the vertex.

Cor. 2.—The axis of y is the tangent at the vertex of the parabola; for if in (325) we put $x' = 0$, $y' = 0$, we get $x = 0$.

Cor. 3.—The equation (324) may be written $y = x\cot\phi' + a\tan\phi'$, from which it is seen that ϕ' is the angle PBY, which the tangent PT at P makes with AY, the tangent at A. Hence we have the following theorem :—

The intrinsic angle of any point of a parabola is equal to the angle which the tangent at that point makes with the tangent at the vertex.

If s denote the length of an arc of any curve measured from some fixed point A to a variable point P; ϕ the inclination of the tangent at the latter point to the tangent at the fixed extremity A; then the equation expressing the relation between s and ϕ has been by Dr. Whewell (*Phil. Trans.*, vol. viii., p. 659) termed the *intrinsic equation* of the curve, a nomenclature which has been adopted by mathematicians. It was this that suggested the propriety of calling ϕ the *intrinsic angle*.

Cor. 4.—Since $TA = x'$, $TS = x' + a = a \sec^2 \phi = SP$, (108, *Cor.* 1); hence $TS = SP$; therefore the angle $SPT = STP = TPN$. Hence PT bisects the angle SPN.

Def.—*If from a fixed point in the plane of a curve perpendiculars be let fall on its tangents, the locus of their feet is called the first positive pedal of the curve with respect to the point. Also the pedal of the first positive pedal is called the second positive pedal, &c.* Conversely, *the curve itself is called, in relation to a positive pedal of any order, the negative pedal of the same order.*

Cor. 5.—If PT meet the tangent at the vertex in B, since $TA = AM$, $TB = BP$; hence the triangles TBS, PBS are equal in every respect; therefore the angle PBS is right, and SB is perpendicular to the tangent. *Hence the pedal of a parabola with respect to the focus is the tangent at the vertex.*

Cor. 6.—If p denote the length of the perpendicular from S on PT,

$$p = \sqrt{a(a + x')}.$$

For since the angle ASB is equal to ϕ', we have

$$AS \div SB = \cos \phi', \text{ that is } \frac{a}{p} = \cos \phi'.$$

Hence
$$p = a \sec \phi' = \sqrt{a\,(a + x')}. \qquad (326)$$

Or thus: the triangles ASB, SBP are equiangular; hence

$$AS : SB :: SB : SP; \text{ that is, } a : p :: p : a + x'.$$

Cor. 7.—The equation of any tangent to a parabola may be written in the form

$$y = mx + \frac{a}{m}, \qquad (327)$$

for equation (324) will reduce to the form if we put $m = \cot \phi'$.

EXAMPLES.

1. The first negative pedal of a right line is a parabola.

2. The circle described about the triangle formed by three tangents to a parabola passes through the focus; for the feet of perpendiculars from the focus on these tangents are collinear.

3. The polar reciprocal of a parabola with respect to the focus is a circle; for the reciprocal is the inverse of the pedal with respect to the focus, which (Cor. 5) is a right line.

4. The polar reciprocal of a circle with respect to a point in its circumference is a parabola.

5. Given four right lines, a parabola can be described to touch them. The focus is the point common to the circumcircles of the triangles formed by the lines.

6. The orthocentre of the triangle formed by any three tangents to a parabola is a point on the directrix. (See Equation (90).)

7. Find the co-ordinates of the intersection of tangents at the points ϕ', ϕ''. *Ans.* $x = a \tan\phi' \tan\phi''$, $y = a(\tan\phi' + \tan\phi'')$. (328).

8. If $\tan\phi''$ bear a given ratio to $\tan\phi'$, the envelope of the chord joining the points ϕ', ϕ'' is a parabola.

9. The area of the triangle formed by three tangents to a parabola is half the area of the triangle formed by joining the points of contact. (Compare Art. 5, Ex. 2, 3.)

10. Three tangents to a parabola form a right-angled triangle ABC, having the angle C right. If D be the point of contact of the side AB with the curve, prove that the points B, D, with one of the Brocard points of the triangle BCD, and the focus of the parabola are concyclic.

11. If a triangle be formed by two tangents to a parabola and their chord of contact, prove that the symmedian line of this triangle, through the vertex, passes through the focus.

12. In the same case, prove that the chord of the circumcircle through the vertex and focus is bisected at the focus.

111. *To find the locus of the middle points of a system of parallel chords.*

Let PP' (see fig. Art. 109) be one of the chords, m its direction tangent; then $m = \dfrac{4a}{y' + y''}$. (See Equation (323).)

Again, if y denote the ordinate of the middle point of PP', we have

$$y = \tfrac{1}{2}(y' + y''); \qquad (329)$$

therefore

$$y = \frac{2a}{m};$$

or, putting $m = \tan\theta$,

$$y = 2a \cot\theta. \qquad (330)$$

Hence the locus of the middle points of a system of parallel chords of a parabola is a line parallel to the axis.

DEF.—*A bisector of a system of parallel chords is called a diameter.*

Cor. 1.—The tangent at the end of a diameter is parallel to the chords which the diameter bisects; for the tangent is a limiting case of a chord of the system.

Or thus:

Let $x'y'$ be the point where the diameter $y = 2a \cot\phi$ meets the curve. Hence $y' = 2a \cot\theta$, and since the tangent at $x'y'$ is

$$yy' = 2a(x + x'), \qquad \text{(Art. 110)}$$

we have

$$y = \tan\theta\,(x + x'),$$

which is parallel to the chords, since its direction tangent is $\tan\theta$.

Cor. 2.—The tangents at the extremities of any chord meet on the diameter which bisects that chord; for the diameter which bisects a system of chords parallel to the join of ϕ', ϕ'' is $y = a(\tan\phi' + \tan\phi'')$ (Equation (329)), which passes through the intersection of tangents at the points ϕ', ϕ''. (See equation (324).)

Cor. 3.—The diameter through the intersection of two tangents bisects their chord of contact.

Cor. 4.—If ϕ be the intrinsic angle of the point where the diameter which bisects the join of ϕ', ϕ'' meets the curve,

$$\tan\phi = \tfrac{1}{2}(\tan\phi' + \tan\phi''). \qquad (331)$$

Cor. 5.—If θ denote the direction angle of the tangent at ϕ, $\theta + \phi = \frac{\pi}{2}$. (Art. 110, *Cor.* 3.) (332)

EXAMPLES.

1. The distance of the focus from the intersection of two tangents is a mean proportional between the focal vectors of the points of contact.

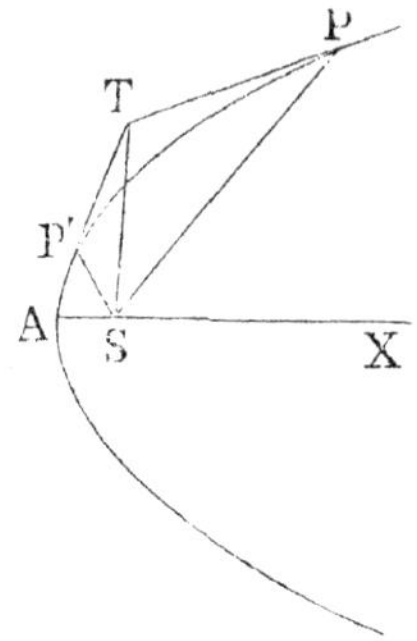

For if ϕ', ϕ'' denote the points of contact, ρ', ρ'', their focal vectors, we have (Art. 108, *Cor.* 1)

$$\rho'\rho'' = a^2 \sec^2\phi' \sec^2\phi''.$$

Again, the co-ordinates of T are $a\tan\phi'\tan\phi''$, $a(\tan\phi' + \tan\phi'')$. Hence the square of the distance of this point from S, whose co-ordinates are a, 0, is $a^2\sec^2\phi'\sec^2\phi''$. Hence

$$ST^2 = \rho'\rho''. \qquad (333)$$

2. If T be the intersection of tangents at ϕ', ϕ'', A the vertex, S the focus, the angle $AST = \phi' + \phi''$. (334)

For, substituting the co-ordinates of T and S in the equation

$$\frac{y' - y''}{x' - x''} = m,$$

which gives the direction tangent of the line through two points, we get

$$\tan XST = \frac{\tan\phi' + \tan\phi''}{\tan\phi' . \tan\phi'' - 1}. \quad \text{Hence } \tan AST = \frac{\tan\phi' + \tan\phi''}{1 - \tan\phi'\tan\phi''}.$$

3. Since $ASP'' = 2\phi''$, $ASP' = 2\phi'$ (Art. 108, *Cor.* 2), $AST = \frac{1}{2}(ASP + ASP'')$. Hence PT bisects the angle $P'SP''$.

4. The triangles $P'ST$, TSP'' are equiangular (Exs. 1 and 3).

5. The angle $P'TP''$ is the supplement of half $P'SP''$.

6. If $P'T$, $P''T$, be two tangents, TM the diameter through T, meeting the chord $P'P''$ in M, TM is bisected by the curve. For, draw the tangent AQ. This is parallel to $P'P''$; and since the diameter through Q bisects AP'' (*Cor.* 3), we have $AN = NP''$. Hence $TQ = QP''$, and therefore $TA = AM$.

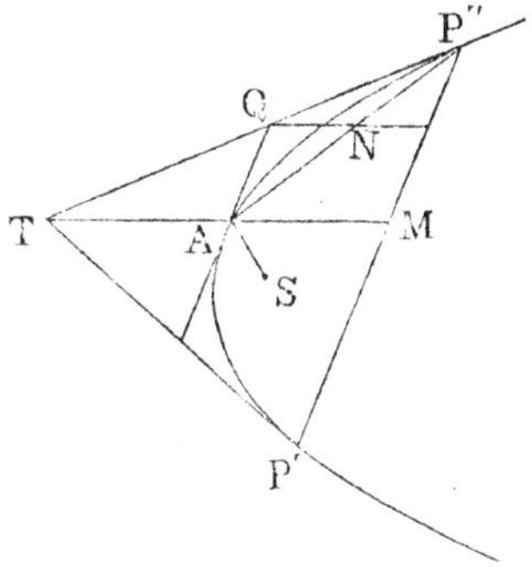

7. Find the co-ordinates of the point A.

$$Ans.\ x = a\left(\frac{\tan\phi' + \tan\phi''}{2}\right)^2, \quad y = a(\tan\phi' + \tan\phi''). \quad (335)$$

8. $$AM = a\left(\frac{\tan\phi' + \tan\phi''}{2}\right)^2. \qquad (336)$$

9. $$AS = a + x' = a\left\{1 + \left(\frac{\tan\phi' + \tan\phi''}{2}\right)^2\right\}.$$

10. If a quadrilateral circumscribe a parabola, the rectangle contained by the distances of the extremities of any of its three diagonals from the focus is equal to the rectangle contained by the distances from the focus of the extremities of either of the remaining diagonals.

112. *To find the equation of the parabola referred to any diameter and the tangent at its vertex as axes.*

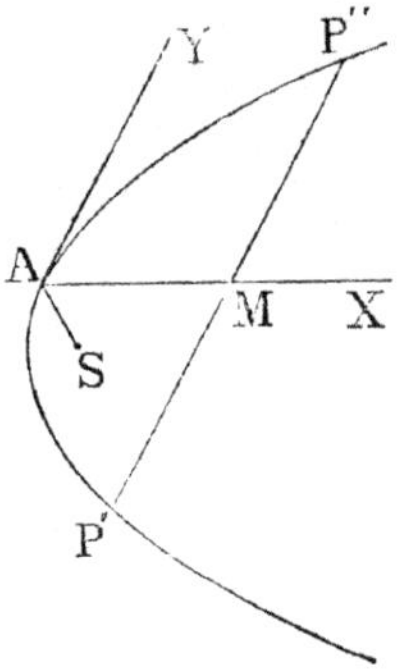

Let $P'P''$ be a double ordinate to the diameter AM; AY the tangent at A; then AY (Art. 111, *Cor.* 1) is parallel to $P'P''$. Let ϕ', ϕ'' be the intrinsic angles of the points P', P''; then (Art. 1)

$$P'P''^2 = a^2(\tan^2\phi - \tan^2\phi'')^2 + 4a^2(\tan\phi' - \tan\phi'')^2;$$

therefore

$$MP''^2 = 4a^2\left(\frac{\tan\phi' - \tan\phi''}{2}\right)^2\left\{1 + \left(\frac{\tan\phi' + \tan\phi''}{2}\right)^2\right\}$$

$$= 4AS \, . \, AM. \qquad \text{(Art. 111, Exs. 8, 9.)}$$

Therefore, denoting AS by a', AM, MP'' by x, y, we have

$$y^2 = 4a'x, \qquad (337)$$

which is the required equation, and identical in form with the old one, $$y^2 = 4ax.$$

Cor. 1.—If the angle between the axes AX, AY be denoted by θ, and if ϕ be the intrinsic angle of the point A, we have, since

$$\theta + \phi = \frac{\pi}{2}, \quad \operatorname{cosec}^2\theta = \sec^2\phi; \text{ but } AS = a\sec^2\phi:$$

therefore $$AS = a\operatorname{cosec}^2\theta. \qquad (338)$$

Cor. 2.—The equation of the tangent to the parabola at any point $x'y'$, referred to the new axes AX, AY, is the same as for rectangular axes, viz.,

$$yy' = 2a(x + x').$$

Examples.

1. From any external point hk can be drawn two tangents to a parabola. For the tangent at a point $x'y'$ of the parabola is $yy' = 2a(x + x')$: if this passes through the point hk, we have

$$ky' = 2a(h + x');$$

but $$y'^2 = 4ax'.$$

Hence $$y'^2 - 2ky' + 4ah = 0. \qquad (339)$$

This quadratic, giving two values for y', proves the proposition.

2. Find the equation of the chord of contact of tangents from hk.

By removing the accents from equation (339), we get

$$y^2 - 2ky + 4ah = 0.$$

This denotes two lines parallel to the axis of x, and passing through the point of contact; and since the parabola is $y^2 - 4ax = 0$, subtracting and dividing by 2, we get the required equation—

$$2a(x + h) - ky = 0. \qquad (340)$$

3. If the chord of contact of two tangents pass through a given point hk, the locus of their intersection is a right line.

For if $\alpha\beta$ be the point of intersection of the tangents, the chord of contact is $2a(x + \alpha) - \beta y = 0$; and since this passes through hk, we have $2a(h + \alpha) - \beta k = 0$, or, putting xy for $\alpha\beta$,

$$2a(x + h) - ky = 0,$$

an equation which is the same in form as (340).

Def.—*The line* $2a(x + h) - ky = 0$ is called the polar of the point hk.

4. If there be two points A, B, and if the polar of A passes through B, the polar of B passes through A.

5. The intercept made on the axis by any two lines is equal to the difference of the abscissae of the poles of these lines.

6. The polar of the focus is the directrix.

7. If any chord pass through the focus, the tangents at the extremities are at right angles. For in the equation of the chord, viz., $2x - (\tan\phi' + \tan\phi'')y + 2a\tan\phi', \tan\phi'' = 0$, substitute the co-ordinates of the focus, and we get $\tan\phi', \tan\phi'' = -1$.

8. The difference between the intrinsic angles of two points being given, to find the locus of the intersection of tangents at these points.

Let $\phi' - \phi'' = \delta$; then $\tan^2\delta = \dfrac{(\tan\phi' + \tan\phi'')^2 - 4\tan\phi'\tan\phi''}{(1 + \tan\phi'\tan\phi'')^2}$; and, substituting $\dfrac{x}{a}$, $\dfrac{y}{a}$ for $\tan\phi' . \tan\phi''$, $\tan\phi' + \tan\phi''$, respectively, we get $(y^2 - 4ax) = (a + x)^2 \tan^2\delta$, which is the required locus.

9. Any line meeting the parabola, and passing through a pole, is cut harmonically by the polar.

10. Find the co-ordinates of the point of intersection of the lines $P'P''$, ST (Art. 111, Ex. 1, fig.).

$$\textit{Ans.}\quad \frac{x}{a} = \frac{\sin^2\phi' + \sin^2\phi''}{\cos^2\phi' + \cos^2\phi''},\qquad \frac{y}{a} = \frac{\sin 2\phi' + \sin 2\phi''}{\cos^2\phi' + \cos^2\phi''}.$$

DEF.—*The normal at any point of a plane curve is the perpendicular to the tangent at that point.*

113. *To find the equation of the normal at the point $x'y'$.*

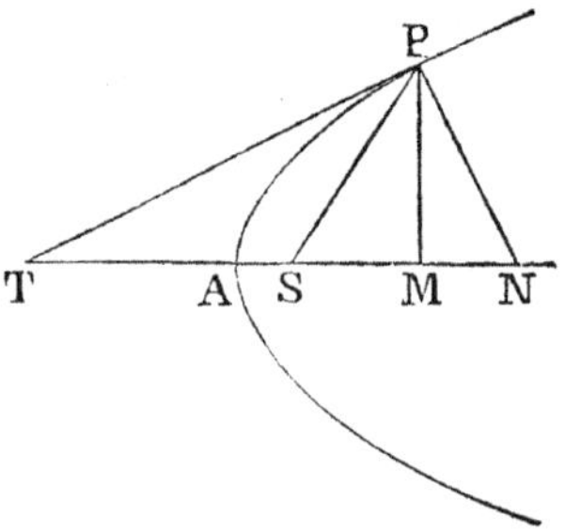

Since the equation of the tangent is

$$yy' = 2a(x + x'),$$

the equation of the normal is

$$y - y' = -\frac{y'}{2a}(x - x'). \quad (341)$$

Cor. 1.—If in the equation of the normal we put $y = 0$, we get $x - x' = 2a$; but in this case $x = AN$, $x' = AM$. Hence $x - x' = MN$; therefore $MN = 2a$.

DEF.—*The line MN intercepted on the axis between the ordinate and the normal is called the* SUBNORMAL. Hence in the parabola the subnormal is constant.

Cor. 2.—Since $SM = x' - a$, and $MN = 2a$, we have $SN = x' + a = SP$.

Cor. 3.—From any point $\alpha\beta$ can be drawn three normals to a parabola; for if the normal (341) passes through $\alpha\beta$, we get, after substituting for $x'y'$ their values in terms of the intrinsic angle,

$$a \tan^3 \phi - (\alpha - 2a) \tan \phi - \beta = 0, \qquad (342)$$

a cubic giving three values for $\tan \phi$.

Cor. 4.—Since the cubic (342) wants its second term, the sum of the three values of $\tan \phi$ must be zero. Hence, if from any point three normals be drawn to a parabola, the sum of the ordinates of their feet is zero.

Cor. 5.—If ϕ, ϕ', ϕ'' be the intrinsic angles of three points on a parabola, the co-ordinates of the centre of a circle pass-through them are (Art. 23, Ex. 4)—

$$x = \frac{a}{2}(\tan^2\phi + \tan^2\phi' + \tan^2\phi'' + \tan \phi \tan \phi' + \tan \phi' \tan \phi'' + \tan \phi'' \tan \phi + 4),$$

$$y = -\frac{a}{4}(\tan\phi + \tan \phi')(\tan\phi' + \tan \phi'')(\tan \phi'' + \tan \phi).$$

Hence, if the three points be consecutive, the co-ordinates of the centre of curvature at the point ϕ are

$$x = a(3 \tan^2 \phi + 2), \quad y = -2a \tan^3 \phi; \qquad (343)$$

and eliminating ϕ between these, we get the locus of the centre of curvature, viz.

$$4(x - 2a)^3 = 27ay^2. \qquad (344)$$

Examples.

1. Find the relation between the co-ordinates of the intersection of normals and the co-ordinates of the intersection of corresponding tangents. The normals at the points ϕ', ϕ'' are

$$y + x \tan \phi' = a(2 \tan \phi' + \tan^3 \phi'),$$

$$y + x \tan \phi'' = a(2 \tan \phi'' + \tan^3 \phi'').$$

Hence the co-ordinates of the point of intersection are

$$x = 2a + a(\tan^2 \phi' + \tan \phi' \tan \phi'' + \tan^2 \phi''),$$

$$y = -a(\tan \phi' + \tan \phi'') \tan \phi' . \tan \phi''.$$

But if α, β denote the co-ordinates of the intersection of corresponding tangents, we have

$$\alpha = a \tan\phi' \tan\phi'', \quad \beta = a(\tan\phi' + \tan\phi'').$$

Hence
$$\left.\begin{aligned} x &= 2a + \frac{\beta^2}{a} - \alpha, \\ y &= -\frac{\alpha\beta}{a} \end{aligned}\right\}. \tag{345}$$

2. If two normals be at right angles, the locus of their points of intersection is a parabola; for if the normals be at right angles the difference between the intrinsic angles is $\frac{\pi}{2}$. Hence, putting $\phi'' = \phi' + \frac{\pi}{2}$, we get for the intersection of the normals—

$$x = 3a + a(\tan\phi' - \cot\phi')^2, \quad y = a(\tan\phi' - \cot\phi').$$

Hence
$$y^2 = a(x - 3a). \tag{346}$$

3. The radius of curvature at the point ϕ is $2a\sec^3\phi$.

4. If PS (see fig., Art. 110) be produced to R until $RS = SP$, the perpendicular at R to PR will meet the normal at P in the centre of curvature.

5. Find the locus of the intersection of normals at the extremities of a chord which passes through a given point.

Since the chord passes through a given point, the intersection of the tangents will be on the polar of the point; and eliminating $\alpha\beta$ between the equation of this polar and the co-ordinates of the point of intersection (see Ex. 1) of normals in terms of the co-ordinates of the point of intersection of tangents, we get the required locus.

6. If normals at $x'y'$, $x''y''$, $x'''y'''$ be concurrent,

$$\frac{x' - x''}{y'''} + \frac{x'' - x'''}{y'} + \frac{x''' - x'}{y''} = 0.$$

7. If four parabolic points be concyclic, the sum of the tangents of their intrinsic angles is zero.

8. If the normals at the points ϕ', ϕ'', ϕ''' be concurrent, the vertex and the points ϕ', ϕ'', ϕ''' are concyclic.

9. If the normal at ϕ meet the curve again in ϕ', then

$$\tan\phi(\tan\phi + \tan\phi') + 2 = 0. \tag{347}$$

10. Find the condition that two of the normals which can be drawn from the point ($\alpha\beta$) are coincident. The condition will evidently be the

vanishing of the discriminant of the cubic in Cor. 3, which is (see Burnside and Panton's *Theory of Equations*),

$$4(\alpha - 2a)^3 - 27a\beta^2 = 0. \qquad \text{(Comp. *Cor.* 5.)}$$

114. *To find the polar equation of the parabola, the focus being pole.*

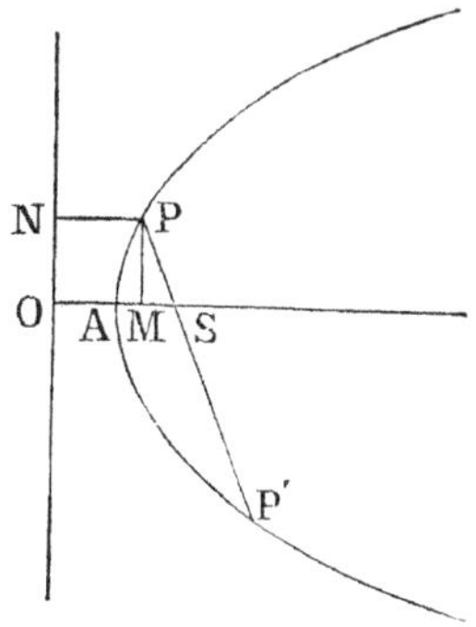

Let S be the focus, P any point in the parabola; then denoting the angle OSP (in Astronomy called the true anomaly) by θ, and SP by ρ. Since $SP = PN = OM = 2a - SM$, we have

$$\rho = 2a - \rho \cos \theta;$$

therefore

$$\rho = \frac{2a}{1 + \cos \theta} = a \sec^2 \tfrac{1}{2}\theta, \qquad (348)$$

which is the required equation.

Cor. 1.—If PS produced meet the curve again in P',

$$PP' = 4a \operatorname{cosec}^2 \theta. \qquad (349)$$

Cor. 2.— $$PS \,.\, SP' = PP' \,.\, a. \qquad (350)$$

Cor. 3.—The polar equation of the tangent at the point whose angular co-ordinate is α, is

$$\frac{2a}{\rho} = \cos \theta + \cos(\theta - \alpha). \qquad (351)$$

For this will be satisfied if we make $\theta = \alpha$; and for other values of θ, the value of ρ derived from this equation is greater than the corresponding value obtained from the equation of the curve. Hence, except at the point α, the line (351) does not meet the curve.

Cor. 4.—The polar equation of the normal at the point α is

$$\frac{a}{\rho} = \cot \frac{\alpha}{2} \cos \frac{\alpha}{2} \sin\left(\theta - \frac{\alpha}{2}\right); \qquad (352)$$

for if we make $\theta = \alpha$, we get $\rho = a \sec^2 \frac{1}{2}\alpha$. Hence the line passes through the point α. Again, if we make $\theta = \pi$, we get the same value for ρ. Now, the focal vector of the foot of the normal is equal to that of the point of contact (Art. 113, *Cor.* 2). Hence the line (352) passes through two points on the normal, and therefore must coincide with it.

EXAMPLES.

1. Find the polar co-ordinates of the intersection of tangents at the points whose angular co-ordinates are $(\alpha + \beta)$, $(\alpha - \beta)$.

2. The equation of the chord joining the points $(\alpha + \beta)$, $(\alpha - \beta)$ is

$$\frac{2a}{\rho} = \cos\theta + \sec\beta \cos(\theta - \alpha).$$

3. If ϕ', ϕ'', ϕ''' be the intrinsic angles of three points on the parabola, prove that $\rho \cos\phi' \cos\phi'' \cos\phi''' = a \cos(\phi' + \phi'' + \phi''' - \theta)$ is the equation of the circumcircle of the triangle formed by the tangents at ϕ', ϕ'', ϕ'''.

4. Find the polar co-ordinates of the point of intersection of the parabolae $y^2 = 4ax$, $x^2 = 4by$, the origin being the pole.

5. Find the locus of the centre of the inscribed circle of the triangle formed by a focal vector, the tangent at its extremity, and the axis of the parabola.

6. Tangents at two points P, P' meet the axis in the points T, T'; prove

$$TT' = SP - SP'.$$

7. If through the focus a perpendicular be drawn to the focal vector, meeting the normal at P in T, the locus of T is

$$\frac{a}{\rho} = \cos^2\left(\frac{\theta}{2} - \frac{\pi}{4}\right) \cot\left(\frac{\theta}{2} - \frac{\pi}{4}\right).$$

8. If l_1, l_2 be the lengths of two tangents to a parabola, ϕ their contained angle, then $l_1^2 + l_2^2 + 2l_1l_2 \cos\phi = \dfrac{(l_1l_2 \sin\phi)^{\frac{4}{3}}}{a^{\frac{2}{3}}}$.

9. If ρ, ρ' be the radii of curvature at the extremities of a focal chord, then

$$\rho^{-\frac{2}{3}} + \rho'^{-\frac{2}{3}} = (2a)^{-\frac{2}{3}}.$$

115. *To find the length of a line drawn from a given point in a given direction to meet the parabola.*

Let O be the given point, OP the given direction, and let the rectangular co-ordinates of O, P be $x'y'$, xy respectively; then, denoting OP by ρ, we have

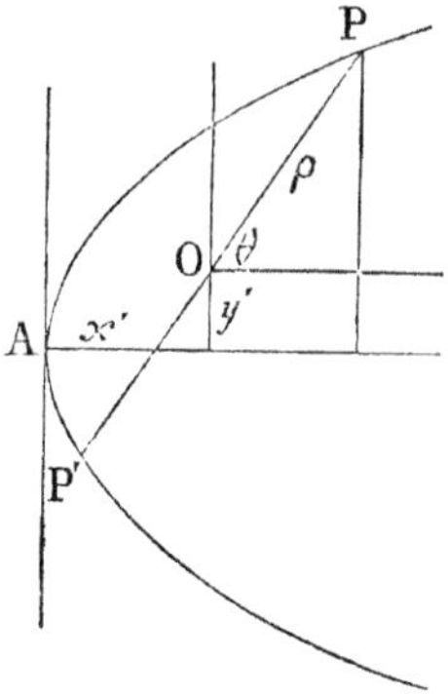

$$x = x' + \rho\cos\theta,\ y = y' + \rho\sin\theta.$$

Substituting these values in the equation $y^2 = 4ax$, we get

$$\rho^2\sin^2\theta + 2(y'\sin\theta - 2a\cos\theta)\rho + y'^2 - 4ax' = 0, \quad (353)$$

a quadratic whose roots are the values required. If the roots of this equation be ρ_1, ρ_2, and if OP meet the curve again in P', we may put $OP = \rho_1$, $OP' = \rho_2$.

Cor. 1.—If PP' be bisected in O, we have $\rho_1 = -\rho_2$, and the coefficient of the second term in (353) is zero. Hence, if θ be constant and y' variable, we see that the locus of the middle points of a system of parallel chords is the line $y = 2a\cot\theta$. (Comp. Art. 111.)

Cor. 2—The prodnct of the roots of equation (353) is $(y'^2 - 4ax')\operatorname{cosec}^2\theta$. Hence

$$OP \,.\, OP' = (y'^2 - 4ax')\operatorname{cosec}^2\theta.$$

Similarly, if another chord QQ' be drawn through O, making an angle θ' with the axis, we have

$$OQ \,.\, OQ' = (y'^2 - 4ax')\operatorname{cosec}^2\theta'.$$

Hence

$$OP \,.\, OP' : OQ \,.\, OQ' :: \operatorname{cosec}^2\theta : \operatorname{cosec}^2\theta'.$$

EXAMPLES.

1. If AX, $A'X'$ be two diameters of a parabola, O, O' any two points in them, PP', QQ' parallel chords through O, O' respectively,

$$AO : A'O :: OP \, . \, OP' : OQ \, . \, OQ'.$$

2. If TR, TV be two tangents, S the focus,

$$TR^2 : TV^2 :: SR : SV.$$

3. If c, c' be the lengths of focal chords parallel respectively to TR, TV,

$$TR^2 : TV^2 :: c : c'.$$

4. If a chord PP' through the point ϕ of a parabola make an angle ψ with the tangent at ϕ, and an angle θ with the axis,

$$\sin\psi = \frac{PP'\cos\phi\sin^2\theta}{4a}.$$

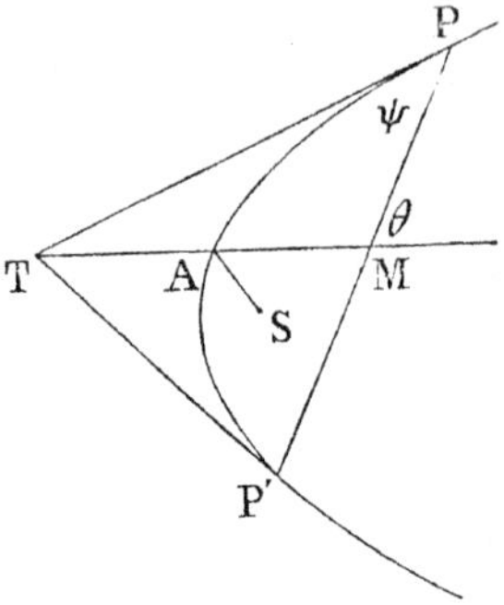

Let PT, $P'T$ be the tangents at PP'; and since the angle MTP is the complement of ψ, we have

$$\sin\psi : \cos\phi :: MT \,(\text{or } 2AM) : MP;$$

therefore $\quad MP\sin\psi = 2AM\cos\phi.$

Again, if S be the focus,

$$4AS \, . \, AM = MP^2; \quad \text{(Art. 112.)}$$

therefore $$2AS \, . \, \sin\psi = MP\cos\phi.$$

But $$AS = a\operatorname{cosec}^2\theta. \qquad \text{(Art. 112, } \textit{Cor. } 1.)$$

Hence $$\sin\psi = \frac{PP'\cos\phi \, . \, \sin^2\theta}{4a}. \qquad (354)$$

5. If through any point ϕ in a parabola be drawn two chords making angles ψ, ψ' with the tangent at ϕ; then, if c, c' be their lengths, θ, θ' their direction angles,

$$\sin\psi : \sin\psi' :: c\sin^2\theta : c'\sin^2\theta'.$$

116. *If λ, μ, ν denote the perpendiculars from the angular points of a circumscribed triangle on any tangent to the parabola, and if ϕ', ϕ'', ϕ''' be the points of contact of its sides,*

$$\frac{\tan\phi' - \tan\phi''}{\lambda} + \frac{\tan\phi'' - \tan\phi'''}{\mu} + \frac{\tan\phi''' - \tan\phi'}{\nu} = 0; \qquad (355)$$

for the equation of any tangent is $x - y \tan\phi + a\tan^2\phi = 0$; and λ being the perpendicular on this from the intersection of tangents at ϕ', ϕ'', we have

$$\lambda = a\cos\phi\,(\tan\phi - \tan\phi')\,(\tan\phi - \tan\phi'');$$

therefore

$$\frac{\tan\phi' - \tan\phi''}{\lambda} = \frac{1}{a\cos\phi}\left\{\frac{1}{\tan\phi - \tan\phi'} - \frac{1}{\tan\phi - \tan\phi''}\right\},$$

with similar values for

$$\frac{\tan\phi'' - \tan\phi'''}{\mu}, \quad \frac{\tan\phi''' - \tan\phi'}{\nu},$$

and these added vanish identically. Hence the proposition is proved.

Cor. 1.—If y', y'', y''' denote the ordinates of the points of contact of the parabola with the sides of the triangle,

$$\frac{y' - y''}{\lambda} + \frac{y'' - y'''}{\mu} + \frac{y''' - y'}{\nu} = 0. \tag{356}$$

Cor. 2.—In like manner, if a polygon of any number of sides be circumscribed to a parabola,

$$\frac{y' - y''}{\lambda} + \frac{y'' - y'''}{\mu} + \frac{y''' - y''''}{\nu} + \ldots \frac{y^{(n)} - y'}{\xi} = 0. \tag{357}$$

Cor. 3.—If the co-ordinates of the angular points be $\alpha'\beta'$, $\alpha''\beta''$, &c., it is easy to to see that

$$\sqrt{\beta'^2 - 4a\alpha'} = a\,(\tan\phi' - \tan\phi'').$$

But $\beta'^2 - 4a\alpha'$ is the power of the point $\alpha'\beta'$ with respect to the parabola. Hence $\sqrt{\beta'^2 - 4a\alpha'}$ may be denoted by $\sqrt{S'}$. Hence we have

$$\frac{\sqrt{S'}}{\lambda} + \frac{\sqrt{S''}}{\mu} + \frac{\sqrt{S'''}}{\nu} + \text{\&c.} = 0, \tag{358}$$

for any circumscribed polygon.

Cor. 4.—If a circumscribed polygon consist of an odd number of sides, y', y'', &c., can be expressed in terms of the ordinates of its angular points; thus, in the case of a triangle, if β', β'', &c., be the ordinates of the angular points, we get, instead of (356), the equation

$$\frac{\beta' - \beta''}{\lambda} + \frac{\beta'' - \beta'''}{\mu} + \frac{\beta''' - \beta'}{\nu} = 0. \qquad (359)$$

Cor. 5.—The perpendiculars from the points ϕ', ϕ'' on the tangent at ϕ are

$$a \cos\phi\,(\tan\phi - \tan\phi')^2, \quad a\cos\phi\,(\tan\phi - \tan\phi'')^2;$$

and the perpendicular from the point of intersection of tangents is

$$a\cos\phi\,(\tan\phi - \tan\phi')\,(\tan\phi - \tan\phi'').$$

Hence we have the following theorem :—*The perpendicular from an external point R on any tangent to the parabola is a mean proportional between the perpendiculars on the same tangent from the points where the polar of R meets the parabola.*

Cor. 6.—From *Cor.* 5 we have immediately the following theorem :—*If a quadrilateral circumscribe a parabola, the product of the perpendiculars from the extremities of one of its three diagonals on any tangent is equal to the product of the perpendiculars on the same tangent from the extremities of either of the remaining diagonals.*

Exercises on the Parabola.

1. Find the polar equation of the parabola, the vertex being the pole.

2. What is the intrinsic angle at either extremity of the latus rectus ?

3. What is the equation of the tangent at an extremity of the latus rectum ?

4. Find the co-ordinates of the centre of a circle passing through the vertex of a parbola, and touching it at the point ϕ.

$$Ans.\ x = \frac{a\tan^4\phi}{4 + 2\tan^2\phi}, \quad y = a\,\frac{(4\tan\phi + 3\tan^3\phi)}{4 + 2\tan^2\phi}.$$

5. Find the equation of the normal at the extremity of the latus rectum.

6. Find the radius of a circle touching a parabola at a point whose abscissa is x.

Ans. $\rho = 2\sqrt{a(a+x)}$.

7. In the figure, Art. 114, prove that the points P', A, N are collinear.

8. If the ordinates of three points on a parabola be in geometrical progression, prove that the pole of the line joining the first and third lies on the ordinate through the second.

9. If from a point O whose abscissa is x a perpendicular be let fall on the polar of O, if this meets the polar in R and the axis in G,

$$SG = SR = x + a.$$

10. If two equal parabolae have a common axis, but different vertices, the tangent to the interior, and bounded by the exterior, is bisected at the point of contact.

11. The tangent at any point of a parabola meets the directrix at equal distances from the focus.

12. If a chord of a parabola subtend a right angle at the vertex, the locus of its pole is $x + 4a = 0$.

13. Prove that the locus of the pole of a chord which subtends a right angle at the point hk is

$$ax^2 - hy^2 + (4a^2 + 2ah)\,x - 2aky + a\,(h^2 + k^2) = 0.$$

14. If from any point in the line $x = a'$ tangents be drawn to a parabola, the product of their direction tangents is $a \div a'$.

15. Find the locus of the intersection of tangents at the points ϕ', ϕ'', if $\tan\phi' = \mu\tan\phi''$. *Ans.* $y^2 = (\mu^{\frac{1}{2}} + \mu^{-\frac{1}{2}})^2\,ax$.

16. Prove that the equation of the chord whose middle point is hk is

$$k(y - k) = 2a(x - h).$$

17. If a chord of a parabola subtend a right angle at the vertex, the locus of its middle point is $y^2 = 2a\,(x - 4a)$.

18. The area of the triangle formed by tangents at the points ϕ', ϕ'' and their chord of contact is

$$\frac{a^2}{2}(\tan\phi' - \tan\phi'')^3.$$

19. If a variable circle touch a fixed circle and a fixed line, the locus of its centre is a parabola.

20. If the difference between the ordinate of two points on a parabola be given, the locus of the intersection of tangents at these points is an equal parabola.

21. If two tangents to a parabola from a variable point P include an angle θ, prove, if S be the focus, PN a perpendicular on the directrix,

$$PN = SP \cos\theta.$$

22. If the points ϕ', ϕ'' subtend a right angle at the vertex, prove

$$\tan\phi' . \tan\phi'' + 4a = 0.$$

23. The area of the triangle formed by the points ϕ', ϕ'' and the focus is

$$a^2(\tan\phi + \tan\phi'')(1 + \tan\phi' \tan\phi'').$$

24. A triangle ABC is inscribed in a parabola whose focus is F; show that one of the circles touching the perpendicular bisectors of FA, FB, FC passes through the circumcentre of the triangle ABC. (R. A. ROBERTS.)

25. The co-ordinates of the centroid of a triangle ABC inscribed in the parabola $y^2 = 4ax$ are α, β; show that the co-ordinates of the centroid of the triangle formed by the tangent at A, B, C are

$$\frac{3\beta^2 - 4a\alpha}{8a}, \ \beta. \qquad \text{(Ibid.)}$$

26. The area of a triangle inscribed in a parabola is

$$\frac{(y' - y'')(y'' - y''')(y''' - y')}{8a}.$$

27. The area of the triangle formed by three tangents is

$$\frac{(y' - y'')(y'' - y''')(y''' - y')}{16a}.$$

28. If a series of circles S, S_1, S_2, S_3, &c., touch each other consecutively along the axis of a parabola; then, if the first be the circle of curvature of the parabola at the vertex, and the others have each double contact with the parabola, prove that their diameters are proportional to the odd numbers 1, 3, 5, &c.

29. If δ be the distance between the centres of curvature of two points at the extremities of a focal chord, which makes an angle θ with the axis; prove $\delta = 16a \cot\theta \operatorname{cosec}^2\theta$.

30. If ρ, ρ' be two radii vectores of a parabola from the vertex at right angles to each other; prove $\rho^{\frac{4}{3}}\rho'^{\frac{4}{3}} = 16a^2(\rho^{\frac{2}{3}} + \rho'^{\frac{2}{3}})$.

31. The perpendicular from the focus on any chord of a parabola meets the diameter which bisects that chord on the directrix.

32. If from any two points ϕ', ϕ'' of a parabola perpendiculars be drawn to the directrix, the intersection of tangents at ϕ', ϕ'' is the centre of a circle through the focus and the feet of the perpendiculars.

33. If from any point P a perpendicular PQ to the axis meet the polar at P in R; find the locus of P, if $PQ.PR$ be constant.

Ans. A parabola.

34. Find the circle whose diameter is the intercept which $y^2 - 4ax = 0$ makes on the line $y = mx + n$.

Ans. $m^2(x^2 + y^2) + 2(mn - 2a)x - 4amy + 4amn + n^2 = 0.$

35. The equation of the circle passing through the feet of normals, from the point hk, is $x^2 + y^2 - (2a + h)x - \frac{1}{2}ky = 0$.

36. If SL be the perpendicular from the focus of a parabola on the normal at any point, find the locus of L.

37. If a chord of a parabola be bisected by a fixed double ordinate to the axis, the locus of the pole of the chord is another parabola.

38. If in the equation $w = z^2$, w and z denote complex variables; prove, if z describe a right line, that w describes a parabola.

39. Two chords from the vertex to points ϕ', ϕ'' of a parabola make an intercept on the directrix, which is bisected by the join of the vertex to the intersection of tangents at ϕ', ϕ''.

40. Two fixed tangents to a parabola are cut proportionally by any variable tangent.

41. Trisect an arc of a circle by means of a parabola.

42. The radical axis of two circles whose diameters are any two chords intersecting on the axis of the parabola passes through the vertex.

43. A coaxal system of circles, having two real points of intersection, are intersected by two chords passing through one of these points. In two systems of points P, P', P'', &c.; Q, Q', Q'', &c.; prove that the chords PQ, $P'Q'$, $P''Q''$, &c., are all tangents to a parabola.

44. LO is the perpendicular at the middle point L of a focal chord, meeting the axis in O. Prove that SO, LO, are the arithmetic and the geometric means of the focal segments of the chord.

45. If ν be the intercept which a tangent to a parabola makes on the axis of y, and ϕ the angle it makes with it, prove that $\nu = a \tan\phi$ is a tangential equation of the parabola.

46. If two circles touch a parabola at the ends of a focal chord, and pass through the focus, they cut orthogonally: also the locus of their second intersection is a circle.

47. Give a geometrical construction for drawing a tangent to a parabola from an external point.

48. If a given parabola roll along a right line, find the locus of its focus.

49. The area of the parabolic segment cut off by any chord is two-thirds of the triangle formed by the chord and the tangents at its extremities.

50. Find the locus of a point P if the perpendicular from it on its polar is constant.

51. Prove that the angle of intersection of $y^2 - 4ax = 0$, $x^2 - 4by = 0$, is

$$\tan^{-1}\left\{\frac{3a^{\frac{1}{3}}b^{\frac{1}{3}}}{2(a^{\frac{2}{3}} + b^{\frac{2}{3}})}\right\}.$$

52. If the base of a triangle be given in position, and its area in magnitude, the locus of its orthocentre is a parabola.

53. If the normal at a point ϕ on a parabola meet the axis in K, the envelope of the parallel through K to the tangent at ϕ is a parabola.

54. If the sum of the abscissæ of two points on a parabola be given, the locus of the intersection of the tangents at the points is a parabola.

55. If from the vertex A of a parabola a perpendicular AP be drawn to any tangent, the locus of the point inverse to P, with respect to a circle whose centre is A, is a parabola.

56. Find the locus of a point P, if the normals corresponding to the tangents from P meet on the line $Ax + By + C = 0$.

Ans. $Ay^2 - Bxy - Aax + 2a^2B + aC = 0$.

57. If $\alpha\beta$ be the co-ordinates of the intersection of two normals; prove that the co-ordinates of the intersection of the corresponding tangents are given by the equations

$$x^3 + ax^2 = a(\beta - 2a)^2, \quad y^3 - a\alpha y = a(2a^2 - \beta^2).$$

58. If normals be drawn from the point $x'y'$ to the parabola; prove that $(x - a)(x + x' - 2a) + y(y + y') = 0$ is the circumcircle of the triangle formed by the corresponding tangents.

59. Two parabolae, S, S', have a common focus, parameter, and axis, their vertices being on opposite sides of the focus; show that if from any point on S two tangents be drawn to S', the circumcircle of the triangle formed by these tangents and their chord of contact touches S'.—

(PROF. F. PURSER.)

60. Two equal parabolae, S, S', have coincident axes, which have the same direction, while the focus F of S is the vertex of S'. Show that if P be a point on S', the chord of S through P, which passes through F, is the minimum chord through P.

(*Ibid.*)

CHAPTER VI.

THE ELLIPSE.

117. Def. I.—*Being given in position a point S, and a line NN′. The locus of a variable point P, whose distance from S has to its perpendicular distance from NN′ a given ratio e, less than unity, is called an* ELLIPSE.

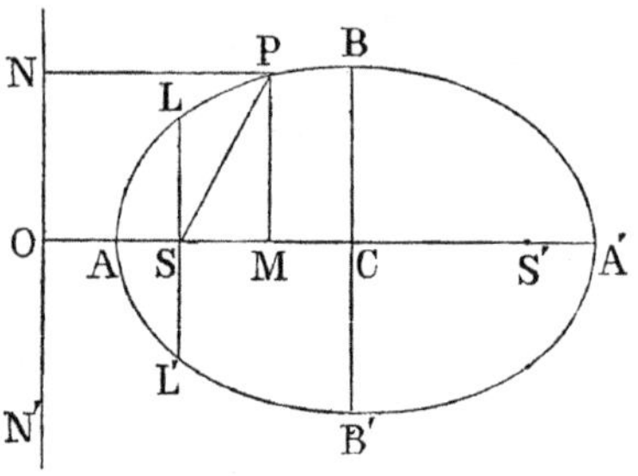

Def. II.—*The point S is called the* FOCUS, *the line NN′ the* DIRECTRIX, *and the ratio e the* ECCENTRICITY *of the ellipse.*

118. *To find the equation of the ellipse.*

1°. Take the focus as origin, and the line through S perpendicular to the directrix as the axis of x, and a parallel to the directrix through S as the axis of y; also denote the perpendicular SO from S on the directrix by f; then, if the co-ordinates SM, MP be xy, we have $SP^2 = x^2 + y^2$, $PN = x + f$; but (Def. I.) $SP \div PN = e$; therefore

$$x^2 + y^2 = e^2(x + f)^2, \tag{360}$$

which is the required equation.

Observation.—It will be seen that equation (360) includes the three conic sections. Thus, when e is less than unity, it represents an ellipse;

when equal to unity, a parabola; and when greater, a hyperbola. Also the general equation $ax^2 + 2hxy + by^2 + 2gx + 2fy + c = 0$ may obviously be written in the form $(x - \alpha)^2 + (y - \beta)^2 = (lx + my + n)^2$; for, by expanding and comparing coefficients, we should obtain a sufficient number of equations to determine α, β, &c., in terms of the coefficients of the general equation. And it is evident that $(x - \alpha)^2 + (y - \beta)^2 = (lx + my + n)^2$ can by transformation be reduced to the form (360).

2°. If in (360) we put $x = x + \dfrac{e^2 f}{1 - e^2}$,

we get
$$x^2 + \frac{y^2}{1 - e^2} = \frac{e^2 f^2}{(1 - e^2)^2}. \qquad \text{(I.)}$$

Hence, if C be the new origin,
$$SC = \frac{e^2 f}{1 - e^2}. \qquad \text{(II.)}$$

Now, putting $y = 0$ in (I.), we get
$$x^2 = \frac{e^2 f^2}{(1 - e^2)^2},$$

giving for x two values, equal in magnitude, but of opposite signs. Hence, denoting the points where the ellipse meets the axis of x by A, A', we have
$$CA' = \frac{ef}{1 - e^2}, \quad CA = -\frac{ef}{1 - e^2};$$

therefore $AC = CA'$, and the line AA' is bisected in C. Hence, denoting AA' by $2a$, we have
$$a = \frac{ef}{1 - e^2}. \qquad \text{(III.)}$$

Again, putting $x = 0$, and denoting the points where the ellipse cuts the axis of y by B, B', we get in the same manner
$$CB = \frac{ef}{(1 - e^2)^{\frac{1}{2}}}, \quad CB' = -\frac{ef}{(1 - e^2)^{\frac{1}{2}}}.$$

Hence BB' is bisected in C; and, denoting BB' by $2b$, we have
$$b = \frac{ef}{(1 - e^2)^{\frac{1}{2}}}. \qquad \text{(IV.)}$$

Now, since equation (I.) may be written

$$\frac{(1-e^2)^2x^2}{e^2f^2} + \frac{(1-e^2)y^2}{e^2f^2} = 1,$$

from (III.) and (IV.) we get

$$\frac{x^2}{a^2} + \frac{y^2}{b^2} = 1. \qquad (361)$$

This is the standard form of the equation of the ellipse.

DEF. III.—*The lines AA', BB' are called, respectively, the* TRANSVERSE *axis and the* CONJUGATE *axis of the ellipse, and the point C the* CENTRE.

DEF. IV.—*The double ordinate LL' through S is called the* LATUS RECTUM *or* PARAMETER.

The name *parameter* is also employed by mathematicians in another and a widely-different signification. Hence, to avoid confusion, it would be better to discontinue its use as a name for the *latus rectum.*

119. The following deductions from the preceding equations are very important:—

1°. $b^2 = a^2(1-e^2)$, from (III.) and (IV.)

2°. If CS be denoted by c, $c = ae$, from (II.) and (III.)

3°. $CO = \frac{a}{e}$, for $CO = CS + f = \frac{e^2 f}{1-e^2} + f = \frac{f}{1-e^2}$.

4°. $b^2 + c^2 = a^2$, from 1° and 2°.

5°. $CS \cdot CO = a^2$, from 2° and 3°.

6°. Latus Rectum $= 2a(1-e^2)$. For in equation (360) put $x = 0$, and we get $SL = ef$; therefore $LL' = 2ef = 2a(1-e^2)$, from (III.)

7°. From 1° and 6°, we infer that the transverse axis AA', the conjugate axis BB', and the latus rectum LL', are continual proportionals.

8°. From the equation (361) it is evident that *the ellipse is symmetrical with respect to each axis.* Hence if we make $CS' = SC$, the point S' will be another focus. Also, if we make $CO' = OC$, and through O' draw MM' perpendicular to the transverse axis, the line MM' will be a second directrix corresponding to the second focus.

Examples.

1. Given the base of a triangle and the sum of the sides, find the locus of the vertex.

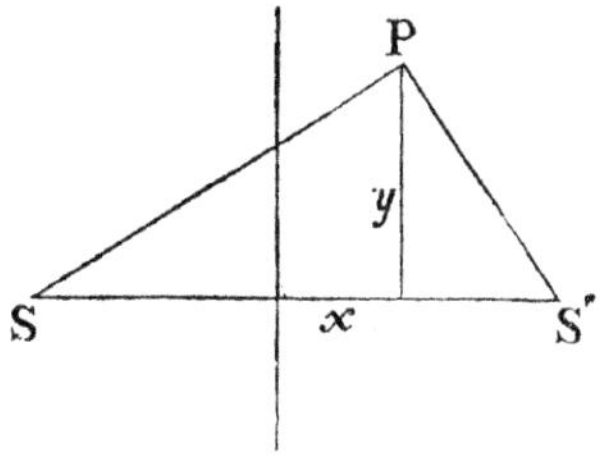

Let $SS'P$ be the triangle, let the sum of the sides equal $2a$, half the base $= c$, and xy the co-ordinates of P; then $SP = \{(c+x)^2 + y^2\}^{\frac{1}{2}}$, $S'P = \{(c-x)^2 + y^2\}^{\frac{1}{2}}$. Hence $\{(c+x)^2 + y^2\}^{\frac{1}{2}} + \{(c-x)^2 + y^2\}^{\frac{1}{2}} = 2a$. (I.)

This cleared of radicals gives

$$(a^2 - c^2)x^2 + a^2y^2 = a^2(a^2 - c^2);$$

or putting $a^2 - c^2 = b^2$,

$$\frac{x^2}{a^2} + \frac{y^2}{b^2} = 1.$$

Hence the locus is an ellipse, having the extremities of the base as foci.

Cor. 1.— $$S'P = a - ex. \quad (362)$$

For in clearing (I.) of radicals, we get

$$a\{(c-x)^2 + y^2\}^{\frac{1}{2}} = a^2 - cx;$$

that is, $a\,S'P = a^2 - aex$; therefore $S'P = a - ex$.

Cor. 2.— $$SP = a + ex. \quad (363)$$

This is also obvious, from Def. I. 117.

2. Given the base of a triangle and the product of the tangents of the base angles, the locus of the vertex is an ellipse.

3. Given the base and the sum of the sides, the locus of the centre of the inscribed circle is an ellipse.

For if xy denote the co-ordinates of the incentre of SPS', we have the perimeter $= 2a + 2c$.

Also $$\tan\tfrac{1}{2}S.\tan\tfrac{1}{2}S' = \frac{s-c}{s} = \frac{a}{a+c} = \frac{1}{1+e}.$$

Now $$\tan\tfrac{1}{2}S = \frac{y}{c+x}, \quad \tan\tfrac{1}{2}S' = \frac{y}{c-x};$$

hence $$\frac{y^2}{c^2 - x^2} = \frac{1}{1+e}.$$

Therefore $$\frac{x^2}{c^2} + \frac{(1+e)\,y^2}{c^2} = 1. \qquad (364)$$

In a similar way it may be proved that the locus of the centre of the escribed circle, which touches the base externally, is the ellipse

$$\frac{x^2}{c^2} + \frac{y^2}{c^2(1+e)} = 1; \qquad (365)$$

and the loci of the centres of the escribed circles which touch the base produced are the directrices of the ellipse which is the locus of the vertex.

4. *MN* is a parallel to the diagonal *AC* of a fixed rectangle *ABCD*. *AE* is made equal to *AD*; and *EM*, *DN* joined; prove that the locus of their intersection *P* is an ellipse.—(POHLKE.)

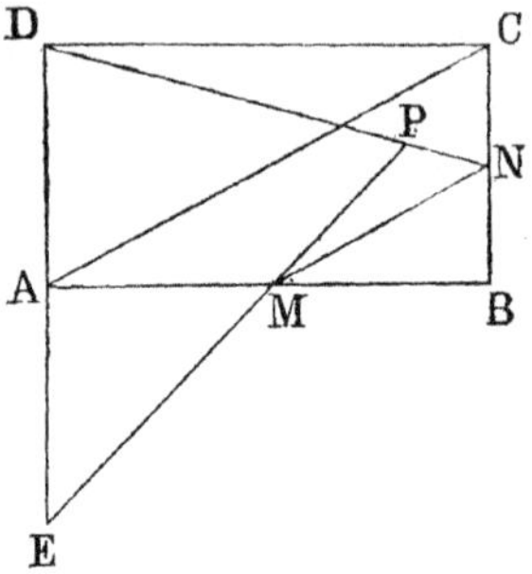

5. If a line *AB* of given length slide between two rectangular lines *OA*, *OB*, the locus of a point *P* fixed in the sliding line is an ellipse. For let $AP = b$, $BP = a$; then, denoting the co-ordinates of *P* by xy, and the angle *OAP* by θ, we have

$$x = a\cos\theta, \quad y = b\sin\theta.$$

Hence, eliminating θ, we get

$$\frac{x^2}{a^2} + \frac{y^2}{b^2} = 1.$$

6. If a fixed point *S*, and a fixed circle, whose centre is *O*, be both at the same side of a fixed line *NN'*, and through *S* any line be drawn meeting the circle in *P*, and *MN* in *R*; then if *RO* be joined, meeting a parallel to *OP*, drawn through *S* in *p*, the locus of *p* is an ellipse.—(BOSCOVICH.)

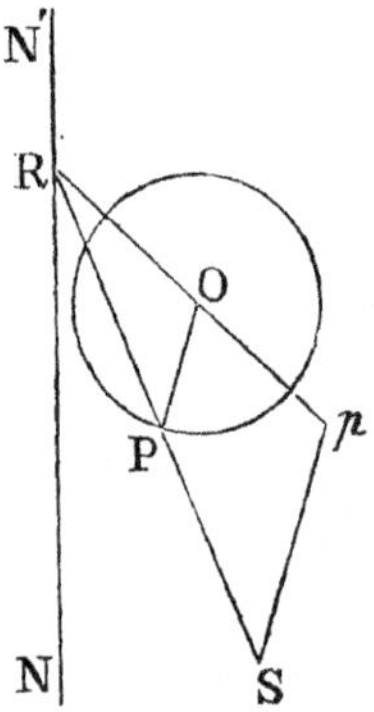

7. Prove that the radius of the *Boscovich Circle*, divided by its distance from the fixed line, is equal to the eccentricity.

8. *CB* is a fixed diameter of a given circle, *A* a fixed point in *CB* produced. Through *A* draw any line meeting the circle in *D* and *E*. Join *CD* and produce to *F*, making $CF = AE$; the locus of *F* is the ellipse

$$\frac{x^2}{AC^2} + \frac{y^2}{AB^2} = 1. \qquad \text{(SIR W. HAMILTON.)}$$

120. *To express the co-ordinates of a point P on an ellipse ABA'B' in terms of a single variable.*

Let AA', BB' be the transverse and the conjugate axes of the ellipse upon AA' as diameter; describe the circle $AP'A'$. Let P be any point of the ellipse, MP its ordinate; produce MP to meet the circle $AP'A'$ in P'. Join OP', and denote the angle MOP' by ϕ; then, since $OM = x$, $OP' = a$, we have $x = a \cos \phi$. This value, substituted in the equation (361) of the ellipse, gives $y = b \sin \phi$: therefore the co-ordinates of P are $a \cos \phi$, $b \sin \phi$.

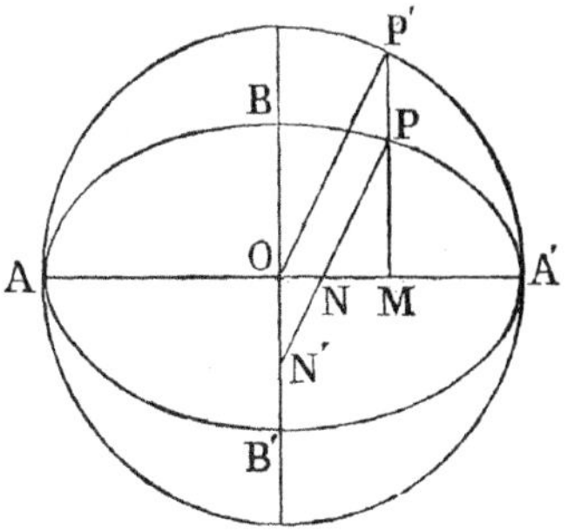

DEF.—*The circle described on AA' as diameter is called the* AUXILIARY *circle of the ellipse, and the angle ϕ the* ECCENTRIC *angle.*

The term eccentric has been taken from Astronomy; the angle ϕ in that science being called the eccentric anomaly.

Cor. 1.—Since $PM = b \sin \phi$, and $P'M = a \sin \phi$,

$$P'M : PM :: a : b. \tag{366}$$

Hence we have the following theorem:—The locus of a point P which divides an ordinate of a semicircle in a given ratio is an ellipse; or again, *If from all the points in the circumference of a circle in one plane perpendiculars be let fall on another plane, inclined to the former at any angle, the locus of their feet is an ellipse* (*called* THE ORTHOGONAL PROJECTION OF THE CIRCLE). For, the diameter of the circle which is parallel to the intersection of the planes is unaltered by projection, and the ordinates of the circle perpendicular to this line are projected into lines having a given ratio to them.

Cor. 2.—If through P the line PN be drawn, making with the transverse axis an angle equal to the eccentric angle, PN is equal to the semi-conjugate axis b.

Cor. 3.—$NN' = a - b.$ (367)

Cor. 4.—If ρ be the radius vector from the centre to any point P of the ellipse, then

$$\rho = a\Delta(\phi), \text{ where } \Delta\phi = \sqrt{1 - e^2 \sin^2 \phi}. \qquad (368)$$

Observation.—If the equation of the ellipse be written in the form

$$\left(1 + \frac{x}{a}\right)\left(1 - \frac{x}{a}\right) = \left(\frac{y}{b}\right)^2,$$

and if

$$\left(1 - \frac{x}{a}\right) = \left(\frac{y}{b}\right)\tan\theta, \quad \left(1 + \frac{x}{a}\right) = \left(\frac{y}{b}\right)\cot\theta,$$

we get

$$2 = \frac{y}{b}(\tan\theta + \cot\theta),$$

or $y = b\sin 2\theta$; hence, if $2\theta = \phi$, we have $y = b\sin\phi$, as before. (Compare Art. 108.)

Examples.

1. The auxiliary circle touches the ellipse at the two points A, A'; hence it has double contact with it.

2. If on the conjugate axis as diameter a circle be described, and ordinates be drawn parallel to the transverse axis, the ordinates of the ellipse are to those of the circle as $a : b$.

3. If a cylinder standing on a circular base be cut by any plane not parallel to the base, the section is an ellipse.

4. If a circle roll inside another of double its diameter, any point invariably connected with the rolling circle, but not on its circumference, describes an ellipse.

121. *The locus of the middle points of a system of parallel chords of an ellipse is a right line.*

Let PP' be a chord of the ellipse, and let the eccentric angles of P, P' be $(\alpha + \beta)$, $(\alpha - \beta)$ respectively; then (Art. 22, Ex. 3) the equation of PP' is

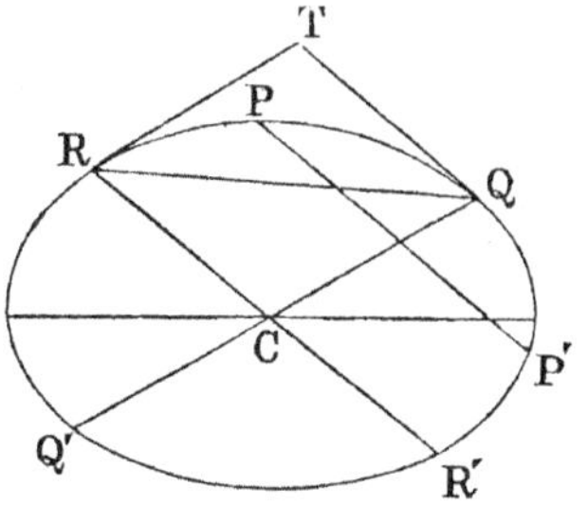

$$b \cos\alpha \,.\, x + a \sin\alpha \,.\, y = ab \cos\beta. \tag{I.}$$

Now it is evident that if α be constant and β variable, PP' will be one of a system of parallel chords.

Let x_1, y_1 be the co-ordinates of the middle point of PP', then we have

$$x_1 = \frac{a}{2}\{\cos(\alpha + \beta) + \cos(\alpha - \beta)\} = a \cos\alpha \cos\beta,$$

$$y_1 = \frac{b}{2}\{\sin(\alpha + \beta) + \sin(\alpha - \beta)\} = b \sin\alpha \cos\beta.$$

Hence $$b \sin\alpha \,.\, x_1 - a \cos\alpha \,.\, y_1 = 0, \tag{369}$$

which is the required equation. This is the line QQ'.

Cor. 1.—Let RR' be the diameter parallel to PP'; then, since RR' passes through the origin, its equation must contain no absolute term. Therefore, from (I.), $\cos\beta = 0$, or $\beta = 90°$; hence the equation of RR' is

$$b \cos\alpha \,.\, x + a \sin\alpha \,.\, y = 0. \tag{370}$$

Cor. 2.—If PP' move parallel to itself until the points P, P' become consecutive, then PP' will become the tangent at Q, and evidently we must have $\beta = 0$; therefore the tangent at Q is

$$b \cos\alpha \,.\, x + a \sin\alpha \,.\, y = ab. \tag{371}$$

Now, if x', y' be the co-ordinates of Q, we have $x' = a\cos\alpha$, $y' = b \sin\alpha$; hence, from (371) we get the tangent at $x'y'$,

$$\frac{xx'}{a^2} + \frac{yy'}{b^2} = 1. \tag{372}$$

Cor. 3.—If the angles which QQ', RR' make with the axis of x be denoted by θ, θ', respectively, we have from (369), (370),

$$\tan\theta = \frac{b}{a}\tan\alpha, \quad \tan\theta' = -\frac{b}{a}\cot\alpha;$$

therefore $$\tan\theta \,.\, \tan\theta' = -\frac{b^2}{a^2}. \qquad (373)$$

Since this remains unaltered by the interchange of θ and θ', it follows that, if two diameters QQ', RR' of an ellipse be such that the first bisects chords parallel to the second, the second also bisects chords parallel to the first.

DEF.—*Two diameters which are such that each bisects chords parallel to the other are called* CONJUGATE *diameters.*

Cor. 4.—Since the eccentric angle of Q is α, and of R $\alpha + \frac{\pi}{2}$ (*Cor.* 1), we see that the differ en ce between theeccentric angles of the extremities of two conjugate semi-diameters is a right angle.

Cor. 5.—If x'', y'' denote the co-ordinates of R, we have

$$x'' = a\cos\left(\alpha + \frac{\pi}{2}\right), \quad y'' = b\sin\left(\alpha + \frac{\pi}{2}\right);$$

but $$x' = a\cos\alpha, \quad y' = b\sin\alpha;$$

therefore $$x'' = -\frac{a}{b}y', \quad y'' = \frac{b}{a}x'. \qquad (374)$$

Cor. 6.—If the conjugate semi-diameters CQ, CR be denoted by a', b' respectively, we have

$$a'^2 = x'^2 + y'^2 = a^2\cos^2\alpha + b\sin^2\alpha;$$

$$b'^2 = x''^2 + y''^2 = a^2\sin^2\alpha + b\cos^2\alpha;$$

therefore $$a'^2 + b'^2 = a^2 + b^2; \qquad (375)$$

hence the sum of the squares of two conjugate semi-diameters is constant.

Cor. 7.—The tangent at Q is parallel to the diameter RR'.

Cor. 8.—The area of the triangle $QCR = \frac{1}{2}(x'y'' - x''y')$,

$$= \frac{1}{2}\begin{vmatrix} a\cos\alpha, & b\sin\alpha, \\ -a\sin\alpha, & b\cos\alpha \end{vmatrix} = \frac{1}{2}ab; \qquad (376)$$

therefore the area of the parallelogram $QCRT$ is equal to ab. Hence it follows that *the area of the parallelogram formed by the tangents at the extremities of any two conjugate diameters of an ellipse is constant.*

Examples.

1. Given any two conjugate semi-diameters OP, OQ of an ellipse, to find the magnitude and direction of its axes.

From P let fall the perpendicular PN on OQ; produce and cut off $PD = OQ$; join OD, and on OD as diameter describe a circle; let C be

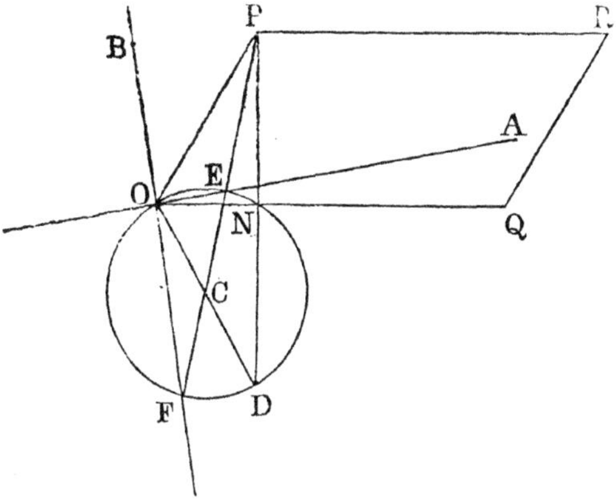

its centre; join PC, cutting the circle in the points E, F; join OE, OF, and make $OB = EP$, and $OA = FP$. Then OA, OB are the semiaxes required.

Dem. $OA^2 + OB^2 = EP^2 + FP^2 = 2CP^2 + 2CE^2 = 2CP^2 + 2OC^2$
$$= OP^2 + PD^2 = OP^2 + OQ^2;$$
that is, equal to the sum of the squares of the semi-conjugate axes.

Again,

$$OA \cdot OB = FP \cdot EP = DP \cdot NP = OQ \cdot NP = \text{parallelogram } OPQR.$$

Hence (*Cors.* 6, 8) OA, OB are the semiaxes required.

The foregoing beautiful construction is due to Mannheim. See *Nouv. An. de Math.*, 1857, p. 188; also Williamson's *Differential Calculus*, fifth edition, p. 374.

2. Being given the transverse and conjugate diameters of an ellipse to construct a pair of equiconjugate diameters.

3. Prove that the acute angle between a pair of equiconjugate diameters is less than the angle between any other pair of conjugate diameters.

122. *To find the equation of an ellipse referred to a pair of conjugate diameters.*

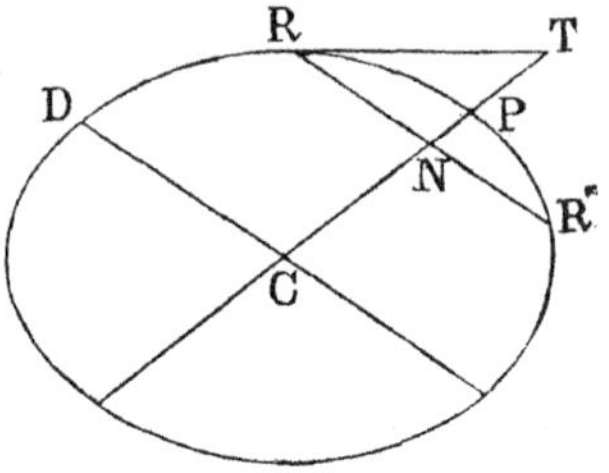

Let CP, CD be two semi-conjugate diameters of lengths a', b'; let RR' be a chord parallel to CD; then RR' is bisected by CP in N. Hence, denoting CN, NR by x, y, and the eccentric angles of R, R' by $(\alpha+\beta)$, $(\alpha-\beta)$ respectively, we have

$$x^2 = \left\{\frac{a\cos(\alpha+\beta)+a\cos(\alpha-\beta)}{2}\right\}^2 + \left\{\frac{b\sin(\alpha+\beta)+b\sin(\alpha-\beta)}{2}\right\}^2$$

$$= (a^2\cos^2\alpha + b^2\sin^2\alpha)\cos^2\beta = a'^2\cos^2\beta. \quad \text{(Art. 121, } Cor. \text{ 6.)}$$

In like manner $$y^2 = b'^2\sin^2\beta;$$

hence

$$\frac{x^2}{a'^2} + \frac{y^2}{b'^2} = 1. \quad \text{(Compare Art. 101, 3°.)} \qquad (377)$$

Cor. 1.—The co-ordinates of any point on an ellipse referred to a pair of conjugate diameters can be represented by

$$a'\cos\beta, \quad b'\sin\beta. \qquad (378)$$

Cor. 2.—The equation of the tangent to an ellipse referred to a pair of conjugate diameters is

$$\frac{xx'}{a'^2} + \frac{yy'}{b'^2} = 1, \quad \text{or} \quad \frac{x\cos\beta}{a'} + \frac{y\sin\beta}{b'} = 1. \qquad (379)$$

Cor. 3.—If the tangent at R meet CP produced in T,

$$CN \,.\, CT = CP^2; \tag{380}$$

for the tangent at R is $\frac{xx'}{a'^2} + \frac{yy'}{b'^2} = 1$; and putting $y = 0$, we get $xx' = a'^2$, or $CN \,.\, CT = CP^2$.

Cor. 4.—The tangents at the extremities of any double ordinates RR' meet its diameter produced in the same point.

Cor. 5.—The line joining the centre to the intersection of two tangents bisects their chord of contact.

Examples.

1. If AB be any diameter of an ellipse, AE, BD tangents at its extremities, meeting any third tangent ED in E and D; prove that $AE \,.\, BD$ = square of semi-diameter conjugate to AB.

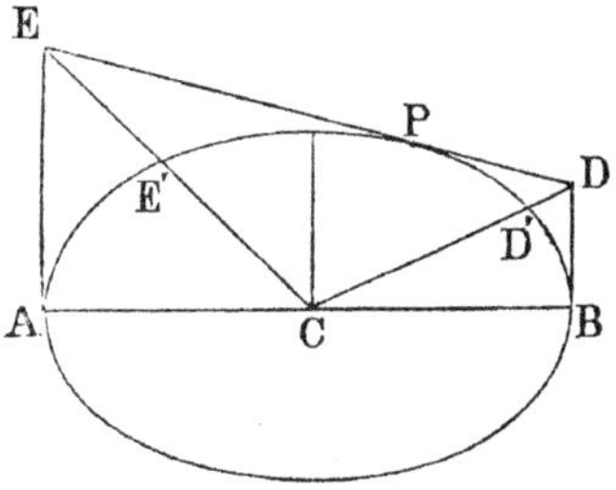

For denoting AC and its conjugate by a', b', the equatian of ED is

$$\frac{x \cos\beta}{a'} + \frac{y \sin\beta}{b'} = 1.$$

(Equation (379).)

Hence, denoting AE, BD by y_1, y_2 respectively, we have, substituting $-a'$, $+a'$, respectively, for x,

$$y_1 \sin\beta = b'(1 + \cos\beta),$$
$$y_2 \sin\beta = b'(1 - \cos\beta);$$

hence

$$y_1 y_2 = b'^2. \tag{381}$$

2. If CD, CE be drawn intersecting the ellipse in D', E'; prove that CD', CE' are conjugate semi-diameters.

3. If P be the point of contact of DE, prove that $DP \,.\, PE$ = square of parallel semi-conjugate diameter. [Make use of Art. 101, 2°.]

4. If AB be the transverse axis, the circle described on DE as diameter passes through the foci.

5. If CP, CD be any two semi-diameters; PT, DE tangents at P and D, meeting CD, CP produced in T and E; prove that the triangle $CPT = CDE$.

6. In the same case, if PN, DM be parallel respectively to DE and PT; prove that the triangle $CPN = CDM$.

Def.—*Two chords, such as AB, BP, joining any point P in the ellipse to the extremities of any diameter AB, are called* SUPPLEMENTAL CHORDS.

7. Diameters parallel to a pair of supplemental chords are conjugates.

8. If a parallel to a fixed line meet a given semicircle in C and its diameter in D; prove that the locus of the point E, which divides CD in a given ratio, is an ellipse.

9. If a line AB of given length slide between two fixed lines; prove that the locus of the point P, which divides AB in a given ratio, is an ellipse.

10. If a given triangle ABC slides with two vertices A, B on two fixed lines OX, OY; prove that the third vertex C describes an ellipse (SCHOOTEN, *Organica Conicorum Descriptio*, 1646, c. 3, Ex. Math. IV.)

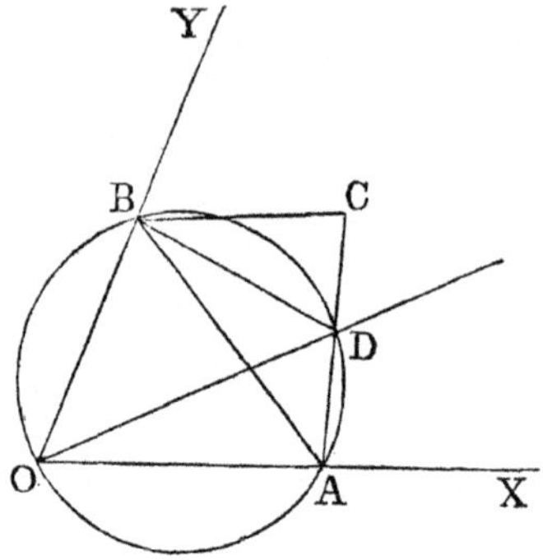

About the triangle OBA describe a circle cutting AC in D; join BD, OD; then, because the angle AOB is given, the angle ADB is given; hence the three angles of the triangle BCD are given; and since BC is given, CD is given; also the angle BOD, being equal to BAC, is given. Hence the line OD is given in position, and the proposition is reduced to the following:—AD, a line of given length, slides between two fixed lines OX, OD, and C is a fixed point in it; therefore (Ex. 9) the locus is an ellipse.

123. *To find the equation of the normal to the ellipse at the point $x'y'$.*

Let α be the eccentric angle of the point $x'y'$; then the equation to the tangent at α (Art. 121, *Cor.* 2) is

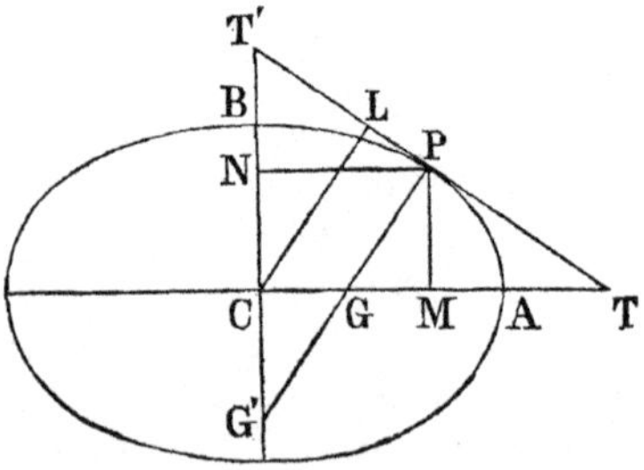

$$b\cos\alpha \,.\, x + a\sin\alpha \,.\, y = ab;$$

hence $a\sin\alpha\,(x - x') - b\cos\alpha\,(y - y') = 0$ is the equation of the normal; and, putting for x', y' their values in terms of α, we get

$$a\sin\alpha \,.\, x - b\cos\alpha \,.\, y = c^2 \sin\alpha\cos\alpha, \qquad (382)$$

or
$$\frac{a^2x}{x'} - \frac{b^2y}{y'} = c^2; \qquad (383)$$
or thus:

The equation of the line bisecting perpendicularly the chord joining the points $(\alpha+\beta)$, $(\alpha-\beta)$ is (equation (44)) $\frac{ax}{\cos\alpha} - \frac{by}{\sin\alpha} = c^2\cos\beta$; and, if the points coincide, the chord becomes a tangent, and $\beta = 0$; thus we get the same equation as before.

Cor. 1.—In equation (382) put $y = 0$, and we get $x = ae^2\cos\alpha$,

or
$$CG = e^2x; \qquad (384)$$
hence
$$MG^2 = (1 - e^2)\,a\cos\alpha.$$

Cor. 2.—$PG^2 = PM^2 + MG^2 = b^2\sin^2\alpha + (1 - e^2)^2a^2\cos^2\alpha$; but $1 - e^2 = \frac{b^2}{a^2}$; therefore $PG^2 = b^2\{\sin^2\alpha + (1 - e^2)\cos^2\alpha\}$ $= b^2(1 - e^2\cos^2\alpha)$; therefore
$$PG = b\sqrt{1 - e^2\cos^2\alpha}. \qquad (385)$$
In like manner,
$$PG' = \frac{a^2}{b}\sqrt{1 - e^2\cos^2\alpha};$$
therefore
$$PG \,.\, PG' = a^2(1 - e^2\cos^2\alpha). \qquad (386)$$

Cor. 3.—If ρ, ρ' be the focal vectors to P, we have
$$\rho = a + ex' = a(1 + e\cos\alpha),$$
$$\rho' = a - ex' = a(1 - e\cos\alpha);$$
therefore
$$PG \,.\, PG' = \rho\rho'. \qquad (387)$$

Cor. 4.—If CR be the semi-diameter conjugate to CP, we have
$$CR^2 = a^2\sin^2\alpha + b^2\cos^2\alpha = a^2(1 - e^2\cos^2\alpha).$$
therefore
$$\rho\rho' = CR^2 = b'^2. \qquad (388)$$
Hence
$$PG \,.\, PG' = b'^2. \qquad (389)$$

Cor. 5.—If CL be perpendicular to the tangent at P,

$$CL^2 = \frac{b^2}{1 - e^2 \cos^2 \alpha}.$$

Therefore $$CL . PG = b^2, \text{ and } CL . PG' = a^2. \qquad (390)$$

Examples.

1. The co-ordinates of the intersection of normals at the points $(\alpha + \beta)$, $(\alpha - \beta)$, are

$$x = \frac{c^2 \cos \alpha . \cos (\alpha + \beta) \cos (\alpha - \beta)}{a \cos \beta}, \quad y = -\frac{c^2 \sin \alpha . \sin (\alpha + \beta) \sin (\alpha - \beta)}{b \cos \beta}. \qquad (391)$$

2. If the normals at α, β, γ be concurrent,

$$\begin{vmatrix} \sec \alpha, & \operatorname{cosec} \alpha, & 1, \\ \sec \beta, & \operatorname{cosec} \beta, & 1, \\ \sec \gamma, & \operatorname{cosec} \gamma, & 1 \end{vmatrix} = 0. \qquad (392)$$

3. The two foci and the points P, G', are concyclic.

4. Find the condition that the normal at the point α on

$$\frac{x^2}{a^2} + \frac{y^2}{b^2} - 1 = 0$$

should pass through the point α on

$$\frac{x^2}{a'^2} + \frac{y^2}{b'^2} - 1 = 0.$$

Ans. $aa' - bb' = c^2$.

5. Find the co-ordinates of the intersection of two consecutive normals. Making $\beta = 0$, in Ex. 1, we get

$$x = \frac{c^2 \cos^3 \alpha}{a}, \quad y = -\frac{c^2 \sin^3 \alpha}{b}. \qquad (393)$$

Or thus :—the cordinates of a point equally distant from α, β, γ (Art 21, Ex. 3) are—

$$\frac{c^2}{a} \cos \tfrac{1}{2}(\alpha + \beta) \cos \tfrac{1}{2}(\beta + \gamma) \cos \tfrac{1}{2}(\gamma + \alpha), \quad -\frac{c^2}{b} \sin \tfrac{1}{2}(\alpha + \beta) \sin \tfrac{1}{2}(\beta + \gamma) \sin \tfrac{1}{2}(\gamma + \alpha);$$

and, supposing the points to become consecutive, we get, for the centre of a circle passing through three consecutive points, the same co-ordinates as before.

Def.—*The circle passing through three consecutive points of a curve is called the* OSCULATING CIRCLE, *or* CIRCLE OF CURVATURE *at the point.*

6. Find the locus of the centre of curvature of all the points of an ellipse. Eliminating α from the equations (393), we get

$$(ax)^{\frac{2}{3}} + (by)^{\frac{2}{3}} = c^{\frac{4}{3}}. \qquad (394)$$

This locus is called the *evolute* of the ellipse.

7. Four normals can in general be drawn from any point to an ellipse. For if hk be the point, the curve of the second degree,

$$\frac{a^2 h}{x} - \frac{b^2 k}{y} = c^2,$$

passes through the feet of the normals.

8. The radius of curvature at α is $= \frac{b'^2}{p}$, where p is the perpendicular from the origin on the tangent.

The radius of curvature is the distance between the points

$$\left(\frac{c^2 \cos^3 \alpha}{a}, \quad - \frac{c^2 \sin^3 \alpha}{b}\right); \ (a \cos \alpha, \ b \sin \alpha),$$

which by an easy reduction can be shown $= \frac{b'^2}{p}$. (395)

9. In the figure, Art. 120, if we complete the rectangle $NON'Q$, prove that the normal at P passes through Q.

10. The equation of the circle, whose diameter is the whole length of the normal intercepted by the ellipse, is

$$(a^4 \sin^2 \alpha + b^4 \cos^2 \alpha)(x^2 + y^2) - 2c^2 \sin \alpha \cos \alpha \, (a \sin \alpha . x - b \cos \alpha . y)$$
$$+ (a^2 + b^2) c^4 \sin^2 \alpha \cos^2 \alpha - a^2 b^2 (a^2 \sin^2 \alpha + b^2 \cos^2 \alpha) = 0.$$

124. *To find the lengths of the perpendiculars from the foci on the tangent at any point ϕ.*

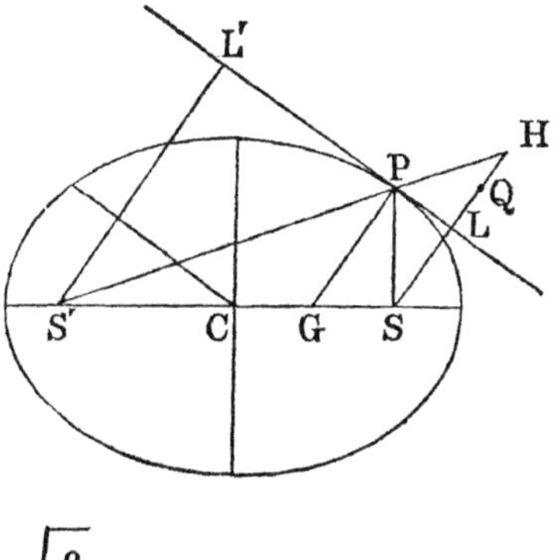

The tangent is

$$b \cos \phi . x + a \sin \phi . y - ab = 0,$$

and the co-ordinates of the focus S are ae, 0. Hence the perpendicular

$$SL = \frac{ab(1 - e \cos \phi)}{a(1 - e^2 \cos^2 \phi)^{\frac{1}{2}}}$$
$$= b\left(\frac{1 - e \cos \phi}{1 + e \cos \phi}\right)^{\frac{1}{2}};$$

or

$$SL = b\sqrt{\frac{\rho}{\rho'}}. \qquad (396)$$

Similarly, $$S'L' = b\sqrt{\frac{\rho'}{\rho}}. \qquad (397)$$

Cor. 1.—$SL \,.\, S'L' = b^2$. (398)

Cor. 2.—$SL \div \rho = \dfrac{b}{\sqrt{\rho\rho'}} = \dfrac{b}{b'} = \dfrac{S'L'}{\rho'} = \sin SPL = S'PL'$. (399)

Cor. 3.—The tangent LL' bisects the external angle at P of the triangle SPS', and the normal PG the internal angle.

Cor. 4.—The first positive pedal (Art. 110) of an ellipse with respect to either focus is the auxiliary circle. For since the angle SPH is bisected by PL, we have $SL = LH$; therefore SH is bisected in L, and SS' is bisected in C; therefore, if CL be joined, $CL = \frac{1}{2}S'H = \frac{1}{2}(S'P + PS) = a$. Hence the locus of L is the auxiliary circle. And conversely, the first negative pedal of a circle with respect to any internal point is an ellipse, having the point for one of its foci.

Cor. 5.—If any point in LL' be joined to S, the circle described on the join will intersect the auxiliary circle in L. Hence may be inferred a method of drawing tangents to an ellipse from an external point. Thus, if Q be the point, join QS, and on QS as diameter describe a circle intersecting the auxiliary circle in L and M. QL, QM are the tangents to the ellipse.

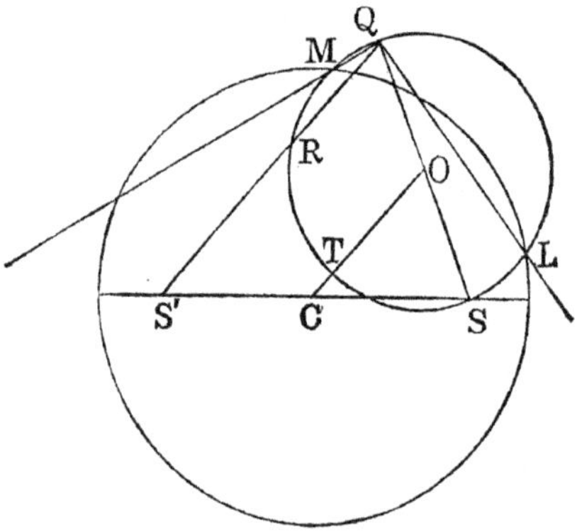

Cor. 6.—The two tangents from Q are equally inclined to the focal vectors QS, QS'. For, join the centres C, O of the circles; then CO is parallel to QS'; therefore it bisects the arc RS, but the line joining the centres also bisects the arc ML. Hence the arc $RM = SL$, and the angle $S'QM = SQL$.

EXAMPLES.

1. Find the relation between the eccentric angles of two points whose joining chord passes through a focus.

If the eccentric angles be $(\alpha + \beta)$, $(\alpha - \beta)$, the chord will be

$$b \cos \alpha . x + a \sin \alpha . y = ab \cos \beta ;$$

and if this passes through the focus $(ae, 0)$, we get

$$e \cos \alpha = \cos \beta. \qquad (400)$$

Hence the equation of any focal chord is

$$\frac{x \cos \alpha}{a} + \frac{y \sin \alpha}{b} = \pm\, e \cos \alpha, \qquad (401)$$

the sign depending on the focus through which the chord passes.

2. The tangents at the extremities of a chord passing through either focus meet on the corresponding directrix. For the tangents at the points $(\alpha + \beta)$, $(\alpha - \beta)$, are $b \cos (\alpha + \beta)\, x + a \sin (\alpha + \beta)\, y = ab$;

$$b \cos (\alpha - \beta)\, x + a \sin (\alpha - \beta)\, y = ab ;$$

and the co-ordinates of the point where these intersect are—

$$\frac{a \cos \alpha}{\cos \beta}, \quad \frac{b \sin \alpha}{\cos \beta}. \qquad (402)$$

Substituting the value of $\cos \beta$ from (400), we get

$$\frac{a}{e}, \quad \frac{b \tan \alpha}{e}, \qquad (403)$$

which are the co-ordinates of a point on the directrix.

3. In the same case the join of the intersection of tangents to the focus is perpendicular to the chord. For the line joining ae, 0 to the point (403) is $a \sin \alpha . x - b \cos \alpha . y - \dfrac{a^2 \sin \alpha}{e} = 0$, which is perpendicular to the chord (401).

4. If the co-ordinates in (402) be denoted by $x'y'$, we get

$$\cos \alpha = \frac{x' \cos \beta}{a}, \quad \sin \alpha = \frac{y' \cos \beta}{b}.$$

Substituting these in the equation of the chord, we get

$$\frac{xx'}{a^2} + \frac{yy'}{b^2} = 1. \qquad (404)$$

Hence the chord of contact of tangents from $x'y'$ is

$$\frac{xx'}{a^2} + \frac{yy'}{b^2} = 1.$$

5. If the chord $b \cos \alpha . x + a \sin \alpha . y = ab \cos \beta$ pass through a fixed point $x'y'$, the locus of the intersection of tangents at its extremities is

$$\frac{xx'}{a^2} + \frac{yy'}{b^2} = 1.$$

For, denoting the co-ordinates (402) by xy, and substituting in $b \cos \alpha . x' + a \sin \alpha . y' = ab \cos \beta$, we get

$$\frac{xx'}{a^2} + \frac{yy'}{b^2} = 1.$$

DEF.—*The line* $\frac{xx'}{a^2} + \frac{yy'}{b^2} = 1$ *is called the* POLAR *of the point* $x'y'$ *with respect to the ellipse.* (Compare Arts. 59, 99.)

Cor.—The directrix is the polar of the focus.

6. If α be variable and β constant, the chord joining the points $(\alpha + \beta)$, $(\alpha - \beta)$ is a tangent to the ellipse

$$\left(\frac{x}{a}\right)^2 + \left(\frac{y}{b}\right)^2 = \cos^2 \beta. \qquad (405)$$

7. In the same case the locus of the intersection of tangents is

$$\left(\frac{x}{a}\right)^2 + \left(\frac{y}{b}\right)^2 = \sec^2 \beta. \qquad (406)$$

8. The equation of the perpendicular from the point (402) on the chord joining the points $(\alpha + \beta)$, $(\alpha - \beta)$ is

$$\frac{ax}{\cos \alpha} - \frac{by}{\sin \alpha} = \frac{a^2 e^2}{\cos \beta} \text{ (compare Art. 123)}, \qquad (407)$$

which meets the axis in the points

$$e^2 \left(\frac{a \cos \alpha}{\cos \beta}\right), \qquad - \frac{a^2 e^2 \sin \alpha}{b \cos \beta};$$

that is, in the points

$$e^2 x', \qquad - \frac{a^2 e^2 y'}{b^2}. \qquad (408)$$

9. Find the condition that the join of $(\alpha + \beta)$, $(\alpha - \beta)$ shall touch the ellipse

$$\left(\frac{x}{a_1}\right)^2 + \left(\frac{y}{b_1}\right)^2 = 1.$$

If ϕ be the point of contact, the equations

$$b_1 \cos \phi . x + a_1 \sin \phi . y - a_1 b_1 = 0,$$

$$b \cos \alpha . x + a \sin \alpha . y - ab \cos \beta = 0$$

must represent the same line; hence, eliminating ϕ from the equations

$$\frac{\cos\phi}{a_1} = \frac{\cos\alpha}{a\cos\beta}, \quad \frac{\sin\phi}{\beta_1} = \frac{\sin\alpha}{b\cos\beta},$$

we get

$$\frac{a_1^2\cos^2\alpha}{a^2} + \frac{b_1^2\sin^2\alpha}{b^2} = \cos^2\beta, \tag{409}$$

which is the required condition.

10. If ϕ denote the angle between the tangents at $(\alpha+\beta)$, $(\alpha-\beta)$, prove

$$\tan\phi = \frac{2ab\sin 2\beta}{(a^2-b^2)\cos 2\alpha - (a^2+b^2)\cos 2\beta}. \tag{410}$$

11. If the angle ϕ be right, we get $(a^2-b^2)\cos 2\alpha = (a^2+b^2)\cos 2\beta$,

or
$$(a^2+b^2)\cos^2\beta = a^2\cos^2\alpha + b^2\sin^2\alpha.$$

Hence, denoting $\dfrac{a\cos\alpha}{\cos\beta}$, $\dfrac{b\sin\alpha}{\cos\beta}$ by x, y, we get the circle

$$x^2 + y^2 = a^2 + b^2 \tag{411}$$

as the locus of the intersection of rectangular tangents.

DEF.—*The circle* (411) *is called the* DIRECTOR *circle of the ellipse.*

12. If in Ex. 9 we put $a_1^2 = a^2 - \lambda^2$, $b_1^2 = b^2 - \lambda^2$, the ellipses will be confocal, and equation (409) reduces, if b' denote the semi-diameter conjugate to that drawn to the point α, to

$$\sin\beta = \frac{\lambda b'}{ab}, \tag{412}$$

which is the condition that the join of the points $(\alpha+\beta)$, $(\alpha-\beta)$ on the ellipse

$$\frac{x^2}{a^2} + \frac{y^2}{b^2} = 1,$$

shall touch the confocal

$$\frac{x^2}{a^2-\lambda^2} + \frac{y^2}{b^2-\lambda^2} = 1.$$

13. If two tangents to an ellipse be at right angles, their chord of contact touches a confocal ellipse (Ex. 11, 12).

14. If from the point $\left(\dfrac{a\cos\alpha}{\cos\beta}, \dfrac{b\sin\alpha}{\cos\beta}\right)$ perpendiculars be drawn to the four focal vectors of the points $(\alpha+\beta)$, $(\alpha-\beta)$, these perpendiculars are equal, their common value being $b\tan\beta$. Hence we have the following theorem:—*The four focal vectors drawn to any two points of an ellipse have one common tangential circle, whose centre is the pole of the chord joining the two points.* The equation of the circle is

$$(x\cos\beta - a\cos\alpha)^2 + (y\cos\beta - b\sin\alpha)^2 = b^2\sin^2\beta. \tag{413}$$

15. The angle ϕ between the tangents to an ellipse can be expressed in terms of the focal vectors to their point of intersection; thus, denoting these by ρ, ρ', we get

$$\rho^2 \cos^2 \beta = b^2 \sin^2 \alpha + a^2 (\cos \alpha + e \cos \beta)^2;$$

then, putting for b^2 the value $a^2 (1 - e^2)$, we get, after an easy reduction,

$$\rho^2 \cos^2 \beta = a^2 \{1 + e \cos(\alpha + \beta)\} \{1 + e \cos(\alpha - \beta)\}.$$

Similarly,

$$\rho'^2 \cos^2 \beta = a^2 \{1 - e \cos(\alpha + \beta)\} \{1 - e \cos(\alpha - \beta)\}.$$

Hence

$$\rho\rho' \cos^2 \beta = a^2 \sqrt{\{1 - e^2 \cos^2(\alpha + \beta)\} \{1 - e^2 \cos^2(\alpha - \beta)\}},$$

and

$$(\rho^2 + \rho'^2 - 4a^2) \cos^2 \beta = (a^2 - b^2) \cos 2\alpha - (a^2 + b^2) \cos 2\beta$$

Now from the value of $\tan \phi$ (Ex. 10) we get

$$\cos \phi = \frac{(a^2 - b^2) \cos^2 2\alpha - (a^2 + b^2) \cos 2\beta}{2a^2 \sqrt{\{1 - e^2 \cos^2(\alpha + \beta)\} \{1 - e^2 \cos^2(\alpha - \beta)\}}}$$

Hence

$$\cos \phi = \frac{\rho^2 + \rho'^2 - 4a^2}{2\rho\rho'}; \tag{414}$$

and putting $\rho + \rho' = 2a'$, we get

$$\cos^2 \tfrac{1}{2}\phi = \frac{a'^2 - a^2}{\rho\rho'}. \tag{415}$$

16. If μ, μ', μ'' be the semi-axes major of three confocal ellipses, and if from any point in the outer, tangents be drawn to the three; then, if $(\widehat{\mu\mu'})$ denote the angle between the confocals μ, μ',

$$\sin^2(\widehat{\mu\mu'}) : \sin^2(\widehat{\mu\mu''}) :: \mu^2 - \mu'^2 : \mu^2 - \mu''^2. \tag{416}$$

17. If tangents at $(\alpha + \beta)$, $(\alpha - \beta)$ intersect on the confocal,

$$\frac{x^2}{a^2 + \lambda^2} + \frac{y^2}{b^2 + \lambda^2} = 1;$$

then

$$\frac{\cos^2 \alpha}{a^2 + \lambda^2} + \frac{\sin^2 \alpha}{b^2 + \lambda^2} = \frac{\sin^2 \beta}{\lambda^2}: \tag{417}$$

If the semi-diameter of $(a^2 + \lambda^2)$, conjugate to that drawn to the point α, be denoted by b', (417) may be written

$$\sin^2 \beta : b'^2 :: \lambda^2 : (a^2 + \lambda^2)(b^2 + \lambda^2); \tag{418}$$

that is, in a *given ratio*.

18. If tangents to two confocals be at right angles, the locus of their intersection is a circle.

19. If c denote the length of the chord joining the points $(\alpha + \beta)$, $(\alpha - \beta)$, we have (Dem. Art. 122) $c^2 = 4b'^2 \sin^2 \beta$, and from Ex. 12,

$$\sin^2 \beta = \frac{\lambda^2 b'^2}{a^2 b^2};$$

therefore

$$c = \frac{2\lambda b'^2}{ab}. \qquad \text{(BURNSIDE.)} \qquad (419)$$

20. If tangents to the confocals

$$\frac{x^2}{a^2} + \frac{y^2}{b^2} - 1 = 0, \quad \frac{x^2}{a_1^2} + \frac{y^2}{b_1^2} - 1 = 0$$

be at right angles to each other, the line joining the point of contact on one to the point of contact on the other is a tangent to a third confocal, the squares of whose semi-axes are

$$\frac{a^2 a_1^2}{a^2 + b_1^2}, \quad \frac{b^2 b_1^2}{a^2 + b_1^2}. \qquad (420)$$

21. If tangents to two confocal ellipses be parallel, the angles subtended at the foci by the points of contact are equal.

125. *The locus of the pole of any tangent to an ellipse, with respect to a circle whose centre is one of the foci, is a circle.*

Dem.—Let S (see fig. Art. 124) be the focus, R the radius of the circle whose centre is S, and with respect to which the poles are taken. Let fall SL perpendicular to the tangent to the ellipse, and make $SL \cdot SQ = R^2$; then L, Q are inverse points with respect to the circle whose radius is R; and since the locus of L is the auxiliary circle, the locus of Q is its inverse, and is therefore a circle; but Q is the pole of LL', and is the point whose locus is required; hence the proposition is proved.

DEF.—*The locus of the poles of all the tangents to any curve with respect to a circle is called the* RECIPROCAL POLAR *of that curve with respect to the circle.*

From this definition we see that the foregoing proposition may be enunciated as follows:—*The reciprocal polar of an ellipse, with respect to a circle whose centre is one of the foci, is a circle.*

Cor. 1.—If we take two consecutive tangents to the ellipse, their poles will be consecutive points on the circle which is the reciprocal polar of the ellipse; but the join of the poles of two lines is the polar of the point of intersection of the line. Hence the locus of the pole of any tangent to a circle is an ellipse. In other words, *The reciprocal polar of a circle with respect to another circle is an ellipse, having the centre of the reciprocating circle for one of its foci.*

Or thus:

Let S be the centre of the reciprocating circle, Q any point on the circle whose reciprocal polar is required; join SQ, and make $SQ \,.\, SL = R^2$, and draw LL' perpendicular to SQ. Now, since $SQ \,.\, SL = R^2$, the locus of L is the circle which is the inverse of that which is to be reciprocated; and since LL' is perpendicular to SL, the envelope of LL' is the first negative pedal of a circle with respect to a given point.

Cor. 2.—Since the auxiliary circles of a system of confocal ellipses is a system of concentric circles, and the inverse of a system of concentric circles is a system of coaxal circles, we have the following theorem:—*The reciprocal polars of a system of confocal ellipses, with respect to a circle whose centre is one of the foci, is a system of coaxal circles, having the focus as one of the limiting points.* Conversely, *The reciprocal polars of a system of coaxal circles, with respect to one of the limiting points, is a confocal system, having that point for one of the foci.*

EXAMPLES.

*1. If a quadrilateral AA', BB' be inscribed in a circle X, and if the diagonals AB, $A'B'$ touch a circle Y of a system coaxal with X, then the sides (*Sequel to Euclid*, p, 126) AA', BB' touch another circle of the same system, and the four points of contact are collinear. Reciprocally, *If a quadrilateral be circumscribed to an ellipse, and if two of its opposite vertices lie on a confocal ellipse, two of the remaining vertices lie on another confocal, and the four tangents at these vertices are concurrent.*.

2. The reciprocal polar of the directrix of an ellipse with respect to a focus is the centre of the circle into which the ellipse reciprocates.

3. If a variable chord of a circle subtend a right angle at a fixed point within the circle, its envelope is an ellipse, having the fixed point for one of its foci.

*4. If L be one of the limiting points of two circles O, O', and LA, LB two radii vectors at right angles to each other, and terminating in those circles, the locus of the intersection of tangents at A and B is a circle coaxal with O, O' (*Sequel to Euclid*, p. 162). Reciprocally, *If two tangents, one to each of two confocal ellipses, be at right angles to each other, the envelope of the line joining the points of contact is a confocal ellipse.*

*5. The envelope of the chord of contact of tangents to a circle which meet at a given angle is a concentric circle. Reciprocally, *the locus of the intersection of tangents to an ellipse, whose chord of contact subtends a given angle at the focus, is an ellipse, having the same focus and directrix.*

126. *The rectangle contained by the segments of any chord, passing through a fixed point in the plane of an ellipse, is to the square of the parallel semidiameter in a constant ratio.* (Compare Art. 100.)

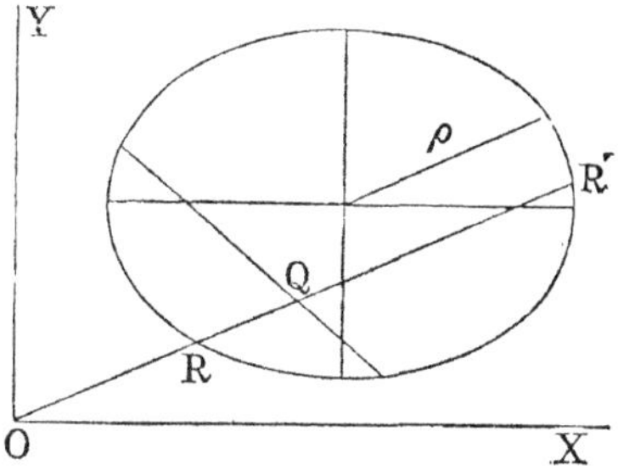

Let O be the fixed point, and take the lines OX, OY parallel to the axes of the ellipse; let the co-ordinates of the centre with respect to OX, OY be x', y'; then transforming to O, as origin, the equation of the ellipse is

$$\frac{(x - x')^2}{a^2} + \frac{(y - y')^2}{b^2} = 1. \qquad (1.)$$

Now, take any point R in the ellipse, join OR, meeting the curve again in R'; then, if r, θ be the polar co-ordinates of R, we have $x = r\cos\theta$, $y = r\sin\theta$. Hence from equation (I.) we get

$$(a^2\sin^2\theta + b^2\cos^2\theta)\,r^2 - 2(a^2y'\sin\theta + b^2x'\cos\theta)r + (b^2x'^2 + a^2y'^2 - a^2b^2) = 0. \quad \text{(II.)}$$

Now the roots of this quadratic in r are OR, OR'.

Hence $$OR\,.\,OR' = \frac{b^2x'^2 + a^2y'^2 - a^2b^2}{a^2\sin^2\theta + b^2\cos^2\theta}.$$

Again, if ρ be the radius vector through the centre parallel to OR, we have

$$\rho^2 = \frac{a^2b^2}{a^2\sin^2\theta + b^2\cos^2\theta};$$

therefore $$\frac{OR\,.\,OR'}{\rho^2} = \frac{x'^2}{a^2} + \frac{y'^2}{b^2} - 1; \quad (421)$$

that is, equal to the power of the point with respect to the ellipse. Hence the proposition is proved.

Cor. 1.—If OS be another line through O cutting the ellipse in S, S', and ρ' the parallel semidiameter,

$$\frac{OS\,.\,OS'}{\rho'^2} = \frac{x'^2}{a^2} + \frac{y'^2}{b^2} - 1.$$

Hence $$\frac{OR\,.\,OR'}{OS\,.\,OS'} = \frac{\rho^2}{\rho'^2}. \quad (422)$$

Cor. 2.—If through another point o two chords be drawn parallel to the chords OR, OS, and cutting the curve in r, r'; s, s', respectively,

$$\frac{OR\,.\,OR'}{OS\,.\,OS'} = \frac{or\,.\,or'}{os\,.\,os'}. \quad (423)$$

Cor. 3.—If the points R, R' coincide, OR becomes a tangent, and if S, S' coincide, OS becomes a tangent; hence, from *Cor.* 1, *Any two tangents to an ellipse are proportional to their parallel semidiameters.*

EXAMPLES.

1. The rectangle $EP.PD$ (see fig., Art. 122, Ex. 1) is equal to the square of the parallel semidiameter.

2. If any tangent meets two conjugate semidiameters of an ellipse, the rectangle under its segments is equal to the square of the parallel semidiameter.

3. If through any point O, in the plane of an ellipse, a secant be drawn meeting the ellipse in two points R, R', the locus of the point Q, which is the harmonic conjugate of O with respect to R, R', is the polar of O. For

$$\frac{2}{OQ} = \frac{1}{OR} + \frac{1}{OR'} = 2\left(\frac{a^2y' \sin\theta + b^2x' \cos\theta)}{a^2y'^2 + b^2x'^2 - a^2b^2}\right).$$

Hence, denoting OQ by ρ, we get, putting $\rho\cos\theta = x$, $\rho\sin\theta = y$,

$$b^2x'(x'-x) + a^2y'(y'-y) = a^2b^2,$$

or, transforming to the centre as origin,

$$\frac{xx'}{a^2} + \frac{yy'}{b^2} + 1 = 0,$$

which is the polar of the point $-x'-y'$ (see Art. 124, Ex. 4).

4. If A, B be any two points, C the centre of the ellipse, and if AG, BH be drawn parallel to CB, CA, intersecting the polars of B, A, respectively, in the points G, H; then $AG\,.\,CB : AC\,.\,BH$: : square of semidiameter through B : square of semidiameter through A.

5. If MN be the polar of the point A; P any point on the ellipse; AF a perpendicular to the tangent at P; PG the portion of the normal intercepted between the curve and the transverse axis; PM a perpendicular from P on MN; then $PG\,.\,AF$ varies as PM. For if the co-ordinates of A be $x'y'$; of P, $x''y''$; then

$$PM\left(\frac{x'^2}{a^4} + \frac{y'^2}{b^4}\right)^{\frac{1}{2}} = \frac{x'x''}{a^2} + \frac{y'y''}{b^2} - 1, \quad AF\left(\frac{x''^2}{a^4} + \frac{y''^2}{b^4}\right)^{\frac{1}{2}} = \frac{x'x''}{a^2} + \frac{y'y''}{b^2} - 1.$$

But
$$\left(\frac{x''^2}{a^4} + \frac{y''^2}{b^4}\right)^{\frac{1}{2}} = \frac{PG}{b^2};$$

therefore
$$PM\left(\frac{x'^2}{a^4} + \frac{y'^2}{b^4}\right)^{\frac{1}{2}} = \frac{PG\,.\,AF}{b^2}.$$

This theorem gives an immediate proof of HAMILTON'S *Law of Force*.—*Proceedings of the Royal Irish Academy*, No. LVII. vol. iii. p. 308. Also *Quarterly Journal of Mathematics*, vol. v. pp. 233–235.

6. Find the equation of the line through the point $x'y'$ parallel to its polar. If $(\alpha+\beta)$, $(\alpha-\beta)$ be the eccentric angles of the points of contact of tangents from $x'y'$, the line required is

$$\frac{x\cos\alpha}{a}+\frac{y\sin\alpha}{b}-\sec\beta=0\equiv L. \qquad (424)$$

7. In the same case the line through the centre and $x'y'$ is

$$\frac{x\sin\alpha}{a}-\frac{y\cos\alpha}{b}=0\equiv M. \qquad (425)$$

8. The equations of the tangents through $x'y'$ to the ellipse are

$$L\cos\beta\pm M\sin\beta=0. \qquad (426)$$

9. The product of the equations of the tangents is

$$\left(\frac{x^2}{a^2}+\frac{y^2}{b^2}-1\right)\left(\frac{x'^2}{a^2}+\frac{y'^2}{b^2}-1\right)\left(\frac{xx'}{a^2}+\frac{yy'}{b^2}-1\right)^2=0. \qquad (427)$$

Compare Articles 55, 99.

*127. *To find the major axis of an ellipse confocal to a given one and passing through a given point.*

Let hk be the given point, $\frac{x^2}{a^2}+\frac{y^2}{b^2}-1=0$ the given ellipse, then, putting $a^2-b^2=c^2$, the equation of the required ellipse will be of the form $\frac{x^2}{a'^2}+\frac{y^2}{a'^2-c^2}=1$, and substituting the given co-ordinates, we get

$$a'^4-(h^2+k^2+c^2)a'^2+c^2h^2=0. \qquad (428)$$

Similarly $$b'^4-(h^2+k^2-c^2)b'^2-c^2k^2=0. \qquad (429)$$

Let the roots of these equations be a'^2, a''^2; b'^2, b''^2, respectively; then

$$a'a''=ch, \quad b'b''=ck\sqrt{-1}. \qquad (430)$$

Hence we have the following theorem:—*Two confocals to the ellipse* $\frac{x^2}{a^2}+\frac{y^2}{b^2}-1=0$ *can be drawn through the point* hk:

* The student is recommended to omit this proposition until he has read the chapter on the hyperbola.

the product of the semiaxes major of these confocals is ch, and of the semiaxes minor, cki; where i denotes, as usual, $\sqrt{-1}$.

It will be seen in Chapter VII. that one of these confocals must be a hyperbola unless $k = 0$, in which case one of them must consist of the two foci.

DEF.—*The semiaxes major* a', a'' *of the two confocals, which can be drawn to a given ellipse through a given point, are called the* ELLIPTIC CO-ORDINATES *of the point* (LAMÉ, 'Co-ordonneés Curvilignes').

Cor. 1.— $$h^2 = \frac{a'^2 a''^2}{c^2}, \quad -k^2 = \frac{b'^2 b''^2}{c^2};$$

therefore $$h^2 + k^2 = \frac{a'^2 a''^2 - b'^2 b''^2}{c^2} = \frac{a'^2(a''^2 - b''^2) + b''^2(a'^2 - b'^2)}{c^2}$$
$$= a'^2 + b''^2 = a''^2 + b'^2. \quad (431)$$

Cor. 2.—The two confocals to a given ellipse which can be drawn through any point cut each other orthogonally. For the tangents are

$$\frac{hx}{a'^2} + \frac{ky}{b'^2} - 1 = 0, \quad \frac{hx}{a''^2} + \frac{ky}{b''^2} - 1 = 0,$$

and these tangents are perpendicular to each other if

$$\frac{h^2}{a'^2 a''^2} + \frac{k^2}{b'^2 b''^2} = 0, \quad \text{or} \quad \frac{1}{c^2} - \frac{1}{c^2} = 0.$$

Cor. 3.—Let p', p'' denote the perpendiculars from the centre on the tangents to the confocals through hk at that point, and β', β'' the semidiameters conjugate to the semidiameter drawn to hk,

$$\beta'^2 + h^2 + k^2 = a'^2 + b'^2; \quad \text{[Equation (375)]}$$

therefore $$\beta'^2 = a'^2 - a''^2. \quad (\textit{Cor. } 1). \quad (432)$$

Similarly, $$\beta''^2 = b''^2 - b'^2. \quad (433)$$

But $\beta' p' = a'b'$ [Art. 121, *Cor.* 8]; $\therefore\ p'^2 = \dfrac{a'^2 b'^2}{a'^2 - a''^2}$. (434)

Similarly, $$p''^2 = \frac{a''^2 b''^2}{b''^2 - b'^2}. \quad (435)$$

Cor. 4.—By means of the values of h^2, k^2, *Cor.* 1, we find, after an easy reduction,

$$\frac{\sqrt{(a'^2-a^2)(a^2-a''^2)}}{(a'^2-a^2)+a''^2-a^2}=\frac{\sqrt{b^2h^2+a^2k^2-a^2b^2}}{h^2+k^2-a^2+b^2};$$

and substituting for hk the values $\dfrac{a\cos\alpha}{\cos\beta}$, $\dfrac{b\sin\alpha}{\cos\beta}$ [Art. 124, Ex. 2], this reduces to $\dfrac{ab\sin 2\beta}{(a^2-b^2)\cos 2\alpha-(a^2+b^2)\cos 2\beta}$. Hence [Art. 124, Ex. 10] we have the following theorem:—*If* ϕ *denote the angle between the tangents to the ellipse* $\dfrac{x^2}{a^2}+\dfrac{y^2}{b^2}-1=0$, *from the point whose elliptic co-ordinates are* a', a'',

$$\tan\phi=\frac{2\sqrt{(a'^2-a^2)(a^2-a''^2)}}{(a'^2-a^2)-(a^2-a''^2)}; \qquad (436)$$

therefore

$$\tan\tfrac{1}{2}\phi=\sqrt{\frac{a^2-a''^2}{a'^2-a^2}}. \qquad (437)$$

Therefore if ψ denote the angle which the tangent at P to the confocal a' makes with the tangent from P to the original ellipse, we have

$$\cot\psi=\sqrt{\frac{a^2-a''^2}{a'^2-a^2}}.$$

Hence

$$\sin\psi=\sqrt{\frac{a'^2-a^2}{a'^2-a''^2}},\quad \cos\psi=\sqrt{\frac{a^2-a''^2}{a'^2-a''^2}}. \qquad (438)$$

Cor. 5.—The results proved give a new demonstration of the propositions, Art. 124, Ex. 16.

The principal theorems in *Cors.* 4 and 5 were first published in a Paper of mine in the *Messenger of Mathematics* in the year 1866, and were extended to sphero-conics, and to curves on confocal quadrics. Corresponding theorems were given by Chasles for geodesic tangents to lines of curvature on the ellipsoid.—Liouville's *Journal*, 1846.

EXAMPLES.

1. The locus of the pole of the line $\mu x + \nu y = 1$, with respect to a system of conics confocal to $\frac{x^2}{a^2} + \frac{y^2}{b^2} - 1 = 0$, is the line

$$\frac{x}{\mu} - \frac{y}{\nu} = c^2. \tag{439}$$

2. The equation of the director circle of an ellipse in elliptic co-ordinates is $a'^2 + a''^2 = 2a^2$.

3. If from the centre of the ellipse $\frac{x^2}{a^2} + \frac{y^2}{b^2} = 1$ a parallel be drawn to the tangent from any point P on $\frac{x^2}{a^2} + \frac{y^2}{b^2} - 1$ to a given confocal (a'), to meet the tangent at P to the first ellipse, the locus of the point of intersection is a circle.

4. If a', a'' be the elliptic co-ordinates of any point, ϕ the angle included between the tangents from this point to $\frac{x^2}{a^2} + \frac{y^2}{b^2} - 1 = 0$; then

$$a'^2 \sin^2 \tfrac{1}{2}\phi + a''^2 \cos^2 \tfrac{1}{2}\phi = a^2. \tag{440}$$

5. If from the intersection of tangents to an ellipse distances be measured along the tangents equal to the focal vectors of the intersection, the length of the join of their extremities $= 2a'$.

6. If a tangent to one confocal be perpendicular to a tangent to another, the chord of contact is trisected by the join of their intersection to the centre.

7. The difference between the squares of the perpendiculars from the centre on parallel tangents to two confocals is constant.

8. The locus of the points of contact of parallel tangents to a system of confocal ellipses is a hyperbola.

9. The locus of the point (α) on a system of confocal ellipses is a confocal hyperbola.

10. If two secants, OR, OS, cut the ellipse in the points R, R'; S, S', respectively, and be tangents to a confocal,

$$\frac{1}{OR} - \frac{1}{OR'} = \frac{1}{OS} - \frac{1}{OS'}. \quad \text{(M. ROBERTS.)} \tag{441}$$

For let $a^2 - \lambda^2$, $b^2 - \lambda^2$ be the semiaxes of the confocal; b', b'' the semidiameters parallel to OR, OS; then

$$\frac{1}{OR} - \frac{1}{OR'} = \frac{RR'}{OR \,.\, OR'} = \frac{2\lambda b'^2}{ab \,.\, OR \,.\, OR'}. \quad \text{[Equation (419)]}$$

In like manner,

$$\frac{1}{OS} - \frac{1}{OS'} = \frac{2\lambda b''^2}{ab\, OS \,.\, OS'}.$$

But $OR : OR' :: b'^2 : b''^2$. [Equation (422)]

Hence the proposition is proved.

128. *To find the polar equation of an ellipse, the focus being pole.*

If the focus be origin the equation of the ellipse is

$x^2 + y^2 = e^2(x+f)^2$. [Art. 118]

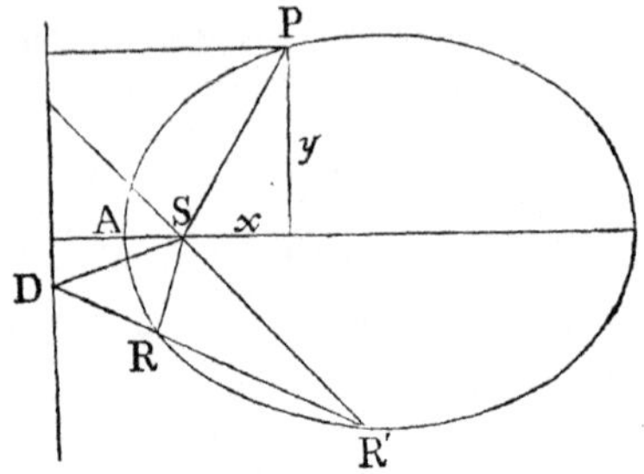

Hence, putting

$x = \rho \cos\theta, \quad y = \rho \sin\theta,$

we get

$$\rho = \frac{ef}{1 - e\cos\theta},$$

that is,

$$\rho = \frac{a(1-e^2)}{1 - e\cos\theta}. \qquad (442)$$

It is usual in Astronomy, when the polar equation is employed, to denote the angle ASP, called the true anomaly, by θ; then the polar equation is

$$\rho = \frac{a(1-e^2)}{1 + e\cos\theta}. \qquad (443)$$

Since $a(1 - e^2) = \frac{1}{2}$ latus rectum $= l$ suppose, the polar equation is

$$\rho = \frac{l}{1 + e\cos\theta}. \qquad (444)$$

Cor. 1.—If the angular co-ordinates of two points on the ellipse be $\alpha + \beta$, $\alpha - \beta$, the equation of their joining chord is

$$\frac{l}{\rho} = e\cos\theta + \sec\beta\cos(\theta - \alpha). \qquad (445)$$

For assuming it to be of the form

$$\frac{l}{\rho} = A\cos\theta + B\cos(\theta - \alpha),$$

and putting in succession for θ the values $\alpha + \beta$, $\alpha - \beta$, we get

$$1 + e\cos(\alpha+\beta) = A\cos(\alpha+\beta) + B\cos\beta,$$
$$1 + e\cos(\alpha-\beta) = A\cos(\alpha-\beta) + B\cos\beta.$$

Hence $A = e, \quad B = \sec\beta.$

Cor. 2.—The equation of the tangent at the point α is

$$\frac{l}{\rho} = e \cos \theta + \cos (\theta - \alpha). \qquad (446)$$

EXAMPLES.

1. If ρ, ρ' denote the segments of a focal chord,

$$\frac{1}{\rho} + \frac{1}{\rho'} = \frac{4}{l}. \qquad (447)$$

2. The rectangle contained by the segments of a focal chord is proportional to the length of the chord.

3. Any focal chord is a third proportional to the transverse axis and the parallel diameter.

4. The sum of the reciprocals of two perpendicular focal chords is constant.

5. If any chord RR' of an ellipse meet the directrix in D, the line SD bisects the external angle of the triangle RSR'.

6. The join of the intersection of two tangents to the focus bisects the angle made by the focal vectors of the points of contact.

7. If any point on an ellipse be joined to the extremities of the transverse axis, the portion of the directrix which the joining lines intercept subtends a right angle at the focus.

8. The angle subtended at the focus by the portion of any variable tangent intercepted by two fixed tangents is constant.

9. If a tangent from a variable point subtend a constant angle δ at the focus, the locus of the point is

$$\frac{l}{\rho} = \cos \delta + e \cos \theta. \qquad (448)$$

10. If a chord PQ subtend a constant angle 2δ at the focus, the locus of the point, where it meets the internal bisector of that angle, is

$$\frac{l}{\rho} = \sec \delta + e \cos \theta. \qquad (449)$$

11. If θ denote the true anomaly, ϕ the supplement of the eccentric angle,

$$\tan \tfrac{1}{2}\theta \,.\, \tan \tfrac{1}{2}\phi = \sqrt{\frac{1 + e}{1 - e}}. \qquad (450)$$

12. If Q be the point on the auxiliary circle corresponding to P on the ellipse, the locus of the intersection of SP and the join of Q to the centre is a circle.

Exercises on the Ellipse.

1. Find the eccentricity of the ellipse $3x^2 + 4y^2 = 1$.

2. If two central vectors of an ellipse be at right angles to each other, the sum of the squares of their reciprocals is constant.

3. Find the equation of the circle through either extremity of the transverse axis and both extremities of the latus rectum.

4. Find the equation of the tangent at either extremity of the latus rectum.

5. The locus of the middle points of chords of an ellipse passing through a given point is an ellipse whose axes are parallel to those of the given ellipse.

6. If from any point in a circle a line be drawn making a given angle with a fixed line, and divided in a given ratio, the locus is an ellipse.

7. Two points P, Q are taken on the minor axis of an ellipse at distances $\frac{ab}{c}$ from the centre, and CV is a perpendicular from the centre on any tangent; prove

$$\frac{PV + QV}{CV} = \frac{2a}{c}. \qquad \text{(Crofton.)} \qquad (451)$$

8. Only one chord of an ellipse can be perpendicular to a given line and bisected by it, unless the line pass through the centre.

9. If a common tangent to the two ellipses

$$\frac{x^2}{a^2} + \frac{y^2}{b^2} - 1 = 0, \qquad \frac{x^2}{a_1^2} + \frac{y^2}{b_1^2} - 1 = 0,$$

touch the first in $x'y'$, and the second in $x''y''$; then $x'x''$ is equal to the square of the abscissa of either of their points of intersection, and $y'y''$ to the square of the corresponding ordinate.

10. If the sum of the tangents drawn from a point to two circles be given, the locus of the point is an ellipse.

11. If a circle described through any point P on the minor axis of an ellipse, and through the two foci intersect the ellipse in the points Q, Q'; prove that PQ, PQ' are either tangents or normals to the ellipse.

12. Tangents are drawn from a fixed point P to a system of confocal ellipses; if T, T' be the lengths of the tangents to any of the ellipses, and θ their included angle, prove

$$\left(\frac{1}{T}+\frac{1}{T'}\right)\cos\tfrac{1}{2}\theta = \text{constant.} \quad \text{(Crofton.)} \quad (452)$$

13. The area of the triangle formed by the tangents from the point

$$\left(\frac{a\cos\alpha}{\cos\beta},\ \frac{b\sin\alpha}{\cos\beta}\right),$$

and their chord of contact, is $ab\sin^2\beta\tan\beta$.

14. If A, A' be the extremities of the transverse axis, P a point in the ellipse, whose eccentric angle is ϕ; prove

$$\tan APA' = -\frac{2b}{ae^2\sin\phi}.$$

15. If from any point T in PT (the tangent at P) a perpendicular TR be drawn to the focal vector SP, and a perpendicular TM on the directrix; then $SR = eTM$.

16. Find the equation of the circle described on the intercept which the ellipse

$$\frac{x^2}{a^2}+\frac{y^2}{b^2}-1=0$$

makes on the line $y = mx + n$; and thence show how to find the length of the normal at any point of an ellipse until it meets the ellipse again.

17. The locus of the intersection of tangents at the extremities of a pair of conjugate diameters is

$$\frac{x^2}{a^2}+\frac{y^2}{b^2}=2,$$

and the envelope of the join of their extremities is

$$\frac{x^2}{a^2}+\frac{y^2}{b^2}=\tfrac{1}{2}.$$

18. Find the co-ordinates of the pole of the normal at the point α, and show that the locus of the pole is

$$\frac{a^6}{x^2}+\frac{b^6}{y^2}=c^4. \quad (453)$$

19. If a tangent at any point P meet the transverse axis in T; then, if S be the focus,

$$\cos SPT = e \cos STP. \qquad (454)$$

20. Find the pedal of the ellipse with respect to its centre.

21. If perpendiculars at a point P of the ellipse to the lines AP, AP' (see Ex. 14) meet the transverse axis in the points Q, Q', prove that QQ' is constant.

22. Prove that two of the normals drawn from the point whose co-ordinates are

$$\frac{c^2 \cos \alpha \cos 2\alpha}{a\sqrt{2}}, \quad -\frac{c^2 \sin \alpha \cos 2\alpha}{b\sqrt{2}}$$

meet the ellipse at the extremities of a pair of conjugate diameters.

23. Find the equation of the pair of lines joining the centre of the ellipse to the points of contact of tangents from $x'y'$.

24. The sum of the eccentric angles of four concyclic points on an ellipse is 2π.

25. If a circle osculate an ellipse at the point α, the co-ordinates of the point where it meets the ellipse again are, $a \cos 3\alpha$, $-b \sin 3\alpha$.

26. The sum of two focal chords of an ellipse parallel to two conjugate diameters is constant.

27. Any two fixed tangents are cut homographically by a variable tangent.

For the angle which the intercept on the variable tangent subtends at the focus is constant.

28. If S be the focus, T any point on the tangent at P, TM a perpendicular on the directrix; then, if $ST = e' \, TM$,

$$\cos PST = \frac{e}{e'}.$$

29. If a chord PP' of an ellipse pass through a fixed point T, and if $ST = e' \, TM$, then

$$\tan \tfrac{1}{2} PST \, . \, \tan \tfrac{1}{2} P'ST = \frac{e - e'}{e + e'}. \qquad \text{(M'Cullagh.)}$$

30. If S, S' be the foci, and if the circle described on SS' as diameter meet the conjugate axis in H, H'; prove that the sum of the squares of the perpendiculars from H, H' on any tangent is constant.

31. If all the tangents to an ellipse be inverted from any internal point, the locus of the centres of all the circles into which they invert is an ellipse.

32. If ν be the intercept which any normal to an ellipse makes on the transverse axis, and ϕ the angle which it makes with it; prove

$$\nu = \frac{c^2}{(a^2 + b^2 \tan^2 \phi)^{\frac{1}{2}}}. \qquad (455)$$

33. If two central vectors of an ellipse be at right angles to each other, the envelope of the join of their extremities is a circle.

34. If the chords joining the pairs of points α, β; γ, δ, respectively, meet the transverse axis in points equally distant from the centre; prove

$$\tan \frac{\alpha}{2} \tan \frac{\beta}{2} \tan \frac{\gamma}{2} \tan \frac{\delta}{2} = 1. \qquad (456)$$

35. The area of the parallelogram formed by the points α, β, and the points diametrically opposite to them $= \dfrac{4ab}{\sin(\alpha - \beta)}$.

36. If the co-ordinates in Ex. 22 be denoted by x, y; prove

$$2(a^2x^2 + b^2y^2)^3 = c^4(a^2x^2 - b^2y^2)^2 \qquad (457)$$

37. If CP, CD be two conjugate semi-diameters, and if the normals at P be produced both ways to Q, Q', making PQ, PQ' each equal to CD; prove that

$$CQ = a + b, \quad CQ' = a - b. \qquad \text{(M'CULLAGH.)} \quad (458)$$

38. The locus of the intersection of normals at points, which have equal eccentric angles, on the ellipses

$$\frac{x^2}{a^2} + \frac{y^2}{b^2} - 1 = 0, \quad \frac{x^2}{a_1^2} + \frac{y^2}{b_1^2} - 1 = 0$$

is an ellipse.

39. If x_1y_1, x_2y_2, x_3y_3 be any three points, and if

$$S \equiv \frac{x^2}{a^2} + \frac{y^2}{b^2} - 1 = 0, \quad S_1 \equiv \frac{x_1^2}{a^2} + \frac{y_1^2}{b^2} - 1, \quad T_{12} = \frac{x_1x_2}{a^2} + \frac{y_1y_2}{b^2} - 1, \ \&c;$$

prove that -4 (area of triangle formed by these points)$^2 \div a^2b^2$ is equal to

$$\begin{vmatrix} S_1, & T_{12}, & T_{13}, \\ T_{12}, & S_2, & T_{23}, \\ T_{13}, & T_{23}, & S_3 \end{vmatrix}. \qquad (459)$$

(PROF. CURTIS, S.J.)

40. If the three points form a self-conjugate triangle, with respect to S, and if Δ denote the discriminant $-\frac{1}{a^2 b^2}$,

$$\text{area} = \tfrac{1}{2}\sqrt{\frac{S_1 S_2 S_3}{\Delta}}. \qquad \text{(BURNSIDE.)} \qquad (460)$$

41. If they form a triangle circumscribed about S,

$$\text{area} = ab\{\sqrt{S_1} + \sqrt{S_2} + \sqrt{S_3}\}. \qquad (461)$$

(PROF. CURTIS, S.J.)

42. If the triangle be inscribed in S,

$$\text{area} = ab\sqrt{\frac{T_{12} T_{23} T_{31}}{2}}. \qquad (\textit{Ibid.}) \qquad (462)$$

43. If PM be an ordinate at any point P of an ellipse, find the locus of the intersection of PM, with the perpendicular from the centre on the tangent at P.

44. Find the locus of the point of bisection of the portion of the tangent to an ellipse which is intercepted by the axes.

45. If a point P whose eccentric angle is θ be joined to the foci, and the joining lines produced meet the ellipse again in Q, R; find the equation of QR, and prove that its polar lies on the normal at θ.

46. If ϕ be the eccentric angle of the point P of an ellipse, Q the point on the auxiliary circle corresponding to P; prove that the area of the parallelogram formed by the points P, Q and the points diametrically opposite to them is $2a(a - b)\sin 2\phi$.

47. In the same case, prove that the area of the parallelogram formed by the tangents at the same points $= \dfrac{8a^2 b}{(a - b)\sin 2\phi}$.

48. If the normal at P meet the transverse and the conjugate axes in the points G, G', respectively; prove that the middle point of CG is the centre of a circle through P and the extremities of the minor axis; and the middle point of CG' the centre of a circle through P and the extremities of the transverse axis.

49. If the product of the direction tangents of two lines touching an ellipse be given, and negative, the locus of their point of intersection is an ellipse.

50. Find the locus of the point of intersection of two normals at right angles to each other.

51. If θ be the angle between a central vector and the normal at the point ϕ; prove

$$\tan\theta = \frac{c^2 \sin 2\phi}{2ab}.$$

52. The lengths of the tangents from the point $x'y'$ to the ellipse

$$S \equiv \frac{x^2}{a^2} + \frac{y^2}{b^2} - 1 = 0$$

are roots of the equation in T,

$$\frac{x'}{a}\sqrt{T^2 - b^2 S'} + \frac{y'}{b}\sqrt{a^2 S' - T^2} = cS' \quad \text{(Crofton.)} \quad (463)$$

53. A circle has double contact with an ellipse at the points P, P'. Prove that the sum of the distances of the points P, P' from either focus is half the sum of the distances from the same focus of the points in which the ellipse is intersected by any circle concentric with the former. (*Ibid.*)

54. If from any point on an ellipse tangents be drawn to the circle on the minor axis, and if the chord of contact meet the major and the minor axes in the points L, M respectively; prove

$$\frac{b^2}{CL^2} + \frac{a^2}{CM^2} = \frac{a^2}{b^2}.$$

55. Find the locus of the middle points—1°. of chords of a given length in an ellipse. 2°. Of the middle points of chords whose distance from the centre is given.

56. If S, S' be the foci, P any point on the ellipse, PQ a normal and a mean proportional between SP, $S'P$, the locus of Q is a circle.

57. The sum of the squares of the perpendiculars from the extremities of any two conjugate semidiameters on any fixed diameter is constant.

58. If CP, CP' be two semidiameters of an ellipse; CD, CD' their conjugates; prove, if PP' pass through a fixed point, that DD' also passes through a fixed point.

59. E, F are the feet of perpendiculars from the centre and focus on any tangent, T the point where the tangent meets the transverse axis; prove $EP.ET = EF^2$.

60. The locus of the points of contact of tangents to a system of confocal ellipses from a fixed point on the transverse axis is a circle.

61. If $x\cos\alpha + y\sin\alpha - p = 0$ be a tangent to $\frac{x^2}{a^2} + \frac{y^2}{b^2} - 1 = 0$;

prove
$$p^2 = a^2\cos^2\alpha + b^2\sin^2\alpha. \quad (464)$$

62. If the circle $x^2 + y^2 + 2gx + 2fy + c = 0$ passes through the extremities of three semidiameters of the ellipse

$$\frac{x^2}{a^2} + \frac{y^2}{b^2} - 1 = 0,$$

prove that the circle

$$x^2 + y^2 + \frac{2fb}{a} x - \frac{2ga}{b} y - (a^2 + b^2 + c) = 0$$

passes through the extremities of the three conjugate semidiameters.—
(R. A. Roberts.) (465)

63. Show that if the first circle in Ex. 62 be orthogonal to $x^2 + y^2 - 2\alpha x - \beta y + c' = 0$, the second is orthogonal to

$$x^2 + y^2 + \frac{2a\beta c}{b} - \frac{2b\alpha y}{a} + a^2 + b^2 - c = 0. \quad (Ibid.) \qquad (466)$$

64. A triangle is inscribed in the ellipse

$$\frac{x^2}{a^2} + \frac{y^2}{b^2} - 1 = 0;$$

prove, if x', y' be the co-ordinates of its centroid, and x, y those of the circumcentre,

$$16(a^2x^2 + b^2y^2) + 9c^4\left(\frac{x'^2}{a^2} + \frac{y'^2}{b^2}\right) - 12c^2(xx' - yy') - c^4 = 0.$$

(*Ibid.*) (467)

65. If normals at $\alpha, \alpha', \alpha''$ be concurrent; prove

$$\sin 2\alpha \sin(\alpha' - \alpha'') + \sin 2\alpha' \sin(\alpha'' - \alpha) + \sin 2\alpha'' \sin(\alpha - \alpha') = 0, \qquad (468)$$

and

$$\sin(\alpha + \alpha') + \sin(\alpha' + \alpha'') + \sin(\alpha'' + \alpha) = 0. \qquad (469)$$

66. If a', b' be conjugate semidiameters, making angles ϕ, ϕ' with the semiaxes, prove

$$\frac{a'^2 - b'^2}{a^2 - b^2} = \frac{\cos(\phi + \phi')}{\cos(\phi - \phi')}. \qquad (470)$$

67. If the rectangle contained by the perpendiculars on a variable line from its pole, with respect to a given ellipse, and from the centre of the ellipse, be constant, the envelope of the line is a confocal ellipse.

68. The normals to an ellipse at the points where the lines

$$\frac{px}{a} + \frac{qy}{b} - 1 = 0, \quad \frac{x}{ap} + \frac{y}{bq} + 1 = 0$$

meet it are concurrent.

69. If PP' be a diameter of an ellipse; find the locus of the intersection of the normal at P with the ordinate at P'.

70. Find the locus of the pole of a chord of given length in an ellipse.

71. The circle whose diameter is any chord, parallel to the conjugate axis of

$$\frac{x^2}{a^2} + \frac{y^2}{b^2} - 1,$$

has double contact with the ellipse

$$\frac{x^2}{a^2 + b^2} + \frac{y^2}{b^2} = 1. \tag{471}$$

72. If focal vectors from any point P meet the ellipse again in Q and R, and if the tangent at P make an angle θ with the transverse axis, and the line QR an angle ϕ; prove

$$\tan\phi = \frac{1 - e^2}{1 + e^2}\tan\theta. \tag{472}$$

73. Being given two confocal ellipses; prove that the distance between the point ϕ on the first and the point ϕ' on the second is equal to the distance between ϕ' on the first and ϕ on the second. (IVORY.)

74. If from an external point O a secant ORR' be drawn, cutting the ellipse in R, R'; then if $OQ^2 = OR \cdot OR'$, the locus of Q is an ellipse.

75. If the angles which any two conjugate diameters subtend at any point of the ellipse be denoted by λ, λ', respectively; then $\cot^2\lambda + \cot^2\lambda'$ is constant.

76. The external angle formed by two tangents to an ellipse is half the sum of the angles subtended by the chord of contact at the foci.

77. If a normal to an ellipse be parallel to one of the equiconjugate diameters, it cuts the ellipse again at a minimum angle.

(PROF. J. PURSER.)

78. Two parallel focal chords of an ellipse meet it in the points G, H, on the same side of the transverse axis; if the join of G, H make intercepts λ, μ on the axes, prove

$$\frac{a^4}{\lambda^2} + \frac{b^4}{\mu^2} = a^2. \tag{473}$$

79. If two normals to an ellipse cut at right angles, the intercepts made on them by the ellipse are divided proportionally at their point of intersection. (PROF. J. PURSER.)

80. Prove that if a parabola be described with a point on an ellipse as focus, and the tangent at the corresponding point on the auxiliary circle as directrix, it passes through the foci of the ellipse. (*Ibid.*)

CHAPTER VII.

THE HYPERBOLA.

129. Def. I.—*Being given in position a point S, and a line NN', the locus of a variable point P, whose distance from S has to its perpendicular distance from NN' a given ratio e greater than unity, is called a* HYPERBOLA.

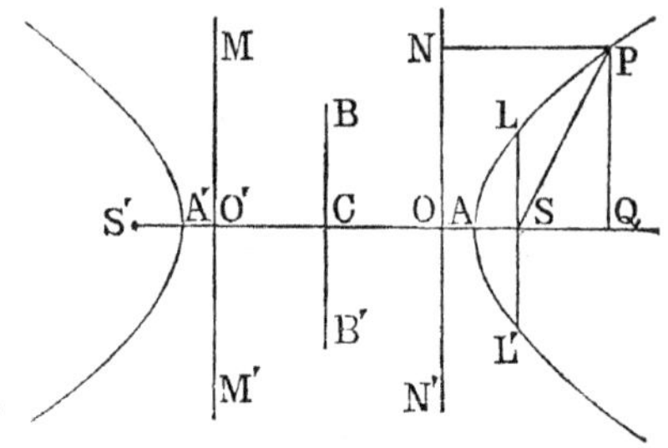

Def. II.—*The point S is called the* FOCUS ; *the line NN' the* DIRECTRIX, *and the ratio e the* ECCENTRICITY *of the hyperbola.*

130. *To find the equation of the hyperbola.*

1°. Take the focus as origin, the line through S, perpendicular to the directrix, as axis of x, and a parallel to the directrix through S as the axis of y; also denote the perpendicular SO from S on the directrix by f; then, denoting the co-ordinates of P by x, y, we have $SP^2 = x^2 + y^2$, and $PN = x + f$; but (Def. I.) $SP \div PN = e$; therefore

$$x^2 + y^2 = e^2 (x + f)^2. \qquad (474)$$

2°. In equation (474) put

$$x = x - \frac{e^2 f}{e^2 - 1},$$

and we get

$$x^2 - \frac{y^2}{e^2 - 1} = \frac{e^2 f^2}{(e^2 - 1)^2}. \qquad (1.)$$

Hence, if C be the new origin, we have

$$CS = \frac{e^2 f}{(e^2 - 1)^2}. \qquad \text{(II.)}$$

Now, putting $y = 0$ in (I.), we get

$$x^2 = \frac{e^2 f^2}{(e^2 - 1)^2},$$

giving for x two values equal in magnitude, but of opposite signs. Hence, denoting the points where the hyperbola cuts the axis of x by A, A', we get $CA = \frac{ef}{e^2 - 1}$, $CA' = -\frac{ef}{e^2 - 1}$. Hence $A'C = CA$; therefore the line $A'A$ is bisected in C, and denoting it by $2a$, we have

$$a = \frac{ef}{e^2 - 1}. \qquad \text{(III.)}$$

Again, putting $x = 0$ in (I.), we get

$$y^2 = -\frac{e^2 f^2}{e^2 - 1}.$$

This gives two imaginary values for y, viz.

$$+\frac{ef\sqrt{-1}}{\sqrt{e^2 - 1}} \text{ and } \frac{-ef\sqrt{-1}}{\sqrt{e^2 - 1}},$$

showing that the hyperbola does not cut the axis of y.

DEF. III.—*The line AA' is called the* TRANSVERSE AXIS *of the hyperbola; and if we make $CB = B'C = \frac{ef}{\sqrt{e^2 - 1}}$, the line BB' is called the* CONJUGATE AXIS, *and the point C the* CENTRE. *The line $B'B$ is denoted by $2b$.*

3°. Since $a = \frac{ef}{(e^2 - 1)}$, $b = \frac{ef}{(e^2 - 1)^{\frac{1}{2}}}$, equation (I.) can be written

$$\frac{x^2}{a^2} - \frac{y^2}{b^2} = 1. \qquad (475)$$

This is the standard form of the equation of the hyperbola.

DEF. IV.—*The double ordinate LL' through S is called the* LATUS RECTUM *of the hyperbola.*

131. The following deductions from the preceding equations are important:—

1°. $b^2 = a^2(e^2 - 1)$.

2°. If CS be denoted by c, $c = ae$.

3°. $CO = \frac{a}{e}$. For $CO = CS - f = \frac{e^2 f}{e^2 - 1} - f = \frac{f}{e^2 - 1}$.

4°. $a^2 + b^2 = c^2$. From 1° and 2°.

5°. $CS \cdot CO = a^2$. From 2° and 3°.

6°. Latus rectum $= 2a(e^2 - 1)$. For in (474) put $x = 0$, and we get $SL = ef$; therefore $LL' = 2ef = 2a(e^2 - 1)$.

7°. The transverse axis : conjugate axis : : conjugate axis : latus rectum. From 1° and 6°.

8°. Since from the form (475) of the equation of the hyperbola each axis is an axis of symmetry of the figure, it follows that, if we make $CS' = SC$, the point S' will be another focus; also, if $CO' = OC$, and through O' a line MM' be drawn perpendicular to the transverse axis, MM' will be a second directrix, corresponding to the second focus S'.

DEF. V.—*If the semiaxes a, b of a hyperbola be equal, the curve is called an* EQUILATERAL HYPERBOLA.

EXAMPLES.

1. Given the base of a triangle and the difference of the sides; find the locus of the vertex.

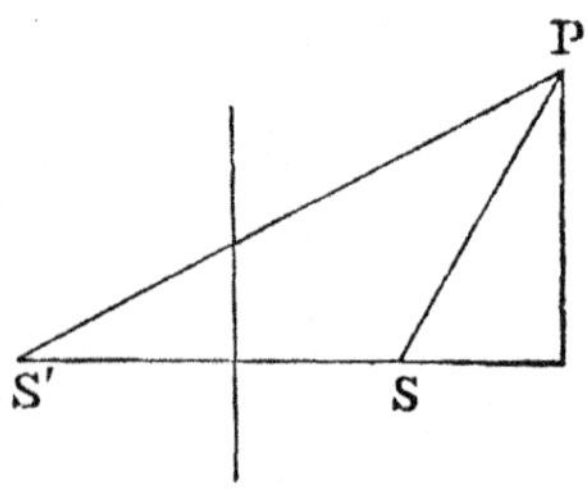

Let $S'SP$ be the triangle; let the base $SS' = 2c$, and the difference of the sides equal $2a$. Let $S'S$ produced be taken as axis of x, and the perpendicular to $S'S$ at its middle point as axis of y; then, if x, y be the co-ordinates of P, we have

$$SP = \{(x + c)^2 + y^2\}^{\frac{1}{2}}, \quad S'P = \{(x - c)^2 + y^2\}^{\frac{1}{2}};$$

therefore $$\{(x+c)^2+y^2\}^{\frac{1}{2}} - \{(x-c)^2+y^2\}^{\frac{1}{2}} = 2a; \qquad \text{(I.)}$$

or cleared of radicals,

$$(c^2-a^2)x^2 - a^2y^2 = a^2(c^2-a^2);$$

or putting $$c^2 - a^2 = b^2, \quad \frac{x^2}{a^2} - \frac{y^2}{b^2} = 1.$$

Cor. 1.— $$SP = ex - a. \qquad (476)$$

For in clearing (I.) of radicals, we get

$$a\{(x-c)^2+y^2\}^{\frac{1}{2}} = cx - a^2;$$

that is, $$a \,.\, SP = aex - a^2.$$

Cor. 2.— $$S'P = ex + a.$$

2. Given the base of a triangle and the difference of the base angles, the locus of the vertex is an equilateral hyperbola.

3. Given the base of a triangle, and the ratio of the tangents of the halves of the base angles, the locus of the vertex is a hyperbola.

4. The locus of the centre of a circle, which passes through a given point and cuts a fixed line at a given angle, is a hyperbola.

5. Trisect a given arc of a circle by means of a hyperbola.

6. If the base of a triangle be given in magnitude and position, and the difference of the sides in magnitude, then the loci of the centres of the escribed circles which touch the base produced are the two branches of a hyperbola; and the loci of the centres of the inscribed circle, and the escribed which touches the base externally, are the directrices of the same hyperbola.

7. If in Ex. 6, Art. 119, the 'Boscovich Circle' cut the line NN', show that the locus of P will be a hyperbola.

8. CB is a fixed diameter of a given circle; and through a fixed point A in CB draw any chord DE of the circle; join CD, and on CD produced, if necessary, take $CF = AE$: the locus of the point F is a hyperbola.

9. $ABCD$ is a lozenge whose diagonals are $2a$, $2b$, respectively; prove, if the diagonals be taken as axes, that the locus of a point P, such that the rectangle $AP \,.\, CP =$ the rectangle $BP \,.\, DP$, is the equilateral hyperbola

$$x^2 - y^2 = \frac{a^2 - b^2}{2}$$

132. *The locus of the middle points of a system of parallel chords of a hyperbola is a right line.*

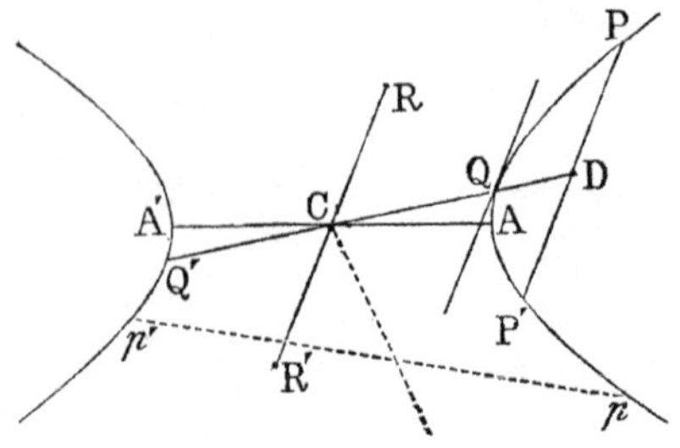

Let the equation of one of the chords be

$$\frac{y}{b} = m\left(\frac{x}{a}\right) + n.$$

Now, if m be constant and n variable, this will represent a line which moves parallel to itself; and eliminating y between it and the equation of the hyperbola, we get

$$(1 - m^2)\,x^2 - 2mnax - a^2n^2 - a^2 = 0.$$

Similarly, by eliminating x, we get

$$(1 - m^2)y^2 - 2nby + b^2n^2 - b^2m^2 = 0.$$

Hence the equation of the circle, whose diameter is the intercept which the hyperbola makes on the line

$$\frac{y}{b} = m\left(\frac{x}{a}\right) + n,$$

is $(1 - m^2)(x^2 + y^2) - 2mnax - 2nby - (a^2 - b^2)\,n^2 - a^2 - m^2b^2 = 0.$ (477)

Now if the co-ordinates of the centre of this circle be x', y', we get

$$x' = \frac{mna}{1 - m^2}, \quad y' = \frac{nb}{1 - m^2}.$$

Hence, eliminating n and omitting accents, the locus of the centre, that is, of the middle point of the chord, is the diameter

$$\frac{y}{b} = \frac{1}{m}\cdot\frac{x}{a}. \qquad (478)$$

This is the line QQ' in the diagram.

Cor. 1.—If a line be drawn through the centre parallel to PP', or, in other words, a diameter conjugate to QQ', its equation must contain no absolute term; hence its equation is

$$\frac{y}{b} = m\left(\frac{x}{a}\right).$$

Hence the product of the tangents of the angles, which two conjugate diameters make with the transverse axis of a hyperbola, is $\frac{b^2}{a^2}$.

Cor. 2.—If the line PP' move parallel to itself until the points P, P' become consecutive, then PP' becomes a tangent such as at Q; and if the co-ordinates of Q be $x'y'$, we must have

$$\frac{y'}{b} = m\left(\frac{x'}{a}\right) + n;$$

and since the line QQ' passes through it, we must have (478)

$$\frac{y'}{b} = \frac{1}{m}\left(\frac{x'}{a}\right).$$

Hence

$$m = \frac{bx'}{ay'}, \quad n = -\frac{b}{y'},$$

which, substituted in

$$\frac{y}{b} = m\left(\frac{x}{a}\right) + n,$$

gives

$$\frac{xx'}{a^2} - \frac{yy'}{b^2} = 1, \qquad (479)$$

which is the equation of the tangent.

Cor. 3.—To find the equation of the chord of contact of tangents from the point hk.

Let $x'y'$, $x''y''$, be the points of contact; then, since the tangent at $x'y'$ passes through hk, we have

$$\frac{hx'}{a^2} - \frac{ky'}{b^2} = 1.$$

Similarly,

$$\frac{hx''}{a^2} - \frac{ky''}{b^2} = 1.$$

Hence it is evident that the line

$$\frac{hx}{a^2} - \frac{ky}{b^2} = 1 \qquad (480)$$

passes through each point of contact, and therefore must be the chord required.

If instead of hk we put $x'y'$, we see that the chord of contact of tangents, from $x'y'$ to the hyperbola, is

$$\frac{xx'}{a^2} - \frac{yy'}{b^2} = 1. \qquad (481)$$

Cor. 4.—If through any point $x'y'$ a chord of the hyperbola be drawn, the locus of the intersection of tangents at its extremities is

$$\frac{xx'}{a^2} - \frac{yy'}{b^2} = 1.$$

Cor. 5.—The line

$$\frac{xx'}{a^2} - \frac{yy'}{b^2} = 1$$

is such that any line passing through $x'y'$ is cut harmonically by it and the hyperbola.

Cor. 6.—If two diameters QQ', RR' of the hyperbola be such that the first bisects chords parallel to the second, the second also bisects chords parallel to the first.

Observation.—It is not necessary that both extremities of the chord PP' should be on the same branch of the hyperbola; the chord may take the position pp', where they are on different branches.

133. DEF. I.—*It has been proved that if we construct the hyperbola*

$$\frac{x^2}{a^2} - \frac{y^2}{b^2} = 1,$$

whose axes are AA', BB', it will be the figure HHHH in the diagram. Again, if we construct the hyperbola, which has BB' for its transverse axis, and AA' for its conjugate axis, it will be the figure $H'H'H'H'$ in the diagram, This second figure is called the CONJUGATE HYPERBOLA.

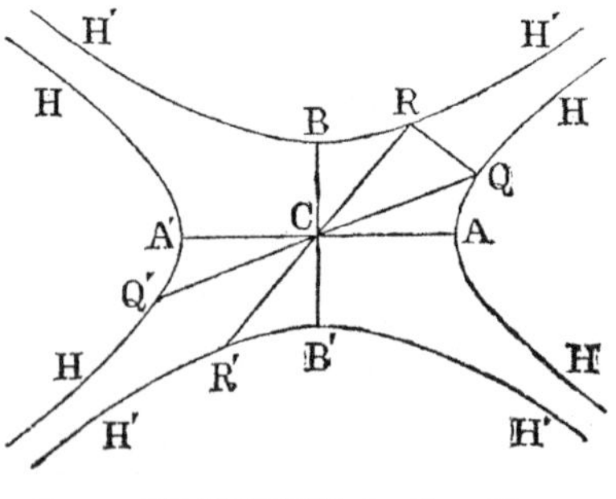

134. *To find the equation of the conjugate hyperbola.*

If the line BB' were the axis of x, and AA' the axis of y; since BB' is the transverse axis and AA' the conjugate axis, the equation of the figure $H'H'H'H'$ would be (Art. 130),

$$\frac{x^2}{b^2} - \frac{y^2}{a^2} = 1.$$

Hence, interchanging x and y, the required equation is

$$\frac{x^2}{a^2} - \frac{y^2}{b^2} = -1. \qquad (482)$$

Cor. 1.—If CQ, CR be conjugate diameters with respect to the hyperbola H, they are conjugate diameters with respect to the hyperbola H'.

For the required condition with respect to H is

$$\tan ACQ \cdot \tan ACR = \frac{b^2}{a^2} \quad \text{(Art. 132, } Cor. \text{ 1)};$$

hence

$$\tan BCR \cdot \tan BCQ = \frac{a^2}{b^2}.$$

Hence the proposition is proved.

Cor. 2.—The tangent at R to the hyperbola H' is parallel to QQ'. For the diameter RR' of H' bisects chords parallel to QQ', and the tangent R is a limiting case of a chord.

Cor. 3.—If the co-ordinates of Q be $x'y'$, the co-ordinates of R are

$$\frac{ay'}{b}, \quad \frac{bx'}{a}. \qquad (483)$$

For these satisfy the equation (482) of the hyperbola H', and the equation of the line RR' is

$$\frac{xx'}{a^2} - \frac{yy'}{b^2} = 0.$$

Cor. 4.—If the conjugate semidiameters CQ, CR be denoted by a', b', respectively, then $a'^2 - b'^2 = a^2 - b^2$. (484)

For $$a'^2 - b'^2 = CQ^2 - CR^2 = x'^2 + y'^2 - \frac{a^2y'^2}{b^2} - \frac{b^2x'^2}{a^2}$$

$$= \left(x'^2 - \frac{a^2y'^2}{b^2}\right) - \left(\frac{b^2x'^2}{a^2} - y'^2\right) = a^2 - b^2, \text{ from (475).}$$

Cor. 5.—Every diameter of an equilateral hyperbola is equal to its conjugate.

Cor. 6.—The area of the triangle $QCR = \frac{1}{2}ab$. (485)

For the area

$$= \frac{1}{2}\left(x' \times \frac{bx'}{a} - y' \times \frac{ay'}{b}\right) = \frac{1}{2}ab\left(\frac{x'^2}{a^2} - \frac{y'^2}{b^2}\right).$$

Hence the area of the parallelogram, whose two adjacent sides are two conjugate semidiameters, is constant.

Cor. 7.—The equation of the line QR is

$$\left(\frac{x'}{a'} - \frac{y'}{b'}\right)\left(\frac{x}{a} + \frac{y}{b}\right) = 1.$$

Hence QR is parallel to the line

$$\frac{x}{a} + \frac{y}{b} = 0.$$

Cor. 8.—The equation of the median, which bisects QR, is

$$\frac{x}{a} - \frac{y}{b} = 0. \qquad (486)$$

135. *To find the equation of an hyperbola referred to two conjugate diameters.*

Let CQ, CR be two conjugate semidiameters (see fig., Art. 132), and take CQ, CR as the new axes of x, y. Let x, y be the old co-ordinates of any point P of the hyperbola, $x'y'$ the new; then, denoting the angles QCA, RCA by α, β, respectively, we have

$$x = x' \cos\alpha + y' \cos\beta, \qquad y = x' \sin\alpha + y' \sin\beta.$$

Substitute these values in the equation $b^2x^2 - a^2y^2 = a^2b^2$; then

$$x'^2(b^2\cos^2\alpha - a^2\sin^2\alpha) - y'^2(a^2\sin^2\beta - b^2\cos^2\beta)$$
$$+ xx'y'(b^2\cos\alpha\cos\beta - a^2\sin\alpha\sin\beta) = a^2b^2;$$

but, since CQ, CR are conjugate semidiameters,

$$\tan\alpha\tan\beta = \frac{b^2}{a^2}$$

(Art. 132, *Cor.* 1). Hence the coefficient of $x'y'$ vanishes, and the equation may be written

$$x'^2\left(\frac{b^2\cos^2\alpha - a^2\sin^2\alpha}{a^2b^2}\right) - y'^2\left(\frac{a^2\sin^2\beta - b^2\cos^2\beta}{a^2b^2}\right) = 1.$$

Now, when $y' = 0$, we have $x' = CQ$. Hence, denoting CQ by a', we have

$$a'^2 = \frac{a^2b^2}{b^2\cos^2\alpha - a^2\sin^2\alpha}.$$

Again, if R be the point where CR meets the conjugate hyperbola (Art. 133), we get

$$CR^2 = \frac{a^2b^2}{a^2\sin^2\beta - b^2\cos^2\beta};$$

and, denoting this by b'^2, we see that the equation can be written

$$\frac{x'^2}{a'^2} - \frac{y'^2}{b'^2} = 1;$$

or, omitting accents on x', y',

$$\frac{x^2}{a'^2} - \frac{y^2}{b'^2} = 1. \qquad (487)$$

This is the same in form as the equation referred to the transverse and conjugate axes. (Compare Art. 101.)

**Or thus*:—Let the co-ordinates of the point P, with respect to the new axes CQ, CR, be denoted by X, Y (see fig., Art. 132), viz., $CD = X$, $DP = Y$; then the power of the origin, with respect to the circle (477) on PP' as diameter, is $X^2 - Y^2$. Hence

$$X^2 - Y^2 = (a^2 - b^2)\frac{n^2}{m^2 - 1} + \frac{a'^2 + m^2b^2}{m^2 - 1}.$$

Now $a^2 - b^2 = a'^2 - b'^2$ (484); and, substituting for m its value, $\frac{bx'}{ay'}$, we get $\frac{a^2 + m^2b^2}{m^2 - 1} = \frac{a^4x'^2 + b^4y'^2}{a^2b^2} = CQ^2$ or b'^2 (483).

Hence $$X^2 - Y^2 = (a'^2 - b'^2)\frac{n^2}{m^2 - 1} + b'^2.$$

Again, if hk denote the co-ordinates of D,

$$X^2 = h^2 + k^2 \quad \text{(Art. 132)}$$

$$= \frac{n^2}{m^2 - 1}\left(\frac{m^2a^2 + b^2}{m^2 - 1}\right) = \frac{n^2}{m^2 - 1}(x'^2 + y'^2) = \frac{n^2a'^2}{m^2 - 1}.$$

Hence $$X^2 - Y^2 = (a'^2 - b'^2)\frac{X^2}{a'^2} + b'^2;$$

or $$\frac{X^2}{a'^2} - \frac{Y^2}{b'^2} = 1.$$

Cor. 1.—The equation of the tangent, when the hyperbola is referred to a pair of conjugate diameters as axes, is

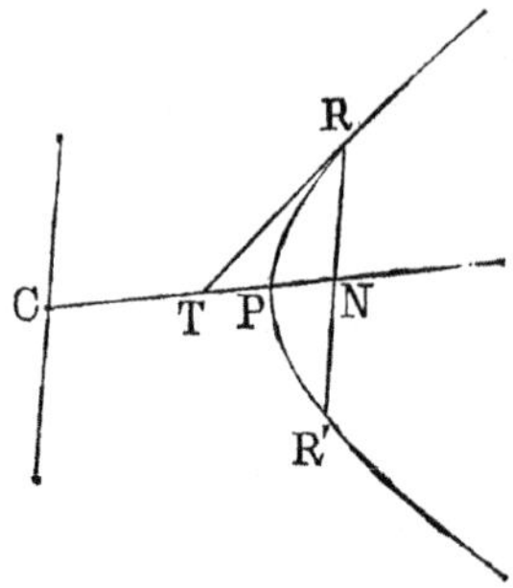

$$\frac{xx'}{a^2} - \frac{yy'}{b^2} - 1 = 0;$$

for, taking two points $x'y'$, $x''y''$ on the hyperbola, the curve

$$\frac{(x - x')(x - x'')}{a^2} - \frac{(y - y')(y - y'')}{b^2} = 0$$

evidently passes through both points. Hence the chord joining both points is

$$\frac{(x - x')(x - x'')}{a^2} - \frac{(y - y')(y - y'')}{b^2} - \left(\frac{x^2}{a^2} - \frac{y^2}{b^2} - 1\right) = 0;$$

and, if the points become consecutive, this reduces to

$$\frac{xx'}{a^2} - \frac{yy'}{b^2} = 1. \qquad (488)$$

Cor. 2.—If the tanget at R meet CP in T, $CN \cdot CT = CP^2$.

Cor. 3.—The tangents at the extremities of any chord meet on the diameter conjugate to that chord.

Cor. 4.—The line joining the intersection of two tangents to the centre bisects the chord of contact.

EXAMPLES.

1. If a chord of a circle be parallel to a line given in position, the locus of a point which divides it into parts, the sum of whose squares is constant, is an equilateral hyperbola.

2. If CP, CD be any two semidiameters of a hyperbola, PN, DM tangents meeting CD, CP in N and M, respectively; triangle $CPN = CDM$.

3. In the same case, if PT, DE be parallels to the tangents meeting CD, CP produced in T and E; the triangle $CDE = CPT$.

4. If a quadrilateral be circumscribed to a hyperbola, the join of the middle points of its diagonals passes through the centre.

5. If AB be any diameter of a hyperbola, AE, BD tangents at its extremities meeting any third tangent in E and D, the rectangle $AE \cdot BD$ is equal to the square of the semidiameter conjugate to AB.

6. If in the fig. of Ex. 5, CD, CE be drawn meeting the hyperbola and its conjugate in D' and E'; CD', CE' are conjugate semidiameters.

7. Diameters parallel to a pair of supplemental chords are conjugate.

8. Find the condition that the line $\lambda x + \mu y + \nu = 0$ shall touch the hyperbola.

Ans. $a^2\lambda^2 - b^2\mu^2 - \nu^2 = 0$, which is the tangential equation of the hyperbola.

9. If AA' be any diameter of an ellipse, PP' a double ordinate to it; if AP, $A'P'$ be produced to meet, the locus of their point of intersection is a hyperbola.

10. Tangents to a hyperbola are drawn from any point in one of the branches of the conjugate hyperbola; prove that the envelope of the chord of contact is the other branch of the conjugate hyperbola.

136. *To find the equation of the normal to the hyperbola at the point $x'y'$.*

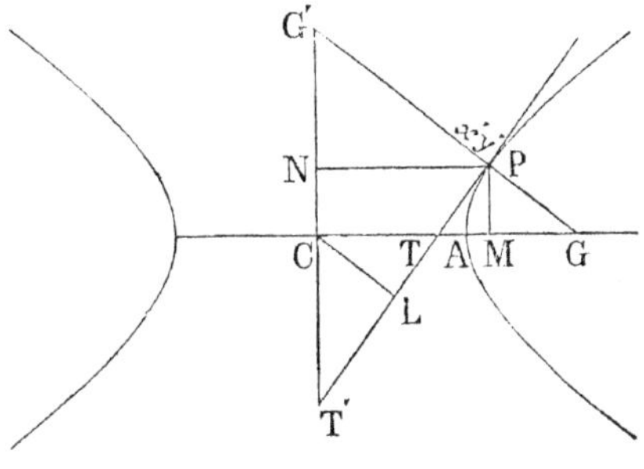

The equation of the tangent at $x'y'$ is

$$\frac{xx'}{a^2} - \frac{yy'}{b^2} = 1.$$

Hence the equation of the perpendicular to this at $x'y'$ is

$$\frac{a^2x}{x'} + \frac{b^2y}{y'} = c^2, \quad (489)$$

which is the equation of the required normal.

Cor. 1.—In equation (489) put $y = 0$, and we get

$$CG = e^2 x'. \tag{490}$$

Hence

$$MG = (e^2 - 1)\, x'. \tag{491}$$

Cor. 2.—$PG^2 = PM^2 + MG^2 = y'^2 + (e^2 - 1)^2 x'^2 =$ (after an easy reduction) to

$$\frac{b^2}{a^2}(e^2 x'^2 - a^2).$$

Hence

$$PG = \frac{b}{a}\sqrt{e^2 x'^2 - a^2}.$$

In like manner,

$$G'P = \frac{a}{b}\sqrt{e^2 x'^2 - a^2}.$$

Hence

$$G'P \,.\, PG = e^2 x'^2 - a^2. \tag{492}$$

Cor. 3.—If ρ, ρ' be the focal vectors to P,

$$G'P \,.\, PG = \rho\rho'. \tag{493}$$

Cor. 4.—In an equilateral hyperbola

$$PG = G'P. \tag{494}$$

Cor. 5.—If CR be the semidiameter conjugate to CP,

$$G'P \,.\, PG = CR^2 = b'^2 = \rho\rho'. \tag{495}$$

Cor. 6.—If CL be perpendicular to the tangent at P,

$$CL \,.\, PG = b^2, \quad CL \,.\, G'P = a^2.$$

EXAMPLES.

1. The points G', P, T' and the two foci are concyclic.

2. A right line parallel to the conjugate axis of a hyperbola meets it and its conjugate in the points M, N; show that normals to these curves at the points M, N intersect on the transverse axis.

3. If the hyperbola be equilateral, and if CL produced meet the curve in L'; prove $CL \,.\, CL' = a^2$.

4. If through the points G, G' parallels be drawn to the axes, the locus of their intersection is a hyperbola.

5. In an equilateral hyperbola half the difference of the base angles of the triangle SPS' is equal to one of the angles which CP makes with SS'.

6. If from any point in a hyperbola perpendiculars be drawn to the axes, the join of their feet is always normal to a hyperbola.

7. If through the point T, where the tangent at P meets the transverse axis, a parallel to the conjugate axis be drawn meeting the join of the points A, P, in J, the locus of J is an ellipse, having the same axes as the hyperbola.

8. If the co-ordinates of a point on the hyperbola

$$\frac{x^2}{a^2} - \frac{y^2}{b^2} = 1$$

be denoted by $a \sec\phi$, $b \tan\phi$; prove that the co-ordinates of the intersection of normals at the points $(\alpha + \beta)$, $(\alpha - \beta)$ are

$$\frac{-}{a} \cdot \frac{\cos\beta}{\cos\alpha \cos(\alpha+\beta) \cos(\alpha-\beta)}, \quad -\frac{c^2}{a} \tan\alpha \,.\, \tan(\alpha+\beta) \,.\, \tan(\alpha-\beta).$$

9. The co-ordinates of the point of intersection of two consecutive normals are

$$\frac{c^2}{a} \sec^3\alpha, \quad -\frac{c^2}{b} \tan^3\alpha. \qquad (496)$$

10. The locus of the centre of curvature of the hyperbola is

$$(ax)^{\frac{2}{3}} - (by)^{\frac{2}{3}} = c^{\frac{4}{3}}. \qquad (497)$$

137. *To find the lengths of the perpendiculars from the foci on the tangent at any point of the hyperbola.*

If the co-ordinates of the point P be $a \sec\phi$, $b \tan\phi$, the equation of the tangent is

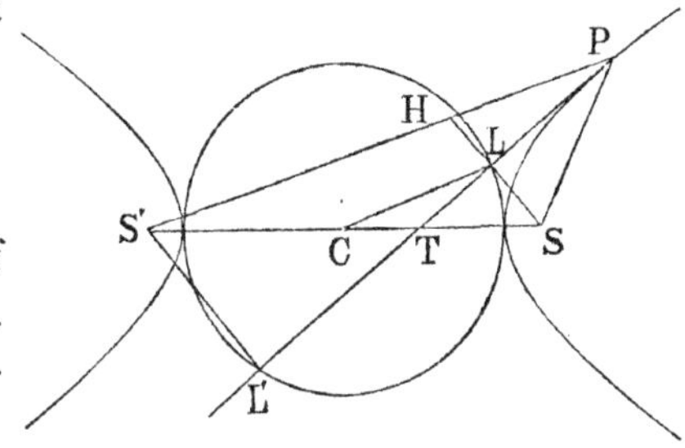

$$\frac{x \sec\phi}{a} - \frac{y \tan\phi}{b} - 1 = 0,$$

and the co-ordinates of the focus S are ae, 0. Hence the perpendicular

$$SL = b\left(\frac{e \sec\phi - 1}{e \sec\phi + 1}\right)^{\frac{1}{2}};$$

or denoting the focal vectors by ρ, ρ',

$$SL = b\sqrt{\frac{\rho}{\rho'}}. \tag{498}$$

Similarly,
$$S'L' = b\sqrt{\frac{\rho'}{\rho}}. \tag{499}$$

Cor. 1.—
$$SL \,.\, S'L' = b^2. \tag{500}$$

Cor. 2.—
$$SL \div \rho = \frac{b}{\sqrt{\rho\rho'}} = \frac{b}{b'}. \quad \text{(Art. 135, } Cor. \text{ 5.)} \tag{501}$$

Cor. 3.—The tangent at P bisects the internal angle at P of the triangle SPS'; and the normal bisects the external angle.

Cor. 4.—Since the angle SPH is bisected by PL, we have $SL = LH$, and $SC = CS'$, because C is the centre. Hence

$$CL = \tfrac{1}{2}S'H = \tfrac{1}{2}(S'P - SP) = a;$$

therefore the locus of L is the auxiliary circle.

Cor. 5.—If a line move so that the rectangle contained by perpendiculars on it from two fixed points on opposite sides is constant, its envelope is a hyperbola.

Cor. 6.—The first positive pedal of a hyperbola, with respect to either focus, is a circle.

Cor. 7.—The first negative pedal of a circle, with respect to any external point, is a hyperbola.

Cor. 8.—The reciprocal of a hyperbola, with respect to either focus, is a circle.

138. *The rectangle contained by the segments of any chord passing through a fixed point in the plane of the hyperbola is to the square of the parallel semidiameter in a constant ratio.*

The proof is the same as that of the corresponding proposition (Art. 126) for the ellipse, and similar inferences may be drawn.

Examples.

1. If an equilateral hyperbola pass through the angular points of a triangle, it passes through the orthocentre.

2. The locus of the centres of all equilateral hyperbolas described about a given triangle is the 'nine-points circle' of the triangle.

3. If P be any point in an equilateral hyperbola whose vertices are A, A'; prove that the normal at P and the line CP make equal angles with the transverse axis.

139. *To find the polar equation of the hyperbola, the centre being pole.*

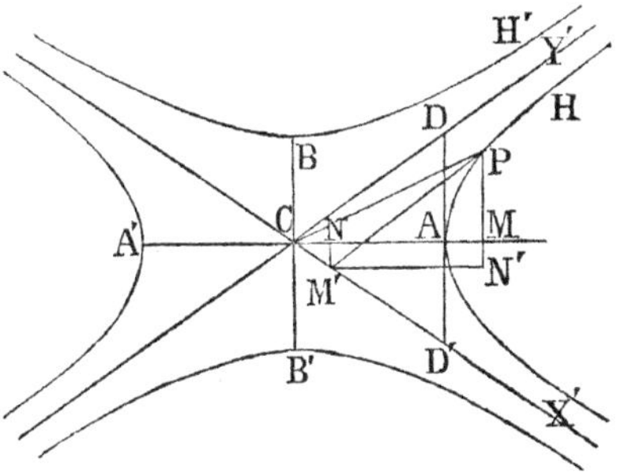

Let H be the hyperbola, $A'A$ its tranverse axis, and $B'B$ its conjugate axis, P any point in the curve; then, if x, y be the rectangular co-ordinates of P, ρ, θ, its polar co-ordinates, we have

$$x = \rho \cos \theta, \; y = \rho \sin \theta;$$

and, substituting these in the equation of the hyperbola, we get

$$\frac{1}{\rho^2} = \frac{\cos^2 \theta}{a^2} - \frac{\sin^2 \theta}{b^2}.$$

Hence
$$\rho^2 = \frac{b^2}{e^2 \cos^2 \theta - 1}, \tag{502}$$

which the polar equation required.

Cor. 1.—The polar equation of the conjugate hyperbola H' is

$$\rho^2 = \frac{b^2}{1 - e^2 \cos^2 \theta}. \tag{503}$$

Cor. 2.—If the hyperbola be equilateral, $b^2 = a^2$, and the polar equation is

$$\rho^2 \cos 2\theta = a^2. \tag{504}$$

Cor. 3.—If in equation (502) the denominator, $e^2\cos^2\theta - 1$, vanish, we get $\rho^2 =$ infinity; therefore $\rho = \pm$ infinity; but if $e^2\cos^2\theta - 1 = 0$, we get $\tan^2\theta = \frac{b^2}{a^2}$ and $\tan\theta = \pm\frac{b}{a}$ Hence, if DD' be erected at right angles to CA, and if AD and $D'A$ be made each equal to b, and CD, CD' joined, these lines produced both ways will each meet the curve at infinity.

Cor. 4.—The equations of the line CD, CD' are respectively

$$\frac{x}{a} - \frac{y}{b} = 0, \quad \frac{x}{a} + \frac{y}{b} = 0. \tag{505}$$

Each of these lines touches the curve at infinity, or, in other words, is an asymptote. (Art. 104.)

For the tangent at $x'y'$ may be written

$$\frac{x}{a^2} - \frac{yy'}{b^2x'} = \frac{1}{x'}.$$

Now, if $x'y'$ be the point where the line $\frac{x}{a} - \frac{y}{b} = 0$ meets the curve, we have $\frac{y'}{x'} = \frac{b}{a}$. Hence the tangent may be written

$$\frac{x}{a} - \frac{y}{b} = \frac{a}{x'}, \text{ or } \frac{x}{a} - \frac{y}{b} = 0, \text{ since } x' \text{ is infinite.}$$

Cor. 5.—Since the product of the equations of the two asymptotes (505) is $\frac{x^2}{a^2} - \frac{y^2}{b^2} = 0$, we see that the equation of the hyperbola differs from the equation of its asymptotes only by the absolute term. (Art. 105, *Cors.* 1, 2.)

Cor. 6.—The asymptotes of an equilateral hyperbola are at right angles to each other. On this account the equilateral hyperbola is also called the rectangular hyperbola.

Cor. 7.—The secant of half the angle between the asymptotes is equal to the eccentricity.

Cor. 8.—The lines joining an extremity of any diameter to the extremities of its conjugate are parallel to the asymptotes.

140. *To find the equation of the hyperbola referred to the asymptotes as axes.*

Let H be the hyperbola, CX', CY' (see last fig.) the asymptotes, P any point in the curve; draw PM' parallel to CY'; then, denoting CM', $M'P$, the co-ordinates of P with respect to the new axes, by $x'y'$, and half the angle between the asymptotes by α, we have, since $CM = CO + M'N'$, and $PM = PN' - M'N$,

$$x = (x' + y')\cos\alpha, \quad y = (y' - x')\sin\alpha;$$

and substituting in the equation

$$\frac{x^2}{a^2} - \frac{y^2}{b^2} = 1,$$

we get

$$\frac{(x' + y')^2\cos^2\alpha}{a^2} - \frac{(y' - x')^2\sin^2\alpha}{b^2} = 1.$$

But $\sec\alpha = e.$ (Art. 139, *Cor.* 7.)

Hence

$$\cos^2\alpha = \frac{a^2}{a^2 + b^2}, \quad \sin^2\alpha = \frac{b^2}{a^2 + b^2};$$

therefore $(x' + y')^2 - (y' - x')^2 = a^2 + b^2,$

or $4x'y' = a^2 + b^2;$

and omitting accents, as being no longer necessary,

$$xy = \frac{a^2 + b^2}{4}, \tag{506}$$

which is the required equation.

Cor. 1.—The area of the parallelogram formed by the asymptotes, and by parallels to them through any point in the curve, is constant.

Cor. 2.—Since the product xy is constant, the larger x is, the smaller y will be, and conversely; hence the hyperbola continually approaches its asymptotes, but never meets them, until it goes to infinity, where it touches them.

Examples.

1. A variable line has its extremities on two lines given in position and passes through a given point; prove that the locus of the point in which it is divided in a given ratio is a hyperbola.

2. From a point P perpendiculars are let fall on two fixed lines; if the area of the quadrilateral thus formed be given; prove that the locus of P is a hyperbola.

3. If any line cuts a hyperbola and its asymptotes; prove that the intercepts on the line between the curve and its asymptotes are equal.

4. If a variable line form with two fixed lines a triangle of constant area, the locus of the point which divides the intercept made on the variable line in a given ratio is a hyperbola.

5. If two sides of a triangle be given in position, and its perimeter given in magnitude, the locus of the point which divides the base in a given ratio is a hyperbola.

6. The equation of a hyperbola passing through three given points, and having its asymptotes parallel to two lines given in position, is

$$\begin{vmatrix} xy, & x, & y, & 1, \\ x'y', & x', & y', & 1, \\ x''y'', & x'', & y'', & 1, \\ x'''y''', & x''', & y''', & 1 \end{vmatrix} = 0, \tag{507}$$

the axes being the lines given in position.

If the lines given in position be denoted by $S \equiv ax^2 + 2hxy + by^2 = 0$, the equation will be

$$\begin{vmatrix} S, & x, & y, & 1, \\ S', & x', & y', & 1, \\ S'', & x'', & y'', & 1, \\ S''', & x''', & y''', & 1 \end{vmatrix} = 0. \tag{508}$$

7. The equation $xy = k^2$, being a special case of the equation $LM = R^2$ (Art. 108), the co-ordinates of a point on the hpperbola can be expressed by a single variable. Thus $x = k \tan \phi$, $y = k \cot \phi$. This will be called the point ϕ.

8. Prove that the equation of the join of the points ϕ', ϕ'' on the hyperbola is

$$\frac{x}{\tan \phi' + \tan \phi''} + \frac{y}{\cot \phi' + \cot \phi''} = k,$$

or

$$\frac{x}{x' + x''} + \frac{y}{y' + y''} = 1. \tag{509}$$

9. The intercepts on the axes are $x' + x''$, $y' + y''$.

10. The tangent at the point ϕ is

$$x \cot \phi + y \tan \phi = 2k, \text{ or } \frac{x}{x'} + \frac{y}{y'} = 2. \tag{510}$$

11. The area of the triangle formed by the asymptotes and any tangent to the hyperbola $= 2k^2$.

12. If a variable point xy on the hyperbola be joined to two fixed points, the intercept on the asymptotes made by the joining lines is constant.

13. The co-ordinates of the point of intersection of tangents at ϕ', ϕ'', are

$$\frac{2k}{\cot \phi' + \cot \phi''}, \qquad \frac{2k}{\tan \phi' + \tan \phi''}. \tag{511}$$

*14. The area of the triangle formed by tangents at the points ϕ', ϕ'', ϕ''' is

$$\frac{2k^2 \{\sin^2 \phi' (\sin 2\phi'' - \sin 2\phi''') + \sin^2 \phi'' (\sin 2\phi''' - \sin 2\phi) + \sin^2 \phi''' (\sin 2\phi' - \sin 2\phi'')\}}{\sin (\varphi' + \phi'') \sin (\phi'' + \phi''') \sin (\phi''' + \phi')}$$

15. The normal at the point ϕ is $x \tan \phi - y \cot \phi = k (\tan^2 \phi - \cot^2 \phi)$.

16. The four normals, from the point $\alpha\beta$ to the hyperbola $xy = k^2$, have the tangents of the parametric angles of their points of meeting the hyperbola connected by the relation $k (\tan^4 \phi - 1) = \alpha \tan^3 \phi - \beta \tan \phi$.

17. The intersection of normals at the points $x'y'$, $x''y''$ are

$$\frac{x'^2 + x'x'' + x''^2 + y'y''}{x' + x''}, \quad \frac{y'^2 + y'y'' + y''^2 + x'x''}{y' + y''}. \tag{512}$$

18. The co-ordinates of the centre of curvature at the point $x'y'$ are

$$\frac{3x'^2 + y'^2}{2x'}, \quad \frac{3y'^2 + x'^2}{2y'}. \tag{513}$$

19. The circle of curvatnre at $x'y'$ meets the curve again in the point, whose co-ordinates are

$$\frac{x'^2}{y'}, \quad \frac{y'^2}{x'}. \tag{514}$$

20. The radius of curvature at $x'y'$ is $(x'^2 + y'^2)^{\frac{3}{2}} \div 2k^2$. (515)

141. *To find the polar equation of the hyperbola, the focus being pole.*

Let $SP = \rho$, the angle $ASP = \theta$. (See fig., Art. 130.)

Then $SP = ePN$ by definition;

that is, $\rho = e(OS + SQ) = ef + e\rho\cos(\pi - \theta)$,

or $\rho = a(e^2 - 1) - e\rho\cos\theta$.

Therefore
$$\rho = \frac{a(e^2 - 1)}{1 + e\cos\theta}. \tag{516}$$

Cor. 1.—If we put $\theta = \frac{\pi}{2}$, we get $\rho = a(e^2 - 1)$; but in this case ρ is half the latus rectum. Hence, denoting it by l, we have

$$\rho = \frac{l}{1 + e\cos\theta}. \tag{517}$$

Cor. 2.—The polar equation of the tangent at the point α is

$$\rho = \frac{l}{\cos(\alpha - \theta) + e\cos\theta}. \tag{518}$$

Examples.

1. The equation of the chord joining the points $(\alpha + \beta)$, $(\alpha - \beta)$, is

$$\rho = \frac{l}{e\cos\theta + \sec\beta\cos(\alpha - \theta)}. \tag{519}$$

2. If α be constant, and β variable, the chord joining the points $(\alpha + \beta)$ $(\alpha - \beta)$, passes through a fixed point.

*142. *To find the area of an equilateral hyperbola, between an asymptote and two ordinates.*

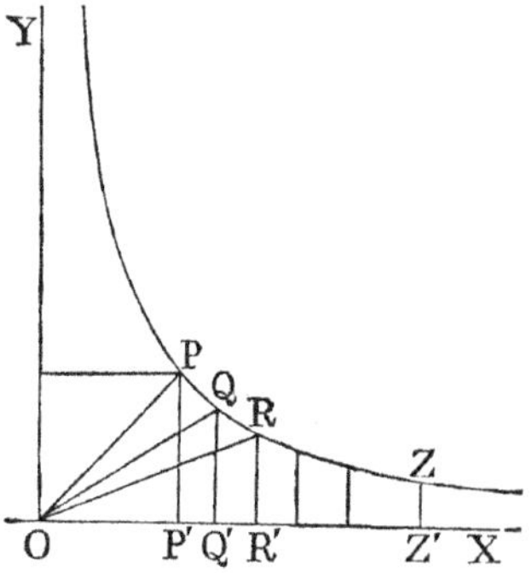

Let PQZ be the hyperbola; OX, OY the asymptotes. Bisect the angle XOY by OP; draw the ordinate PP' and ZZ'; then denoting OP' by unity, and $P'Z'$ by x the area enclosed by PP', ZZ', $P'Z'$, and the hyperbola, $= \log_e(1 + x)$.

Demonstration.—Divide $P'Z'$ into any number of parts n, in the points Q', R', &c.; so that OP', OQ', QR', &c., are in geometrical progression, and draw the ordinates $Q'Q$, $R'R$, &c. Join PQ, QR, &c.; also join OQ, QR. Now, denoting the co-ordinates of the points P, Q, R by $x'y'$, $x''y''$, $x'''y'''$, we have area of the triangle OQR

$$\tfrac{1}{2}(x''y''' - x'''y'') = \tfrac{1}{2}\left(\frac{x''y''^2}{y'} - \frac{x''^2 y''}{x'}\right);$$

since $$y''' = \frac{y''^2}{x'} \text{ and } x''' = \frac{x''^2}{x'}.$$

Hence area of triangle OQR

$$= \tfrac{1}{2}\frac{x''y''}{x'y'}(x'y'' - x''y') = \tfrac{1}{2}(x'y'' - x''y'),$$

or equal area of triangle OPQ. But it is easy to see that the triangle OPQ is equal to the trapezium $PP'Q'Q$, and OQR equal to the trapezium $QQ'R'R$. Hence the trapeziums are equal; and therefore the whole rectilineal figure $PP'Z'Z$ is equal to n times the trapezium $PP'Q'Q$. Again, we have $OZ' = OP' + P'Z = 1 + x$; and $OQ' = OP' + P'Q' = 1 + P'Q'$; and since OP', OQ', . . . OZ' are in geometrical progression, and there are n terms, we have $(1 + P'Q')^n = 1 + x$; therefore

$$P'Q' = (1 + x)^{\frac{1}{n}} - 1, \text{ and } PP' = 1.$$

Hence, when n is indefinitely large, the area of the trapezium $PP'Q'Q = (1+x)^{\frac{1}{n}} - 1$. Therefore the hyperbolic area $PP'Z'Z$ is equal to the limit of

$$n\{(1+x)^{\frac{1}{n}} - 1)\} = \log_e(1+x). \text{ (See *Trig.*)} \quad (520)$$

Cor. 1.—The hyperbolic sector $OPZ = \log_e(1+x)$. (521)

Cor. 2.—If AZ be an equilateral hyperbola, whose equation is $x^2 - y^2 = 1$, and if the co-ordinates OZ', $Z'M$ of a point Z be xy, the sectorial area

$$OAZ = \tfrac{1}{2}\log(x+y).$$

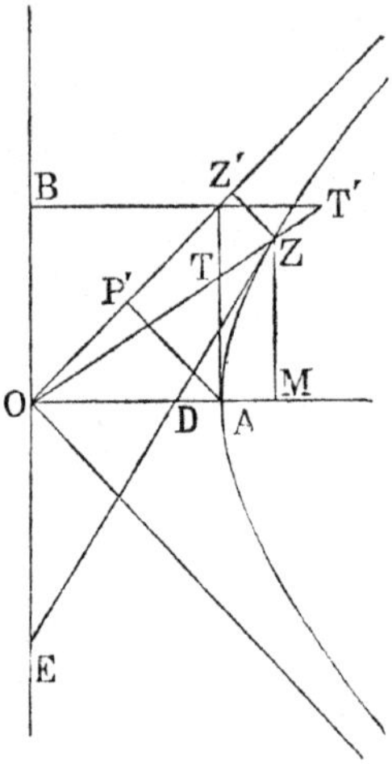

Dem—In the foregoing proof OP' is taken to be the linear unit, but in the general case it is evident that the proposition proved is that the sectorial area $= OP'^2 \times \log_e(OZ' \div OP')$; but it is easy to see that $OZ' \div OP' = (OM + MZ) \div OA$, and $OP'^2 = \frac{1}{2}OA^2$. Hence the area of the hyperbolic sector $OAZ = \frac{1}{2}a^2 \log_e \frac{(x+y)}{a}$.

Hence, when a is unity, sectorial area $= \frac{1}{2}\log_e(x+y)$. (522)

Cor. 3.—If u denote twice the sectorial area OAZ, then

$$x = \frac{e^u + e^{-u}}{2}, \quad y = \frac{e^u - e^{-u}}{2}. \quad (523)$$

For $\log_e(x+y) = u$; therefore $e^u = x+y$; and

$$e^{-u} = \frac{1}{x+y} = x - y.$$

DEF.—x, y *are called, respectively, the* HYPERBOLIC COSINE *and* HYPERBOLIC SINE *of* u, *and are denoted by the notation* *Chu, Shu.* (See *Trigonometry.*)

Cor. 4.—If $\sqrt{-1}$ be denoted by i, $Chu = \cos(ui)$, $Shu = \frac{\sin(ui)}{i}$. These follow from the values of x, y, and the trigonometric expansions of $\cos(ui)$, $\sin(ui)$.

143. The other hyperbolic functions are defined as follows, thus :—OD = hyperbolic secant $u \equiv \sec hu$, AT = hyperbolic tangent $u \equiv Thu$, BT' = hyperbolic cotangent $u \equiv \cot hu$, OE = hyperbolic cosecant $u \equiv \operatorname{cosec} hu$.

From the known properties of the hyperbola we have immediately the following relations :—

$$\sec hu = \frac{1}{Chu}, \quad Thu = \frac{Shu}{Chu}, \quad \cot hu = \frac{Chu}{Shu}, \quad \operatorname{cosec} h = \frac{1}{Shu},$$

corresponding to the known relations of circular functions ; and from them can be constructed a theory of these functions. (See Author's *Trigonometry*.)

144. From the values $Chu = \cos(ui)$, $\sin hu = \dfrac{\sin(ui)}{i}$, we see that, if we put $ui = \phi$, we have $x = \cos\phi$, $y = \dfrac{\sin\phi}{i}$; so that the co-ordinates of any point on the equilateral hyperbola can be denoted by the circular functions of an *imaginary angle* ϕ. In like manner, the co-ordinates of a point on the hyperbola

$$\frac{x^2}{a^2} - \frac{y^2}{b^2} = 1$$

can be expressed in a manner analogous to the method of the eccentric angle for the ellipse. Thus we can put

$$\frac{x}{a} = \cos\phi, \quad \frac{y}{b} = \frac{\sin\phi}{i};$$

and by these substitutions we could give proofs analogous to those of the ellipse for the corresponding propositions of the hyperbola.

The following exercises can be solved by using the imaginary eccentric angle.

EXAMPLES.

1. If the chord joining the points $(\alpha+\beta)$, $(\alpha-\beta)$ pass through the focus; prove

$$e\cos\alpha = \cos\beta. \quad (524)$$

2. The tangents at the extremities of a focal chord meet on the directrix.

3. In the same case, the line joining the intersections is perpendicular to the chord.

4. Prove that the eccentric angles of the two points which are the extremities of a pair of conjugate semidiameters differ by $\frac{\pi}{2}$.

5. Apply the method of the eccentric angle to the proof of the proposition, that the locus of the middle points of a system of parallel chords is a right line.

6. Find the equation of the hyperbola, referred to a pair of conjugate diameters by means of the eccentric angle.

7. The co-ordinates of the point of intersection of tangents at the points $(\alpha+\beta)$, $(\alpha-\beta)$, are

$$\frac{a\cos\alpha}{\cos\beta}, \quad \frac{bi\sin\alpha}{\cos\beta}. \quad (525)$$

8. If α be variable and β constant, the chord joining the points $(\alpha+\beta)$, $(\alpha-\beta)$ is a tangent to the hyperbola

$$\frac{x^2}{a^2} - \frac{y^2}{b^2} = \cos^2\beta. \quad (526)$$

9. In the same case, the locus of the intersection of tangents at the extremities of a chord is

$$\frac{x^2}{a^2} - \frac{y^2}{b^2} = \sec^2\beta. \quad (527)$$

10. If ϕ be the angle between the tangents at $(\alpha+\beta)$, $(\alpha-\beta)$,

$$\tan\phi = \frac{2abi\sin\beta}{(a^2+b^2)\cos 2\alpha - (a^2-b^2)\cos 2\beta}. \quad (528)$$

11. Find the locus of the pole of a chord which subtends a right angle at a fixed point hk.

Let $(\alpha+\beta)$, $(\alpha-\beta)$ be the eccentric angles at the extremities of the chord; then the equation of the circle which has the chord for diameter is

$$(x-a\cos\alpha\cos\beta)^2+(y-bi\sec\alpha\cos\beta)^2=(a^2\sin^2\alpha-b^2\cos^2\alpha)\cos^2\beta,$$

and evidently hk is a point on this circle; hence

$$(h-a\cos\alpha\cos\beta)^2+(k-bi\sin\alpha\cos\beta)^2=(a^2\sin^2\alpha-b^2\cos^2\alpha)\cos^2\beta,$$

or

$$h^2+k^2-2(a\cos\alpha\cos\beta)h-2(bi\sin\alpha\cos\beta)k+a^2(\cos^2\beta-\sin^2\alpha)$$
$$+b^2(\cos^2\alpha-\cos^2\beta)=0.$$

Now, if x, y be the co-ordinates of the pole of the chord joining $(\alpha+\beta)$, $(\alpha-\beta)$, we have

$$a\cos\alpha=x\cos\beta,\quad bi\sec\alpha=y\cos\beta;$$

therefore

$$h^2+k^2-(2hx+2ky-a^2+b^2)\cos^2\beta-a^2\sin^2\alpha+b^2\cos^2\alpha=0;$$

or, eliminating α,

$$h^2+k^2-(2hx+2ky-a^2+b^2)-\frac{a^2y^2}{b^2}+\frac{b^2x^2}{a^2}\cos^2\beta=0.$$

But $$\left(\frac{x^2}{a^2}-\frac{y^2}{b^2}\right)\cos^2\beta=1.\qquad \text{(Ex. 9.)}$$

Hence, eliminating β, we get

$$\left(\frac{h^2+k^2+b^2}{a^2}\right)x^2-\left(\frac{h^2+k^2-a^2}{b^2}\right)y^2-2(hx+ky)+a^2-b^2=0,\qquad (529)$$

which represents a hyperbola, a parabola, or an ellipse, according as the point hk is outside the auxiliary circle, on it, or inside it.

12. The discriminant of this equation (529) is the product of the two factors

$$b^2h^2-a^2k^2-a^2b^2 \text{ and } h^2+k^2-(a^2-b^2).$$

Hence we infer that the locus will break up into two lines if the co-ordinates hk satisfy the equation of the hyperbola. In other words, if a chord of a hyperbola subtend a right angle at any fixed point on the curve, the locus of its pole consists of two right lines.

From the factor $h^2+k^2-(a^2-b^2)=0$ we infer that, if the chord subtend a right angle at any point on the director circle, its pole will be the same point.

Exercises on the Hyperbola.

1. The perpendicular from the focus on either asymptote is equal to the semiconjugate diameter.

2. If e, e' be the eccentricities of a hyperbola and its conjugate; prove

$$\frac{1}{e^2} + \frac{1}{e'^2} = 1. \qquad (530)$$

3. The equations of the asymptotes, with the focus as origin, are

$$\frac{x}{a} \pm \frac{y}{b} = e. \qquad (531)$$

4. If SP be parallel to an asymptote, P being a point on the curve; prove

$$SP = \frac{l}{2}. \qquad (532)$$

5. If from a point K in the transverse axis a perpendicular KL be drawn to an asymptote, and a normal KM to the curve; prove that LM is perpendicular to the transverse axis.

6. An ellipse referred to the equal conjugate diameters being

$$x^2 + y^2 = \frac{a^2 + b^2}{2};$$

prove that it is confocal with the hyperbola

$$xy = \frac{a^2 - b^2}{4}. \qquad \text{(Crofton.)} \qquad (533)$$

7. Also, this hyperbola cuts orthogonally all conics passing through the ends of the major and minor axes of the ellipse in Ex. 6. The general equation of these conics is

$$x^2 \cos^2 \alpha + y^2 \sin^2 \alpha = \frac{a^2 + b^2}{4}. \qquad (\textit{Ibid.}) \qquad (534)$$

8. The chord of contact of two tangents to a hyperbola is parallel to, and half way between, the lines joining the intersections of tangents with the asymptotes.

9. The locus of the centre of a variable circle which makes given intercepts on two given lines is a hyperbola.

10. If from any point P on a given line tangents be drawn to the ellipses

$$\frac{x^2}{a^2} + \frac{y^2}{b^2} = 1, \quad \frac{x^2}{a_1^2} + \frac{y^2}{b_1^2} = 1,$$

the locus of the intersection of their chords of contact is an equilateral hyperbola.

11. If ϕ, ϕ', ϕ'', ϕ''' be the parametric angles of four concyclic points on the hyperbola $xy = k^2$; prove

$$\tan\phi \,.\, \tan\phi' \,.\, \tan\phi'' \,.\, \tan\phi''' = 1. \tag{535}$$

12. The product of the perpendiculars from four concyclic points of a hyperbola on one asymptote is equal to the product of the perpendiculars on the other asymptote.

13. If the extremities of a chord of an ellipse which is parallel to the transverse axis be joined to the centre and to one extremity of that axis, the locus of the intersection of the joining lines is a hyperbola.

14. Parallels drawn from any system of points on a hyperbola to the asymptotes divide the asymptotes homographically; prove this, and thence infer the following theorem:—

If x', x'', x'''; y', y'', y''', denote the distances of two triads of points on two lines given in position from two fixed points O, O' on these lines; prove, if x, y be the distances of two variable points on the same lines from O, O', that x, y will divide the lines homographically if the determinant

$$\begin{vmatrix} xy, & x, & y, & 1, \\ x'y', & x', & y', & 1, \\ x''y'', & x'', & y'', & 1, \\ x'''y''', & x''', & y''', & 1, \end{vmatrix} = 0. \tag{536}$$

15. Prove that the sum of the eccentric angles of four concyclic points on a hyperbola is 2π.

16. If p, p', π be the perpendiculars from the points $\alpha + \beta$, $(\alpha - \beta)$, and the point of intersection of their tangents on any third tangent to the hyperbola; prove

$$pp' = \pi^2 \cos^2\beta. \tag{537}$$

17. If a circle osculates the hyperbola $xy = k^2$ at the point ϕ, the common chord of the circle and the hyperbola is

$$x\tan\phi + y\cot\phi + k(\tan^2\phi + \cot^2\phi) = 0. \tag{538}$$

18. A, B are two fixed points; if from A a perpendicular AP be drawn to the polar of B with respect to an equilateral hyperbola, and from B a perpendicular BQ to the polar of A; then, if C be the centre,

$$CA : AP :: CB : BQ.$$

19. An ellipse circumscribes a fixed triangle so that two of the vertices are at the extremities of a pair of conjugate diameters; prove that the locus of its centre is a hyperbola.

20. The polar of any point on an asymptote is parallel to that asymptote.

21. The points where any tangent meets the asymptotes, and the points where the corresponding normal meets the axes, are concyclic.

22. The two foci and the points of intersection of any tangent with the asymptotes are concyclic.

23. The angles which the intercept, made by the asymptotes on any tangent, subtends at the foci are constant.

24. Given in magnitude and position any two conjugate semidiameters OP, OQ of a hyperbola, to find the axes.

25. If P, P' be the extremities of two conjugate semidiameters of a hyperbola; and if S, S' be the interior foci of the branches of the hyperbola and its conjugate, on which are the points P, P', prove that

$$SP - S'P' = AC - BC. \qquad (539)$$

26. If an ellipse and a confocal hyperbola intersect in any point, the corresponding point on the auxiliary circle of the ellipse lies on one of the asymptotes of the hyperbola.

27. If a system of hyperbolas have the same asymptotes, the normals drawn to them at the points where a parallel to either axis meets them are concurrent.

28. A hyperbola, whose eccentricity is e, has a focus at the centre of the circle $x^2 + y^2 = a^2$; prove that the envelope of the tangents to the hyperbola at the points where it meets the circle is the hyperbola.

29. The chord of contact of two tangents to a parabola subtends a constant angle at the vertex; show that the locus of their intersection is a hyperbola.

30. If two hyperbolas have the same asymptotes, and if from any point in one tangents be drawn to the other, the envelope of their chord of contact is a hyperbola, having the same asymptotes.

31. If a variable circle touch each branch of a hyperbola it subtends a constant angle at either focus, and makes intercepts of constant lengths on the asymptotes.

32. The centre of mean position of the points of intersection of a circle and an equilateral hyperbola bisects the distance between their centres.

33. If PQ be the chord of an equilateral hyperbola which is normal at P; prove

$$3CP^2 + PQ^2 = CQ^2. \qquad (540)$$

34. The area of the triangle formed with the asymptotes by the normal of the hyperbola $x^2 - y^2 = a^2$, at the point $x'y'$, is

$$\frac{(x'^2 - y'^2)^2}{a^2}. \qquad (541)$$

35. The locus of the pole of any tangent to the circle whose diameter is the distance between the foci of $\frac{x^2}{a^2} - \frac{y^2}{b^2} = 1$, with respect to $\frac{x^2}{a^2} - \frac{y^2}{b^2} = 1$, is the ellipse

$$\frac{x^2}{a^4} + \frac{y^2}{b^4} = \frac{1}{a^2 + b^2}. \qquad (542)$$

36. Two circles described through two points on the same branch of an equilateral hyperbola, and through the extremities of any diameter, are equal.

37. If ϕ, ϕ', ϕ'', ϕ''' be the parametric angles of four points on an equilateral hyperbola, such that either is the orthocentre of the remaining three,

$$\tan\phi \tan\phi' \tan\phi'' \tan\phi''' + 1 = 0. \qquad (543)$$

38. If the normal at the point ϕ of the hyperbola $xy = k^2$ meet it again at the point ϕ'; prove

$$\tan^3\phi \,.\, \tan\phi' + 1 = 0. \qquad (544)$$

39. If four points on an equilateral hyperbola be concyclic, prove that the parametric angle of any point and of the orthocentre of the remaining points are supplemental.

40. If the osculating circle of an equilateral hyperbola, at the point whose parametric angle is ϕ, meet it again at the point ϕ'; prove

$$\tan^3\phi \,.\, \tan\phi' = 1. \qquad (545)$$

41. If the eccentric angle of the point $(k \tan\phi,\ k \cot\phi)$ be θ; prove

$$\cot\theta = \cos\phi + i \sin\phi. \qquad (546)$$

42. If two sides AB, AC of a fixed triangle be chords of two equal circles, show that the locus of the second intersection of the circles is an equilateral hyperbola.

43. If the point $(k \tan \alpha, k \cot \alpha)$ be the centroid of an inscribed triangle; prove that the ellipse $(3 \cot \alpha \,.\, x + 3 \tan \alpha \,.\, y)^2 = 4xy$ touches the three sides of the triangle.

44. If θ, θ', θ'', θ''' be the eccentric angles of four points on a hyperbola, and if the join of θ, θ' be perpendicular to the join of θ'', θ'''; prove

$$e^{\iota(\theta + \theta' + \theta'' + \theta''')} + 1 = \{e^{\iota(\theta + \theta')} + e^{\iota(\theta'' + \theta''')}\} \cos \omega, \qquad (547)$$

where e is the Napierian constant 2·718281, and ω the angle between the asymptotes. .

45. Prove that the ellipse in Ex. 43 touches the hyperbola, and that the tangents to it at the remaining points of meeting the hyperbola are parallel to the asymptotes.

46. Show that the polar circle of the triangle formed by three tangents to an equilateral hyperbola touches the 'Nine-points Circle' of the triangle formed by the points of contact at the centre of the curve.

(R. A. ROBERTS).

47. If two vertices of a triangle circumscribed about an ellipse move along confocal hyperbolae, prove that the locus of the centre of the inscribed circle is a concentric ellipse. (*Ibid.*)

48. Two circles, whose centres A, B are points on the transverse axis of a given ellipse, have each double contact with the ellipse, and intersect in a point P; if the difference of the angles ABP, BAP be given, the locus of P is an equilateral hyperbola. (*Ibid.*)

49. The circle inscribed in the triangle formed by the asymptotes and any tangent to the auxiliary circle of a hyperbola intersects the hyperbola in the point where it touches the tangent to the auxiliary circle.

50. The circle on GG' as diameter (see fig., Art. 136) passes through the points where the tangent PT meets the asymptotes.

51. If α, α' be the eccentric angles of two points P, Q on a hyperbola, such that the normal at P passes through the pole of the normal at Q; prove

$$4a^4 \sin \alpha \sin \alpha' + 4b^4 \cos \alpha \cos \alpha' = c^4 \sin 2\alpha \sin 2\alpha'. \qquad (548)$$

52. If three points on an equilateral hyperbola be concyclic with the centre, the angular points of the triangle formed by tangents at these points are concyclic with the centre.

53. The angular points of a self-conjugate triangle of an equilateral hyperbola are concyclic with the centre.

54. P, Q are points on an equilateral hyperbola, such that the osculating circle at P passes through Q; the locus of the pole of PQ is

$$(x^2 + y^2)^2 = 4k^2xy.$$

55. In the same case the envelope of PQ is

$$4(4k^2 - xy)^3 = 27k^2(x^2 + y^2)^2. \qquad (549)$$

56. The hyperbola $\frac{x^2}{a^2} - \frac{y^2}{b^2} = \frac{a^2 - b^2}{a^2 + b^2}$ cuts orthogonally all the conics passing through the extremities of the axes of the ellipse

$$\frac{x^2}{a^2} + \frac{y^2}{b^2} = 1. \qquad \text{(CROFTON.)}$$

57. ACB is a given right-angled triangle, having the angle C right; from any point D of AC a perpendicular DE is let fall on the hypotenuse; prove that the locus of the intersection of BD and CE is an equilateral hyperbola.

58. CA, CB are two lines given in position; AB is a variable line intersecting them; if $CA + CB \pm AB$ be given, prove that the locus of a point which divides AB in a given ratio is a hyperbola whose asymptotes are parallel to CA, CB.

59. If from any point in the hyperbola $x^2 - y^2 = a^2 + b^2$ a pair of tangents be drawn to the hyperbola $\frac{x^2}{a^2} - \frac{y^2}{b^2} = 1$; prove that the four points where they cut the axes are concyclic.

60. If through the point α on an ellipse a line be drawn bisecting the angle formed by the joins of α to the point $(\alpha + \beta)$, $(\alpha - \beta)$; prove, if α be constant and β variable, that the locus of its intersection with the join of the points $(\alpha + \beta)$, $(\alpha - \beta)$ is a hyperbola.

CHAPTER VIII.

MISCELLANEOUS INVESTIGATIONS.

SECTION I.—CONTACT OF CONIC SECTIONS.

145. *If* $S = 0$, $S' = 0$ *be the equations of two curves, then* $S - kS' = 0$ *represents a curve passing through every point of intersection of the curves* S and S'.

This proposition is a simple case of the evident principle that the points of intersection of two curves S and S' must satisfy the equations $S = 0$ and $S' = 0$, and, therefore, must satisfy the equation $S - kS' = 0$. (Compare Art. 19, *Cor.* 4.)

146. The following are special cases of this general theorem :—

1°. If $S = 0$ be any conic section, and $S' = 0$ the product of two lines, $S - k^2S' = 0$ denotes a conic section through the four points, where S is intersected by the two lines denoted by S'; for example, $S - k^2\alpha\beta = 0$ denotes a conic passing through the points where S is intersected by the lines $\alpha = 0$, $\beta = 0$.

2°. If the lines denoted by S' become indefinitely near, S' may be denoted by L^2, where $L = 0$ represents a line; then $S - k^2L^2 = 0$ denotes a conic, touching S in each point where L intersects S; in other words, having double contact with S. By giving different values to k, we get different conics, each having double contact with S, and having a common chord of contact, namely, $L = 0$. If the line $L = 0$

intersect S in two real points, $S - k^2L^2 = 0$ will have real double contact with S. If the line L meet S in two imaginary points—in other words, if it does not meet it in real points, $S - k^2L^2 = 0$ will have imaginary double contact with S. This form of equation may also be written $S^{\frac{1}{2}} - kL = 0$, or $S^{\frac{1}{2}} + kL = 0$; for either equation cleared of radicals gives $S - k^2L^2 = 0$. In conic sections there are many instances of imaginary double contact.

3°. If $S = 0$ denote the product of two lines, say MN; then $MN - k^2L^2 = 0$ will denote a conic section, touching the lines $M = 0$, $N = 0$, and having the line $L = 0$ as the chord of contact.

4°. By supposing one of the three lines L, M, N to be at infinity, we get three different cases. Thus: 1°. Let L be at infinity, then L becomes a constant; and if M, N be real, the equation $MN = k^2L^2$ will denote a hyperbola, of which M, N are the asymptotes. 2°. Let L be at infinity, and let M, N denote the two conjugate imaginary factors $x + y\sqrt{-1}$, $x - y\sqrt{-1}$, the equation $MN = k^2L^2$ will represent a circle. From this it follows that *all circles pass through the same two imaginary points on the line at infinity*. For the circle $x^2 + y^2 = r^2$ passes through the points where the line at infinity meets the lines $x + y\sqrt{-1} = 0$, $x - y\sqrt{-1} = 0$, and the circle $(x - a)^2 + (y - b)^2 = r^2$ passes through the points where infinity meets the lines $(x - a) + (y - b)\sqrt{-1} = 0$, $(x - a) - (y - b)\sqrt{-1} = 0$, which, since parallel lines meet at infinity, will be the same points. 3°. Let one of the factors M, N be a constant, and let $L = 0$ denote a finite line, the equation will be of the form $px = y^2$, and the curve denotes a parabola. Hence we have the important theorem that *every parabola touches the line at infinity*.

5°. If $S = 0$ be the product of two lines, viz., $\alpha\gamma = 0$, and S' the product of two others, namely, $\beta\delta = 0$, then $S - kS'$ becomes $\alpha\gamma - k\beta\delta = 0$. Hence $\alpha\gamma - k\beta\delta = 0$ denotes a conic section passing through the four points $\alpha\beta$, $\alpha\delta$, $\beta\gamma$, $\gamma\delta$; in other words, it denotes a circumconic of the quadrilateral formed by the lines α, β, γ, δ, taken in order.

147. In the equation $S - k^2\alpha\beta = 0$ (Art. 146, 1°.), if the lines $\alpha = 0$, $\beta = 0$, intersect on S, the curve $S - k^2\alpha\beta = 0$ touches S in the point $\alpha\beta$, and will intersect it in the points where the lines $\alpha = 0$, $\beta = 0$ meet S again. For evidently the curves have two consecutive points common at the intersection of the lines α, β. This is called *contact of the first order*.

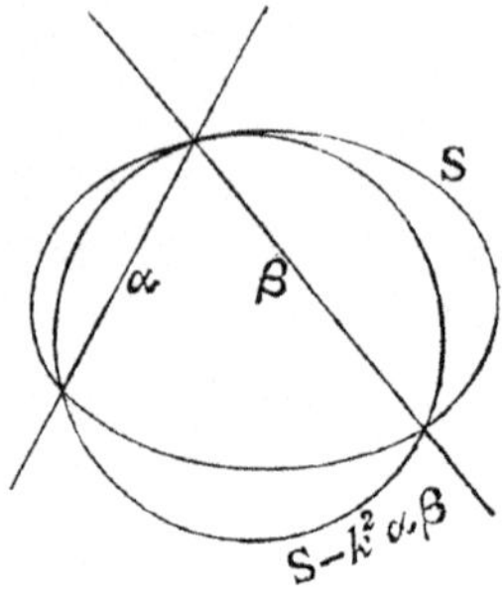

Again, if one of the lines $\alpha = 0$, $\beta = 0$—say $\alpha = 0$—touch S at the intersection of α, β, the second point in which α meets S coincides with the point $\alpha\beta$, and the curve $S - k^2\alpha\beta$ will have at the point $\alpha\beta$ three consecutive points common with S, and will intersect it in the second point, in which β meets S. The contact of S and $S - k^2\alpha\beta$ in this case is called *contact of the second order*, and $S - k^2\alpha\beta$ is said to *osculate* S.

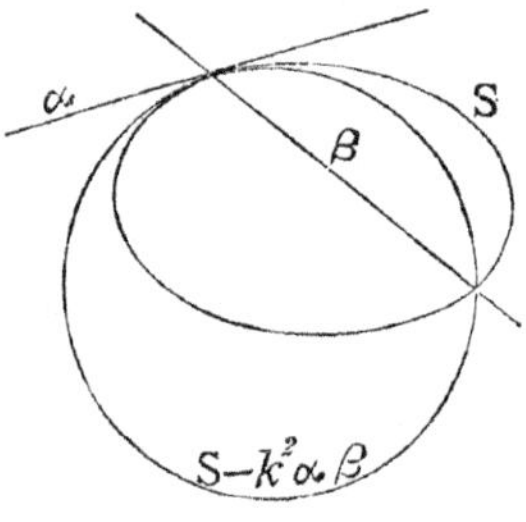

Lastly—Let the lines $\alpha = 0$, $\beta = 0$ coincide with each other, and with the tangent to S; then the product $\alpha\beta$ becomes α^2, and the two conics will have four consecutive points common, which is the highest order of contact that two conics can have. This is called *contact of the third order;* and the equations of two conics which have this species of contact will be of the forms $S = 0$, $S - k^2\alpha^2 = 0$, where α is a tangent to S. It is evident, from

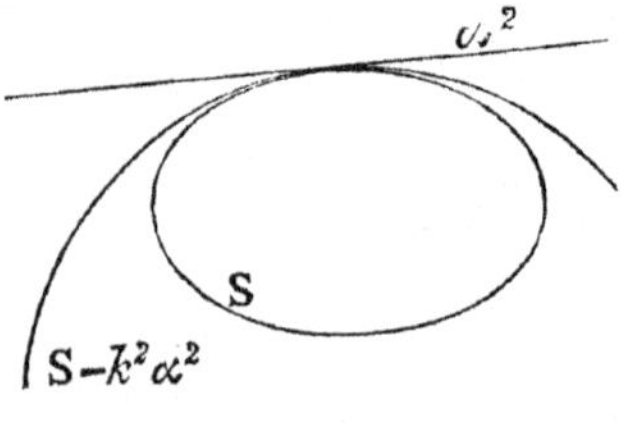

Art. 146, 2°, that the equation of a conic, having double contact with another S, and the equation of one having four-pointic contact, are the same in form, and that one changes into the other, when the chord of contact becomes a tangent.

148. The following examples will illustrate the foregoing principles :—If $S \equiv ax^2 + 2hxy + by^2 + 2gx = 0$, $S - k^2\alpha\beta \equiv a'x^2 + 2h'xy + b'y^2 + 2g'x = 0$, the lines $\alpha = 0$, $\beta = 0$, will be the two factors of the expression $(ag' - a'g)x^2 + 2(hg' - h'g)xy + (bg' - b'g)y^2 = 0$, got by eliminating the terms of the first degree. Now if one of these lines coincide with the tangent at the origin, we must have x as a factor, which requires that the coefficient of y^2 vanish. Hence, if the conics $ax^2 + 2hxy + by^2 + 2gx = 0$, $a'x^2 + 2h'xy + b'y^2 + 2g'x = 0$ osculate at the origin, $bg' = b'g$. Thus, if the circle $x^2 + y^2 + 2xy\cos\omega - 2rx\sin\omega = 0$ osculate $ax^2 + 2hxy + by^2 + 2gx = 0$, we must have $r = -\dfrac{g}{b\sin\omega}$, and this is the value of the radius of curvature of S at the origin. If the condition $bg' = b'g$ be fulfilled, the fourth point common to the two conics will be the point distinct from the origin, in which the line $(ag' - a'g)x + 2(hg' - h'g) = 0$ meets S. This will also coincide with the tangent at the origin if, in addition to the condition $bg' = b'g$, the coefficients of S, S' fulfil the condition $hg' = h'g$, and the conics will have at the origin contact of the third order. Thus the parabola $h^2x^2 + 2bhxy + b^2y^2 + 2hgx = 0$ has contact of the third order at the origin with S.

149. If in the equation $S - k^2L^2 = 0$ (Art. 146, 2°.) S denote a circle, we get the following theorem :—*The locus of a point, such that the tangent from it to a fixed circle is in a constant ratio to its distance from a fixed line, is a conic having double contact with the circle;* the contact will be *real* when the line L cuts S; *imaginary* when it does not. In this case, if we suppose S to reduce to a point, we get, evidently,

the focus and directrix. Hence we have the following definition :—*The focus of a conic is an infinitely small circle, having imaginary double contact with the conic, the directrix being the chord of contact.*

150. If the focus be made the origin, the equation (Articles 118, 130) is of the form $x^2+y^2=k^2L^2$, or $(x+y\sqrt{-1})(x-y\sqrt{-1}) = (kL)^2$, showing that the imaginary lines $x+y\sqrt{-1}=0$, $x-y\sqrt{-1}=0$ are tangents to the curve. But $x+y\sqrt{-1}=0$, $x-y\sqrt{-1}=0$, are (Art. 146, 4°) the lines from the origin to the circular points at infinity. If we denote these points by I and J, we see that the joins of either focus to I and J are tangents to the curve. Hence *all confocal conics are inscribed in the same imaginary quadrilateral, the six vertices of which are the two circular points at infinity, the two real foci, and two imaginary points on the conjugate axis, called* ANTI-FOCI.

Examples.

1. The chord of curvature through the centre of an ellipse is

$$= \frac{2b'^2}{a'}, \tag{550}$$

a' being the semi-diameter, terminating in the point of osculation. For, draw an ordinate to the semi-diameter, terminating in the point of osculation indefinitely near the point of intersection, and through its extremities, and the extremity of the semidiameter a' describe a circle, this will be the circle of curvature, and the proposition is evident.

2. The radius of curvature at the extremity of a' is

$$= \frac{b'^3}{ab}. \tag{551}$$

3. The focal chord of curvature at any point of a conic is equal to the focal chord of the conic, parallel to the tangent at the point.

Describe a circle through the point, and through the extremities of a chord, parallel to and indefinitely near the tangent.—(Townsend.)

4. If ϕ be the eccentric angle at any point of an ellipse, then the equation of the common chord of the ellipse and the circle osculating at ϕ is

$$\frac{x}{a}\cos\phi - \frac{y}{b}\sin\phi = \cos 2\phi.$$

5. The equation of the circle osculating the ellipse

$$\frac{x^2}{a^2} + \frac{y^2}{b^2} - 1 = 0,$$

at the point ϕ, is

$$\left(x - \frac{c^2\cos^3\phi}{a}\right)^2 + \left(y + \frac{c^2\sin^3\phi}{b}\right)^2 = \frac{b'^3}{ab};$$

(Art 123, Ex. 5.)

or
$$x^2 + y^2 - \frac{2c^2x'^3x}{a^4} + \frac{2c^2y'^3y}{b^4} + a'^2 - 2b'^2 = 0. \qquad (552)$$

6. Through any point in the plane of a conic can be described six osculating circles of the conic.

7. In the last exercise, show that if the point be on the conic, there can be described only three circles distinct from the osculating circle at the point.

Observation.—The theorem Ex. 7 is STEINER'S (see SALMON'S *Conics*, p. 229). The extension of it in Ex. 6 was first given in the Author's memoir on *Bicircular Quartics*, 1869.

8. If S, S' be two conics; α, β a pair of their chords of intersection, such that $S - S' = \alpha\beta$; then

$$k^2\alpha^2 - 2k(S + S') + \beta^2 = 0 \qquad (553)$$

represents a conic having double contact with S, S'. For it may be written either

$$(k\alpha + \beta)^2 - 4kS = 0, \quad \text{or} \quad (k\alpha - \beta)^2 - 4kS' = 0.$$

(SALMON.)

9. If a variable ellipse have four-pointic contact with a fixed ellipse at the extremity of its minor axis, the locus of its foci is a circle whose radius is equal to half the radius of curvature.

10. The general equation of a conic, having double contact with S and $S + L^2 + M^2$, is

$$S - (L\cos\alpha + M\sin\alpha)^2 = 0. \qquad (554)$$

(CROFTON.)

11. If a conic have double contact with two others which have the same focus and directrix, the chords of contact pass through the focus, and are perpendicular to each other. (*Ibid.*)

12. From any point P on an outer confocal tangents are drawn to an inner; prove that the ellipse passing through P, which has the points of contact as foci, has four-pointic contact with the outer confocal.

(CROFTON.)

13. The latus rectum of the variable ellipse, Ex. 12, is constant.

14. The equation of a conic having double contact with $A^2x^2 + B^2y^2 = C$, and $A^2x^2 + B^2y^2 = D$ is

$$Ax^2 + By^2 - C = \left(1 - \frac{C}{D}\right)(Ax\cos\theta + By\sin\theta)^2. \qquad (555)$$

15. If $\alpha \pm \beta \pm \gamma = 0$ be the sides of a quadrilateral, the conic

$$\alpha^2 \sec^2\phi + \beta^2 \operatorname{cosec}^2\phi - \gamma^2 = 0$$

touches its four sides; or again, the conic

$$k^2\alpha^2 - 2k(\alpha^2 + \beta^2 + \gamma^2) + \beta^2 = 0 \text{ touches the four sides.} \qquad (556)$$

16. To find the foci of a conic given by the general equation.

Let x', y' be the co-ordinates of a focus, then the imaginary line $(x - x') + (y - y')\sqrt{-1}$ will (Art. 150) be a tangent. Hence, comparing this with $\lambda x + \mu y + \nu$, we have to substitute for λ, μ, ν, in equation (320), $1, \sqrt{-1}$, and $-(x' + y'\sqrt{-1})$; and we get, after omitting accents, and equating real and imaginary parts to zero, the two equations—

$$(Cx - G)^2 - (Cy - F)^2 = \Delta(a - b), \qquad (557)$$

$$(Cx - G) \cdot (Cy - F) = \Delta(h), \qquad (558)$$

which determine the foci. (SALMON.)

17. If S, S' represent two circles, prove that $S^{\frac{1}{2}} \pm S'^{\frac{1}{2}} - k = 0$ has double contact with each.

18. If two conics have each double contact with a third, their chords of contact with the third conic and a pair of their chords of intersection with each other form a harmonic pencil.

19. The diagonals of a quadrilateral inscribed in a conic, and the diagonals of the quadrilateral formed by tangents at its angular points, form a harmonic pencil.

20. If three conics Σ, Σ', Σ'' have each four-pointic contact with a given conic, and contact of the first order with each other, taken two by two; prove that the triangle formed by the points of contact of Σ, Σ', Σ'' with each other is inscribed in the triangle formed by their points of contact with S, and in perspective with it, and also with the triangle formed by the tangent at the points of contact on S. (CROFTON.)

21. If three conics have each double contact with a fourth, six of their chords of intersection are, three by three, concurrent.

22. If a hexagon be described about a conic, the three lines joining opposite angular points are concurrent. (BRIANCHON.)

23. A conic is described touching a fixed conic at P, and passing through its foci S, S'; prove that the pole of SS' with respect to this conic will be on the normal at P, and will coincide with the centre of curvature if the conics osculate.

24. If a parabola have double contact with a given ellipse, and have its axis parallel to a given line, the locus of its focus is a hyperbola confocal with the ellipse, and having one asymptote in the given direction.

25. If a variable conic having double contact with a fixed conic pass through two given points, the chord of contact passes through one or other of two given points. (SALMON.)

26. Three conics which have double contact are met by three of their non-concurrent common chords in six points, which lie on a conic.

(*Ibid.*)

27. If an ellipse have double contact with each of two confocals, the tangents at the points of contact form a rectangle.

28. If the asymptotes of a hyperbola coincide in direction with the equiconjugate diameter of an ellipse; prove that the hyperbola cuts orthogonally all conics passing through the ends of the axes of the ellipse.

29. Two parabolae osculate a circle, and meet it again in two points P, P'; prove that the angle between their axes is one-fourth of the angle subtended by PP' at the centre of the circle.

30. The centre of curvature at any point of an ellipse is the pole of the tangent at the same point with respect to the confocal hyperbola passing through it.

31. The focal chord of curvature at any point of a conic is double the harmonic mean between the focal radii at the same point.

(SALMON.)

32. The locus of the centre of an equilateral hyperbola, having contact of the third order with a given parabola, is an equal parabola.

33. If two tangents be drawn to an ellipse from any point of a confocal ellipse, the excess of the sum of these two tangents over their intercepted arc is constant. (GRAVES.)

34. Find the lengths of the axes of a conic given by the general equation

$$ax^2 + 2hxy + by^2 + 2gx + 2fy + c = 0.$$

Transforming to the centre as origin, we get (Art. 94, *Cor.* 4),

$$ax^2 + 2hxy + by^2 + \frac{\Delta}{C} = 0.$$

Now, if the auxiliary circle be $x^2 + y^2 - r^2$, it has double contact with the conic. Hence, putting c' for $\frac{\Delta}{C}$, and eliminating the constants, we get

$$(ar^2 + c')x^2 + 2hr^2xy + (br^2 + c')y^2 = 0,$$

which must be a perfect square; therefore the roots of the equation

$$(ab - h^2)r^4 + (a + b)c'r^2 + c'^2 = 0,$$

or

$$C^3r^4 + (a + b)C\Delta r^2 + \Delta^2 = 0 \tag{559}$$

are the squares of the roots of the semi-axes.

35. If the conic given by the general equation be an ellipse, its area is

$$\frac{\pi\Delta}{C^{\frac{3}{2}}}. \tag{560}$$

36. If two tangents TP TQ be drawn to an ellipse from any point T in a confocal hyperbola, which cuts the intercepted arc of the ellipse in K; prove that the difference of the arcs PK, KQ is equal to the difference between the tangents PT, TQ. (M'CULLAGH.)

37. If a variable conic has four-pointic contact with a fixed conic, and also touch its directrix; prove that the chord of contact passes through the focus of the fixed conic. (CROFTON.)

38. The co-ordinates of a point in an elliptic quadrant which divides it into parts, whose difference is equal to the difference of the semi-axes, are

$$\left(\frac{a^3}{a + b}\right)^{\frac{1}{2}}, \quad \left(\frac{b^3}{a + b}\right)^{\frac{1}{2}}. \tag{561}$$

39. Show that the locus of points on a system of confocal conics, the circles of curvature at which pass through one of the foci, is a circle of which the foci are inverse points. (MR. F. PURSER, F.T.C.D.)

40. Prove that for a system of conics, having the same focus and directrix, this locus is a parabola. (*Ibid.*)

41. From a fixed point O a tangent OT is drawn to one of a system of confocal conics, and a point P taken on the tangent, such that $OP \cdot OT$ is constant; prove that the locus of P is an equilateral hyperbola.

(PROF. J. PURSER.)

42. If a polygon circumscribe a conic, and if the loci of all the vertices but one be confocal conics, the locus of the remaining vertex is a confocal conic.

It will be sufficient to prove this proposition in the case of a triangle, as the proof for the triangle can be extended to the polygon.

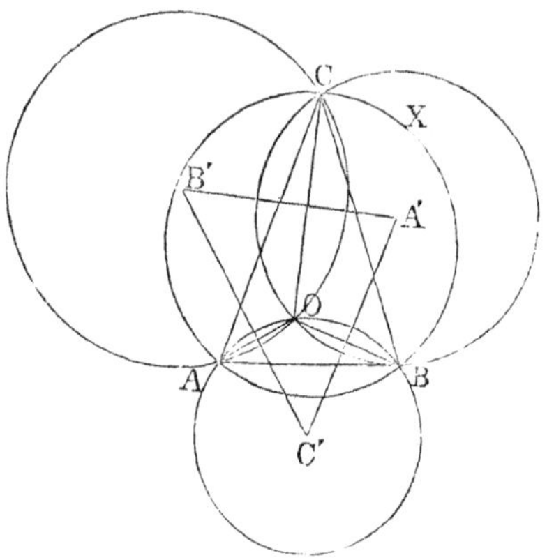

Let ABC be a triangle inscribed in a circle X; then (Sequel, VI., Sect. V., Prop. 12) if the envelopes of two sides of ABC be coaxal circles, the envelope of the third side is a coaxal circle. Now let O be one of the limiting points, and describe circles about the triangles OAB, OBC, OCA; let their centres be C', A', B'; then (Sequel, VI., Sect. V., Prop. 8, *Cor.* 4) the envelopes of these circles are circles concentric with X, and the loci of their centres A', B', C' are conics whose foci are O, and the centre of X; that is, they are confocal conics. Also, since the lines OA, OB, OC are bisected perpendicularly by the sides of the triangle $A'B'C'$, that triangle is circumscribed to a triangle whose foci are O, and the centre of X. Hence the proposition is proved.

43. If the base of a triangle and its vertical angle be given, the locus of its symmedian point is an ellipse having double contact with the circumcircle.

44. If the conic $\alpha\beta - k\gamma^2 = 0$ touch the circle $\alpha\beta \sin C + \beta\gamma \sin A + \gamma\alpha \sin B = 0$, the point of contact is on one of the symmedians of the triangle ABC.

(BROCARD.)

Section II.—Similar Figures.

Def.—*If from the circumcentre O of a triangle ABC three perpendiculars be drawn to its sides, the points A′, B′, C′, in which they meet the circle described on the join of O to the symmedian point K as diameter, called the Brocard circle, form a triangle, which we shall call* Brocard's first triangle.

151. *Brocard's first triangle is inversely similar to the triangle ABC.*

Dem.—Since OA' is perpendicu'ar to BC and OB' to AC, the angle $A'OB'$ is equal to BCA; but $A'OB'$ is equal to

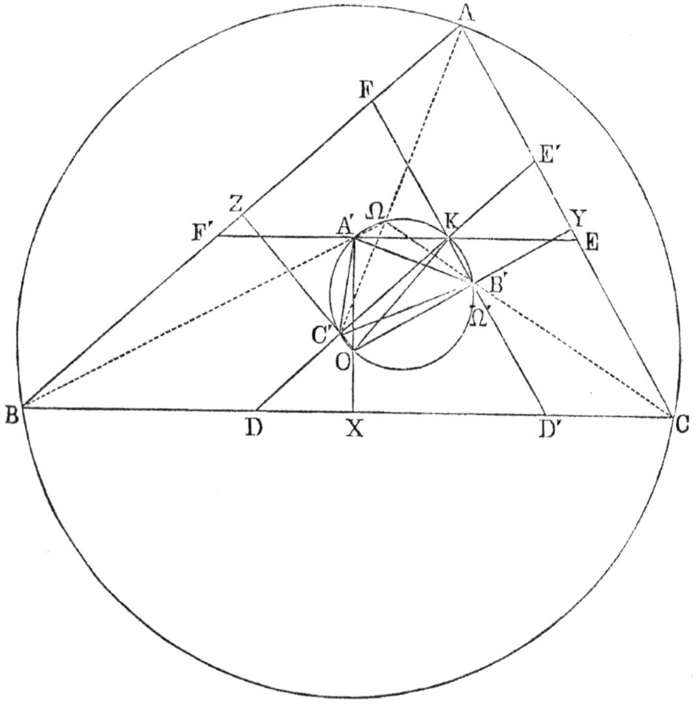

$A'C'B'$. Hence the angle $A'C'B'$ is equal to ACB. In like manner, the other angles are equal, and the triangles have evidently different aspects. Hence they are inversely similar.

Cor. 1.—If K be the symmedian point, the lines $A'K$, $B'K$, $C'K$, are parallel to the sides of the triangle.

Cor. 2.—The three lines $A'B$, $B'C$, $C'A$ are concurrent, and meet on the Brocard circle of the triangle.

For, produce BA', CB' to meet in Ω; then, since K is the symmedian point, the perpendiculars from K on the sides of the triangle ABC are proportional to its sides; but these perpendiculars are equal to $A'X$, $B'Y$, $C'Z$, respectively. Hence $A'X : B'Y :: BX : CY$, and the triangles $A'BX$, $B'CY$ are equiangular; therefore the angle $BA'X$ is equal to $CB'Y$, that is equal to $OB'\Omega$. Hence (*Euclid*, III. XXII.) the points A', O, B', Ω are concyclic; therefore $A'B$, $B'C$ meet on the Brocard circle. Hence the proposition is evident.

Cor. 3.—It may be shown in a similar way that the lines AB', BC', CA' are concurrent, and meet in another point Ω' on the Brocard circle.

Cor. 4.—The six angles ΩAB, ΩBC, ΩCA, $\Omega' BA$, $\Omega' CB$, $\Omega' AC$ are equal.

DEF.—*If the common value of the angles* ΩAB, *&c., be denoted by* ω, ω *is called the* BROCARD ANGLE *of the triangle, and* Ω, Ω' *the* BROCARD POINTS.

Cor. 5.—To find the value of the Brocard angle. Since the lines $A\Omega$, $B\Omega$, $C\Omega$ are concurrent, we have

$$\sin(A-\omega)\sin(B-\omega)\sin(C-\omega)=\sin^3\omega.$$

Hence $$\cot\omega=\cot A+\cot B+\cot C;$$

or $$\operatorname{cosec}^2\omega=\operatorname{cosec}^2 A+\operatorname{cosec}^2 B+\operatorname{cosec}^2 C.$$

(HYMER'S *Trigonometry*, p. 141.)

DEF.—*If the Brocard circle of the triangle* ABC *meet its symmedian lines again in the points* A'', B'', C'', $A''B''C''$ *is called* BROCARD'S SECOND TRIANGLE.

152. *If three figures directly similar be described on the sides of the triangle* ABC, *their centres of similitude* (*Euclid*, VI. XX., Ex. 2), *taken in pairs, are the vertices of Brocard's second triangle.*

Dem.—Join OA'', $A''B$, $A''C$, and let AA'' meet the circum-circle in T. Now, since OK is the diameter of the Brocard circle, the angle $OA''K$ is right. Hence AT is bisected in

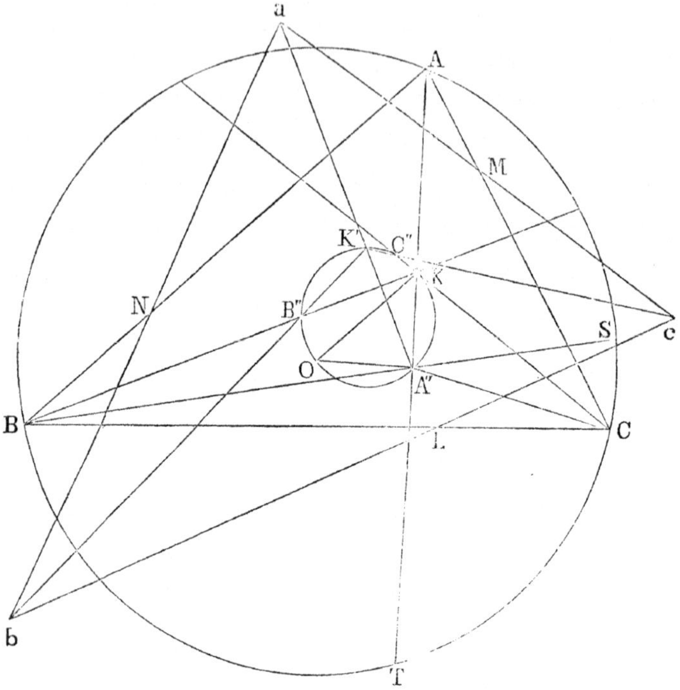

A''; therefore A'' is the focus of the parabola touching AB, AC in the points B, C (Exs. 11, 12, p. 144); the triangle $BA''A$ is directly similar to the triangle $AA''C$. Hence the proposition is proved.

Cor. 1.—If figures directly similar be described on the sides of the triangle ABC, the symmedian lines of the triangle (abc) formed by any three corresponding lines pass respectively through the vertices of Brocard's second triangle. For since A'' is the centre of similitude of the figures described on BA, AC, and that ba, ac are corresponding lines in these figures, A''a divides the angle bac into parts equal to those into which $A''A$ divides BAC; therefore $A''a$ is a symmedian line of the triangle bac. Hence the proposition is proved.

Cor. 2.—The symmedian point of the triangle bac is on the Brocard circle of BAC (M'Cay.)

For, from the similarity of figures, the angle of intersection of the lines $A''a$ and $B''b$ is equal to the angle of intersection $A''A$, $B''B$. Hence the angle $A''K'B'' = A''KB''$. Hence K' is on the Brocard circle.

Cor. 3.—The lines through K' parallel to the sides of abc pass through the vertices of Brocard's first triangle.

Cor. 4.—If the area of abc be given, the envelope of each side is a circle, the centres being the vertices of Brocard's first triangle. This follows at once from the similarity of the triangles abc, ABC, and *Cor.* 3.

Cor. 5.—The centre of similitude of ABC, abc is a point on the Brocard circle of ABC; for, since AK, aK', corresponding lines of the two figures, meet in A'', their centre of similitude is the point of intersection of the circumcircles of the triangles $A''KK'$, $A''A$a (*Euclid*, VI. xx., Ex. 2.)

Cor. 6.—In the same manner, it may be shown that the centre of similitude of any two triangles, each formed by three corresponding lines of figures directly similar, described on the sides of the triangle ABC, is a point on its Brocard circle.

153. *Properties of corresponding points of similar figures.*

1°. *If figures directly similar be described on the sides of the triangle ABC, and if the join of two homologous points A', B' of these figures pass through a given point hk, the locus of each point is a circle.*

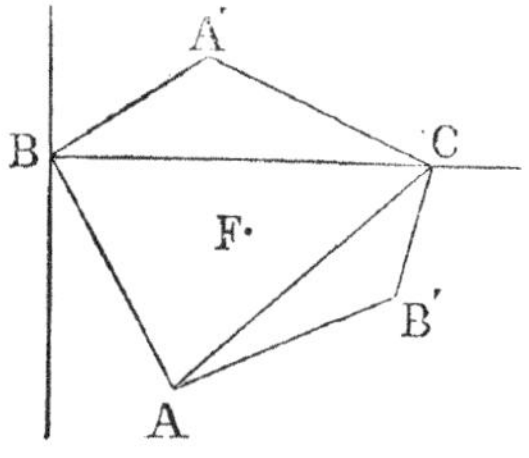

Dem.—Taking B as origin, BC as axis of x, and a perpendicular to it as axis of y; then, from the hypothesis, the triangles $BA'C$, $CB'A$ are directly similar;

therefore the angles CBA', ACB' are equal. Hence, denoting each by θ, and BA', CB' by ρ, ρ' respectively, from the conditions of the question we get,

$$\begin{vmatrix} \rho\cos\theta, & \rho\sin\theta, & 1, \\ a-\rho'\cos(C+\theta), & -\rho'\sin(C+\theta), & 1, \\ h, & k, & 1 \end{vmatrix} = 0;$$

or, expanded, and reduced by putting $\rho' = \frac{b}{a}\rho$, and turning into Cartesian co-ordinates,

$$b\sin C(x^2+y^2)+(ak+bk\cos C-bh\sin C)x+(a^2-ah$$
$$-bh\cos C-bk\sin C)y-a^2k=0.$$

This circle passes through hk; for if we put $\rho\cos\theta$, $\rho\sin\theta$ for hk, the determinant will have two rows alike.

Cor.—If hk be the centroid of the triangle ABC, the circle will be

$$x^2+y^2-ax+\frac{a\cot\omega}{3}y+\frac{a^2}{3}=0, \qquad (562)$$

where ω is the Brocard angle of the triangle.

2°. *If the lines joining the points A, A', B, B' be parallel, the loci of A', B' are circles.* (NEUBERG.)

The equations of the lines AA', BB' are respectively

$$(\rho\sin\theta+C\sin B)x-(\rho\cos\theta-C\cos B)y-\rho C\sin(B+\theta)=0,$$

$$\rho'\sin(C+\theta)x+(a-\rho'(\cos C+\theta)y=0.$$

Hence, from the condition of parallelism, and reducing as before, we get

$$x^2+y^2-ax+a\cot\omega\,.\,y+a^2=0.$$

This circle is the locus of the point A', when the Brocard angle of the triangle BCA' is equal to that of BCA. (Compare Chap. III., Ex. 74.)

In the same manner, it may be shown that the loci of A', B' are circles, if AA', BB' be inclined to each other at a given angle.

3°. *Upon the sides of ABC are described three triangles directly similar, viz. ABC', BCA', CAB'; it is required to investigate in what cases the triangles ABC, A'B'C' are in perspective.*

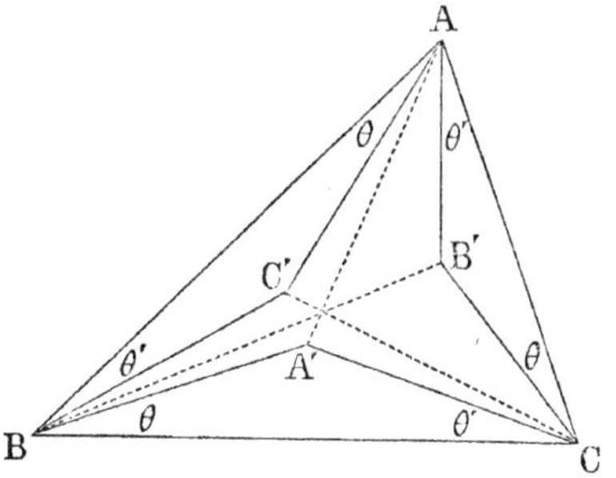

Let α, β, γ be the trilinear co-ordinates of the centre of perspective, θ, θ' the base angles of the triangles; then we have, evidently,

$$\alpha : \beta :: BC' \sin(B-\theta') : AC' \sin(A-\theta) :: \sin\theta \, . \, \sin(B-\theta') : \sin\theta' \, . \, \sin(A-\theta);$$

therefore

$$\alpha \sin A \cot\theta - \beta \sin B \cot\theta' - (\alpha \cos A - \beta \cos B) = 0.$$

Hence, eliminating θ, θ' from this and two similar equations, we get

$$\begin{vmatrix} \alpha \sin A, & \beta \sin B, & \alpha \cos A - \beta \cos B, \\ \beta \sin B, & \gamma \sin C, & \beta \cos B - \gamma \cos C, \\ \gamma \sin C, & \alpha \sin A, & \gamma \cos C - \alpha \cos A \end{vmatrix} = 0.$$

Substituting for the second column the difference between the first and second, and then adding the second and third rows to the first, we get a result which may be written

$$(\alpha \sin A + \beta \sin B + \gamma \sin C) \begin{vmatrix} 1, & 0, & 0, \\ \beta \sin B, & \beta \sin B - \gamma \sin C, & \beta \cos B - \gamma \cos C, \\ \gamma \sin C, & \gamma \sin C - \alpha \sin A, & \gamma \cos C - \alpha \cos A \end{vmatrix} = 0;$$

or

$$(\alpha \sin A + \beta \sin B + \gamma \sin C)(\alpha\beta \sin(A-B) + \beta\gamma \sin(B-C) + \gamma\alpha \sin(C-A)) = 0. \qquad (563)$$

Hence, if the triangle ABC and that formed by three corresponding points be in perspective, the locus of the centre of perspective is either the line at infinity or the equilateral hyperbola

$$\alpha\beta \sin (A - B) + \beta\gamma \sin (B - C) + \gamma\alpha \sin (C - A) = 0,$$

called KIEPERT'S HYPERBOLA, *Nouv. Annales, t.* VIII., 1869, pp. 40–42.

In the former case the lines AA', BB', CC' are parallel, and the locus (2°) of each point A', B', C' is a Neuberg circle.

Again, if we add the equation

$$\alpha \sin A \cot \theta - \beta \sin B \cot \theta' - (\alpha \cos A - \beta \cos B),$$

and the two similar ones got by interchanging letters, we get

$$(\cot \theta - \cot \theta')(\alpha \sin A + \beta \sin B + \gamma \sin C) = 0.$$

Hence, if the triangles be in perspective, either

$$\alpha \sin A + \beta \sin B + \gamma \sin C = 0,$$

as found before, or $\cot \theta = \cot \theta'$, and the three similar triangles will be isosceles; and we have the following theorem, due to KIEPERT:—*If upon the three sides of a triangle ABC similar isosceles triangles be described, the triangle formed by their vertices is in perspective with ABC, and the locus of their centre of perspective is an equilateral hyperbola.*

4°. *If the distance $A'B'$ of two corresponding points be given, the locus of each point is a circle.*

If m be the length of the line $A'B'$, the conditions of the question give us

$$(\rho \cos \theta + \rho' \cos (C + \theta) - a)^2 + (\rho \sin \theta + \rho' \sin (C + \theta))^2 = m^2.$$

Hence, reducing, &c., we get

$$(x^2 + y^2)(a^2 + b^2 + 2ab \cos C) - 2a^2(a + b \cos C)x$$
$$+ 2a^2 b \sin Cy + a^2 (a^2 - m^2) = 0.$$

If m vanish, the two points will coincide, and the circle will be a point, viz.,

$$(x^2+y^2)(a^2+b^2+2ab\cos C)-2a^2(a+b\cos C)x$$
$$+\,2a^2b\sin C\,.\,y+a^4=0.$$

This will be the point circle, which is the centre of similitude A'' (Art. 152) of figures described on the lines BA, AC.

5°. *If the ratio of $B'A' : A'C'$ be given, the locus of each point is a circle.*

If the ratio be $k:1$, since the co-ordinates of the point C' are evidently $c\cos B-\rho''\cos(B-\theta)$, $\rho''\sin(B-\theta)-c\sin B$, where ρ'' denotes AC', as in 4°, we get

$$\frac{(x^2+y^2)(a^2+b^2+2ab\cos C)-2a^2(a+b\cos C)x+2a^2b\sin C\,.\,y+a^4}{(x^2+y^2)(a^2+c^2+2ac\cos B)-2ac(c+a\cos B)x+2a^2c\sin B\,.\,y+a^2c^2}=\frac{k^2}{1};$$

or denoting the equations of the point circles A'', B'', by S, S',

$$S-k^2S'=0,$$

which, if k vary, denotes a coaxal system whose limiting points are two of the vertices of Brocard's second triangle.

6°. *If the area of the triangle formed by three corresponding points be given, the locus of each point is a circle.*

Dem.—Denoting the area of the triangle by Δ', the conditions of the question give us

$$\begin{vmatrix} \rho\cos\theta, & \rho\sin\theta, & 1, \\ a-\rho'\cos(C+\theta), & -\rho'\sin(C+\theta), & 1, \\ c\cos B-\rho''\cos(B-\theta), & \rho''\sin(B-\theta)-c\sin B, & 1 \end{vmatrix}=-2\Delta';$$

or reduced,

$$(ab\sin C+bc\sin A+ac\sin B)(x^2+y^2-ax)$$
$$+(ab\cos C+bc\cos A+ac\cos B)y=2a^2(\Delta'-ac\sin B).$$

Hence, if the area of the triangle ABC be Δ, the locus of A is

$$x^2+y^2-ax+\frac{a\cot\omega}{3}y+\frac{a^2}{3}\left(1-\frac{\Delta'}{\Delta}\right)=0. \qquad (564)$$

Cor. 1.—From the foregoing values of the co-ordinates of A', B', C', it follows at once that the centroid of $A'B'C'$ coincides with that of ABC.

Cor. 2.—Since the area of the triangle $A'B'C'$ may be taken either as positive or negative, the locus will consist of two concentric circles.

Cor. 3.—If the area of the triangle $A'B'C'$ be zero, the points A', B', C' will be collinear; the line of collinearity will pass through the centroid of ABC. The locus of A' will be

$$x^2 + y^2 - ax + \frac{a \cot \omega}{3} y + \frac{a^2}{3} = 0.$$

(Compare 1°, *Cor.*) (565)

Cor. 4.—If the point A in the diagram (Art. 153, 1°) were on the positive side of BC, the equation would be

$$x^2 + y^2 - ax - \frac{a \cot \omega}{3} + \frac{a^2}{3} = 0.$$

154. M'Cay's Circles.—

When three corresponding points are collinear, the circles which are their loci possess several interesting properties. Their most complete discussion, due to Mr. M'Cay, F.T.C.D., is published in the *Transactions of the Royal Irish Academy*, vol. xxviii., pp. 453–470. They have been also studied by Neuberg, who, in a letter to the author, has called them M'Cay's circles. We shall give here a few of their most important properties.

1°. *If in the equation* (565) *we substitute the co-ordinates of the centroid of the triangle, the equation is satisfied. Hence each of* M'Cay's *circles passes through the centroid of* ABC.

2°. Denoting the three circles by α, β, γ, and their centres by X, Y, Z, we have, from the equation,

$$OX = \frac{a \cot \omega}{6}, \quad \text{and } OA' = \frac{a \tan \omega}{2}.$$

Hence $OX \,.\, OA = \frac{a^2}{12}$, equal to the power of the point O with respect to α.

Therefore $A'K$ is the polar of Q with respect to α, and A' is the polar of BC with respect to α. Hence we have the

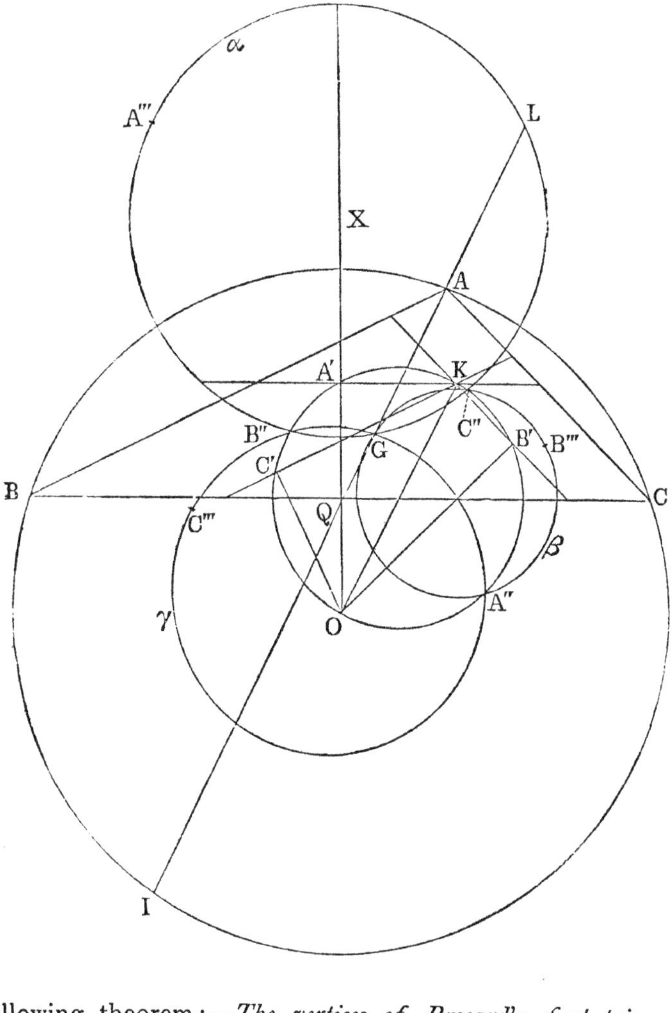

following theorem:—*The vertices of Brocard's first triangle are, with regard to* M'CAY'S *circles, respectively, the poles of the sides of the original triangle.*

DEF.—*If* $A''G$, $B''G$, $C''G$ *be produced to meet* M'CAY'S *circles in the points* A''', B''', C''', *respectively,* $A'''B'''C''$ *is* BROCARD'S THIRD TRIANGLE. *Its linear dimensions are evidently equal to twice those of* $A''B''C''$.

3°. *If through the point* G *a tangent be drawn to any of* M'CAY'S *circles, the intercepts made on it by the other two circles are equal.*

This and the preceding theorem are immediate inferences, from the fact that G is the mean centre of the points in which any transversal through G meets the circles.

4°. If through G we draw any transversal, cutting the circles in three points, the tangents to the circles at these points will be corresponding lines, and will meet the sides of the triangle ABC in corresponding points. Hence we infer the following theorem :—*The polars of corresponding points on the sides of the triangle* ABC, *taken respectively with regard to* M'CAY'S *circles, are three concurrent lines, and the locus of their point of concurrence is the Brocard circle of the triangle.*

5°. If we take the middle point O of BC as origin, the equation of the circle α will be

$$x^2 + y^2 - \frac{a \cot \omega}{3} + \frac{a^2}{12} = 0 ;$$

and invert this with respect to the circle on BC as diameter, and we get NEUBERG'S circle

$$x^2 + y^2 - a \cot \omega \cdot y + \frac{3a^2}{4}$$

(Chap. III., Ex. 75). *Hence the circle described on any side of the triangle as diameter, and the corresponding* M'CAY'S *and* NEUBERG'S *circles, are coaxal.*

6°. *The circle* α *is the locus of the centroids of all triangles described on* BC, *having their Brocard angles equal to that of the triangle* ABC.

7°. *If the median* OA *meet the circle* α *in* L, *and the circumcircle in* I, OL *is equal to* OI, *and* L *is the foot of the perpendicular let fall from the orthocentre on the median.*

8°. *The two rectangular lines, joining G to the highest and lowest points of the circle* α, *are the rectangular axes of the maximum ellipse inscribrd in ABC.*

For these axes are the rectangular pair of the pencil in involution determined by GA, GB, GC, and parallels through G to the sides. The intersection of this pencil with the circle α may be seen to be in perspective at A', whence the above result.

155. DEF.—*Two figures are said to be homothetic when radii vectores from any point of the first figure are proportional to parallel vectores from the homologous point in the second.*

Two conics being given by their general equations, it is required to find the condition that they may be homothetic. If the equations of both conics be transferred to their centres as origins, they will be, respectively, of the forms

$$ax^2 + 2hxy + by^2 + c = 0, \quad a'x^2 + 2h'xy + b'y^2 + c' = 0;$$

or in polar co-ordinates,

$$\rho^2 = -\frac{c}{a\cos^2\theta + 2h\sin\theta\cos\theta + b\sin^2\theta},$$

$$\rho'^2 = \frac{c'}{a'\cos^2\theta + 2h'\sin\theta\cos\theta + b'\sin^2\theta}.$$

Now, in order that ρ, ρ' may have a ratio independent of θ, it is evident that we must have

$$\frac{a}{a'} = \frac{h}{h'} = \frac{b}{b'},$$

which is the required condition.

156. If the conics be similar, but not homothetic, it is plain that if the axes of co-ordinates for one be turned round through a certain angle, the new coefficients a, h, b will be proportional to the old coefficients a', h', b'. Suppose this done, and that they become ka', kh', kb'; then, from the property of invariants, we have, for rectangular co-ordinates,

$$(a + b) = k(a' + b'), \quad ab - h^2 = k^2(a'b' - h'^2).$$

Hence, eliminating k, the required condition is

$$\frac{(a+b)^2}{ab-h^2} = \frac{(a'+b')^2}{a'b'-h'^2}; \qquad (566)$$

and, for oblique co-ordinates,

$$\frac{(a+b)-2h\cos\omega)^2}{ab-h^2} = \frac{(a'+b'-2h'\cos\omega')^2}{a'b'-h'^2}.$$

EXAMPLES.

1. The intercepts made on the sides of the triangle by Lemoine's parallels are proportional to $\sin(A-\omega)$, $\sin(B-\omega)$, $\sin(C-\omega)$.

2. The lines AA', BB', CC' (see figure, Art. 151) are concurrent; the co-ordinates of the point of concurrence are a^{-3}, b^{-3}, c^{-3}.

3. Prove that Brocard's triangles $A'B'C'$, $A''B''C''$ are in perspective; that their centre of perspective is the centroid of ABC, and the axis of perspective its polar with respect to the Brocard circle.

4. If the locus of the point a (fig. Art. 152) be a right line, the envelopes of the lines ba, ac are parabolae.

5. If $A,' B'$, C' be three corresponding points, and if A' describe the line $\lambda x + \mu y + 1 = 0$, prove that the points B', C' describe respectively the lines

$$(\lambda a\cos C - \mu a\sin C)x + (\lambda a\sin C + \mu a\cos C)y + \mu a^2\sin C - \lambda a^2\cos C - b = 0,$$

$$(\lambda a\cos B + \mu a\sin B)x + (\lambda a\sin B - \mu a\cos B)y - C - \lambda ac = 0.$$

6. If m_a, m_b, m_c denote the medians of the triangle ABC, Δ its area; prove that the parameters of the three parabolae which can be described, each touching two sides, and having the third as chord of contact, are respectively

$$\frac{2\Delta^2}{m_a{}^3}, \quad \frac{2\Delta^2}{m_b{}^3}, \quad \frac{2\Delta^2}{m_c{}^3}.$$

7. The vertices A'', B'', C'' of Brocard's second triangle are the foci respectively of three parabolae whose directrices are the medians of ABC, and which are inscribed in the quadrilateral formed by the internal and external bisectors of an angle, and the perpendiculars at the middle points of its sides. (ARTZT.)

8. The parameters of the three parabolae of Ex. 7 are respectively

$$\frac{\Delta(b^2-c^2)}{2m_a{}^3}, \quad \frac{\Delta(c^2-a^2)}{2m_b{}^3}, \quad \frac{\Delta(a^2-b^2)}{2m_c{}^3}. \qquad (Ibid.)$$

9. If through the Brocard point Ω we describe three circles, each passing respectively through the angular points of ABC, the triangle formed by their centres has the circumcentre of ABC for one of its Brocard points. (DEWULF.)

10. If by means of the other Brocard point Ω' we describe in the same manner another triangle, the two triangles will be in perspective; the circumcentre of ABC will be their centre of perspective. (*Ibid.*)

11. If the sum of the squares of the sides of the triangle $A'B'C'$, formed by three corresponding points (Art. 153), be given, the locus of each is a circle.

12. In the same case, if any angle of the triangle $A'B'C'$ or the difference of the squares of any two sides be given, the locus of each point is a circle.

13. Upon a given line, and on the same side of it BC, can be described six triangles equiangular to a given triangle: prove that their six vertices are concyclic. (NEUBERG.)

14. If h_1, h_2 be the altitudes of the highest and lowest points of M'Cay's circle (α); and if we put

$$h_1 = \tfrac{1}{2}a \tan \phi_1, \quad h_2 = \tfrac{1}{2}a \tan \phi_2;$$

prove that ϕ_1, ϕ_2 are the roots of the quadratic

$$\sin (2\phi + \omega) = 2 \sin \omega. \qquad \text{(M'CAY.)}$$

15. In the same case, prove

$$\frac{\sin A}{\cos (A + \phi)} + \frac{\sin B}{\cos (B + \phi)} + \frac{\sin C}{\cos (C + \phi)} = 0. \qquad (\textit{Ibid.})$$

16. Prove that the Brocard angle (ω) satisfies the equation

$$\sin A \cos (A + \omega) + \sin B \cos (B + \omega) + \sin C \cos (C + \omega) = 0.$$

(NEUBERG.)

17. A circle touching an ellipse passes through its centre: prove that the locus of the foot of the perpendicular from the centre of the ellipse on the chord of intersection is a concentric and homothetic ellipse.

18. All parabolae are similar figures.

19. If two conics be homothetic and concentric; prove that they have double contact at infinity.

20. A variable triangle of given species is described with its vertices lying on three given lines: show that its circumcircle has double contact with a given conic.

21. If $S = 0$ be the equation of a conic, $\beta = 0$ the equation of a line, k any constant; then $S - k\beta = 0$ is the equation of a homothetic conic.

22. If three figures be homothetic, two by two, their homothetic centres are collinear.

23. Pairs of tangents drawn to a conic S are parallel to pairs of conjugate diameters of a conic S': prove that the locus of their point of intersection is a conic homothetic with S'.

24. If ϕ_1, ϕ_2 have the same signification as in Ex. 14, prove that the angles $\omega + 2\phi_1$, $\omega + \phi_2$ are supplemental. (M'CAY.)

25. Prove that M'Cay's circles are respectively the inverses of the three sides of Brocard's first triangle, with respect to the circle whose centre is G (fig. Art. 154), and which cuts Brocard's circle orthogonally.

26. If δ be the diameter of Brocard's circle; prove that the diameters of Neuberg's circles are respectively

$$2\delta \sin A \cot \omega, \quad 2\delta \sin B \cot \omega, \quad 2\delta \sin C \cot \omega.$$

27. If L, L' be the limiting points of M'Cay's circle α and the side a of the triangle ABC; M, M' of the circle β and the side b; N, N' of the circle γ and the side c; prove that the triangles LMN, $L'M'N'$ are equilateral.

28. If $S = 0$ be the equation of any conic; prove that the equation of a homothetic conic passing through three given points is

$$\begin{vmatrix} S, & \alpha, & \beta, & \gamma, \\ S', & \alpha', & \beta', & \gamma', \\ S'', & \alpha'', & \beta'', & \gamma'', \\ S''', & \alpha''', & \beta''', & \gamma''' \end{vmatrix} = 0. \qquad (567)$$

SECTION III.—THE GENERAL EQUATION—TRILINEAR CO-ORDINATES.

157. ARONHOLD'S NOTATION.—1°. In this notation, a point is denoted by a single letter, and its trilinear co-ordinates by the same letter, with suffixes. Thus the point x is the point whose co-ordinates are x_1, x_2, x_3.

2°. The trilinear equation of a right line, viz., $a_1x_1 + a_2x_2 + a_3x_3 = 0$, is denoted by $a_x = 0$, the x being a suffix to a.

3°. The general equation of the nth degree is denoted by $a_x^n = 0$; that is, by $(a_1x_1 + a_2x_2 + a_3x_3)^n$, where after the involution a_1^n is replaced by the coefficient of x_1^n in the given equation, $na_1^{n-1}a_2$ by the coefficient of $x_1^{n-1}x_2$, &c. Thus the conic $a_{11}x_1^2 + a_{22}x_2^2 + a_{33}x_3^2 + 2a_{12}x_1x_2 + 2a_{23}x_2x_3 + 2a_{31}x_3x_1 = 0$ is denoted by $(a_1x_1 + a_2x_2 + a_3x_3)^2$, or $a_x^2 = 0$. It is evident that in this notation the symbols a_1, a_2, a_3 have no meaning of themselves for curves of the second or higher degree, until the involution is performed.—SALMON'S *Algebra*, p. 267; CLEBSCH, *Theorie der Binären Algebraeschen Formen.*

4°. Any non-homogeneous equation in two co-ordinates may be transformed into a homogeneous equation by the substitutions $x_1 \div x_3$, $x_2 \div x_3$ for the variables and the clearing of fractions.

158. Several well-known results assume a very simple form when expressed in ARONHOLD'S notation. We shall merely state them here, as they present no difficulty.

1°. JOACHIMSTHAL'S equation (297), which gives the ratio in which the join of the points y, z is divided by the conic $a_x^2 = 0$ is

$$a_y^2 + 2ka_y \,.\, a_z + k^2a_z^2 = 0. \qquad (568)$$

2°. The equation of the polar of the point y, with respect to $a_x^2 = 0$, is

$$a_x \,.\, a_y = 0. \qquad (569)$$

3°. The condition that y and z may be conjugate points, with respect to $a_x^2 = 0$, is

$$a_y \,.\, a_z = 0. \qquad (570)$$

4°. The equation of the pair of tangents, from the point y to $a_x^2 = 0$, is

$$a_y^2 \,.\, a_x^2 - (a_x \,.\, a_y)^2 = 0. \qquad (571)$$

5°. The discriminant of a_x^2 is

$$\begin{vmatrix} a_{11}, & a_{12}, & a_{13}, \\ a_{21}, & a_{22}, & a_{23}, \\ a_{31}, & a_{32}, & a_{33} \end{vmatrix} = 0; \qquad (572)$$

and the minors of this are denoted by A_{11}, A_{12}, &c.

6°. The tangential equation of $a_x^2 = 0$; that is, the condition that the line $\lambda_1 x_1 + \lambda_2 x_2 + \lambda_3 x_3 = 0$ or λ_x may be a tangent, is

$$A_\lambda^2, \text{ or } (A_1\lambda_1 + A_2\lambda_2 + A_3\lambda_3)^2 = 0. \qquad (573)$$

7°. The co-ordinates of the pole of λ_x, with respect to a_x^2, are found thus:—Let y be the pole; then, comparing the equations $\lambda_x = 0$, and $a_x \cdot a_y$, we get the identities

$$\left| \begin{array}{l} a_{11}y_1 + a_{12}y_2 + a_{13}y_3 = \lambda_1, \\ a_{21}y_1 + a_{22}y_2 + a_{23}y_3 = \lambda_2, \\ a_{31}y_1 + a_{32}y_2 + a_{33}y_3 = \lambda_3. \end{array} \right|$$

Hence, denoting the discriminant by Δ, we get

$$\Delta y_1 = \lambda_1 A_{11} + \lambda_2 A_{12} + \lambda_3 A_{13}, \text{ or } \Delta y_1 = A_1\lambda_A.$$

Similarly, $$\Delta y_2 = A_2\lambda_A, \quad \Delta y_3 = A_3\lambda_A. \qquad (574)$$

8°. The condition that two lines $\lambda_x = 0$, $\mu_x = 0$, may be conjugate lines, with respect to the conic $a_x^2 = 0$, is found by substituting in either the co-ordinates of the pole of the other. Thus, we get

$$\lambda_A \cdot \mu_A = 0. \qquad (575)$$

9°. If $a_x^2 = 0$, $b_x^2 = 0$, be two conics, it is required to find the locus of the poles with respect to $a_x^2 = 0$, of tangents to b_x^2.

The polar of the point y, with respect to a_x^2, is

$$(a_1x_1 + a_2x_2 + a_3x_3)(a_1y_1 + a_2y_2 + a_3y_3);$$

or putting $$Y_1 = a_{11}y_1 + a_{12}y_2 + a_{13}y_3, \text{ \&c.,}$$

$$Y_1x_1 + Y_2x_2 + Y_3x_3.$$

And the condition that this should be tangential to $b_x^2 = 0$ is

$$(B_1Y_1 + B_2Y_2 + B_3Y_3)^2 = 0, \text{ or } B_Y^2 = 0. \qquad (576)$$

159. *In the general trilinear equation* $a\alpha^2 + 2h\alpha\beta + b\beta^2 + 2f\beta\gamma + 2g\gamma\alpha + c\gamma^2 = 0$, *to explain the geometrical signification of the vanishing of a coefficient.*

1°. The vanishing of the coefficients of the squares of the variables has been fully explained in Art. 78.

2°. *When the coefficients of the products vanish.*

Suppose the coefficient h, for example, to vanish, then the equation becomes $a\alpha^2 + b\beta^2 + c\gamma^2 + 2f\beta\gamma + 2g\gamma\alpha = 0$. Now this will meet the line $\gamma = 0$ in the two points where the lines $a\alpha^2 + b\beta^2 = 0$ meet $\gamma = 0$; that is, in two points which are harmonic conjugates to the points where the lines $\alpha = 0$, $\beta = 0$, meet γ. Hence we have the following theorem:—*If in the general equation the coefficient of the product of any two variables vanish, the third side of the triangle of reference is cut harmonically by the other sides and the conic.*

Cor. 1.—If the coefficients of all the products vanish, each side of the triangle of reference is cut harmonically by the conic. In other words, the triangle of reference is self-conjugate with respect to the conic.

This may be shown otherwise. Let the conic be

$$l^2\alpha^2 + m^2\beta^2 - n^2\gamma^2 = 0,$$

then we have

$$(n\gamma + l\alpha)(n\gamma - l\alpha) = m^2\beta^2.$$

Hence $n\gamma + l\alpha$, $n\gamma - l\alpha$ are tangents, and β is the chord of contact, which proves the proposition.

Cor. 2.—Any point on the conic $l^2\alpha^2 + m^2\beta^2 - n^2\gamma^2 = 0$ will be common to the lines denoted by the system of determinants

$$\left\| \begin{matrix} l\alpha, & m\beta, & n\gamma, \\ \cos\phi, & \sin\phi, & 1, \end{matrix} \right\| \qquad (577)$$

each equated to zero, which may be called the point ϕ on the conic.

Cor. 3.—The equation of the join of the points $\psi + \psi'$, $\psi - \psi'$ is

$$\begin{vmatrix} l\alpha, & m\beta, & n\gamma, \\ \cos(\psi + \psi'), & \sin(\psi + \psi'), & 1, \\ \cos(\psi - \psi'), & \sin(\psi - \psi'), & 1 \end{vmatrix} = 0,$$

or $$l\alpha \cos \psi + m\beta \sin \psi - n\gamma \sin \psi' = 0. \qquad (578)$$

Hence the equation of the tangent at the point ψ is

$$l\alpha \cos \psi + m\beta \sin \psi - n\gamma = 0. \qquad (579)$$

Cor. 4.—The co-ordinates of the point of intersection of tangents at $\psi + \psi'$, $\psi - \psi'$, are

$$\frac{\cos \psi}{l}, \quad \frac{\sin \psi}{m}, \quad \frac{\cos \psi'}{n}. \qquad (580)$$

Cor. 5.—The equation of a conic referred to a focus and directrix is $x^2 + y^2 = (e\gamma)^2$, where $\gamma = 0$ denotes the directrix. Hence it is a special case of

$$l^2\alpha^2 + m^2\beta^2 - n^2\gamma^2 = 0.$$

EXAMPLES.

1. Find the values of l, m, n, in order that $l^2\alpha^2 + m^2\beta^2 + n^2\gamma^2 = 0$ may represent a circle.

Ans. $l^2 = \sin 2A, \quad m^2 = \sin 2B, \quad n^2 = \sin 2C.$

2. If the conic $l^2\alpha^2 + m^2\beta^2 + n\gamma^2 = 0$ passes through a fixed point, three other points on it are determined.

3. Find the condition that the join of the points $\psi + \psi'$, $\psi - \psi'$ should touch the conic $l'^2\alpha^2 + m'^2\beta^2 + n'^2\gamma^2 = 0$.

$$\textit{Ans.} \quad \frac{l^2 \cos^2 \psi}{l'^2} + \frac{m^2 \sin^2 \psi}{m'^2} + \frac{n^2 \cos^2 \psi'}{n'^2}. \qquad (581)$$

4. Find the co-ordinates of the pole of the line $\lambda_x = 0$, with respect to the conic

$$\sqrt{lx_1} + \sqrt{mx_2} + \sqrt{nx_3} = 0.$$

From equation (574) it is seen that the co-ordinates of the pole are the differentials of the tangential equation of the conic, with respect to $\lambda_1, \lambda_2, \lambda_3$, respectively. But the tangential equation of the given conic is

$$l\lambda_2\lambda_3 + m\lambda_3\lambda_1 + n\lambda_1\lambda_2 = 0.$$

Hence the required co-ordinates are

$$x_1' = m\lambda_3 + n\lambda_2, \quad x_2' = n\lambda_1 + l\lambda_3, \quad x_3' = l\lambda_2 + m\lambda_1.$$

5. Find the locus of the pole of $\lambda_x = 0$ with respect to the conic

$$\sqrt{lx_1} + \sqrt{mx_2} + \sqrt{nx_3},$$

being given that the conic fulfils another condition, such as to touch a given line, say $L_x = 0$.—(HEARN.)

Solving the equations in Ex. 4, l, m, n are proportional to

$$\lambda_1 (\lambda_2 x_2' + \lambda_3 x_3' - \lambda_1 x_1'),\ \lambda_2 (\lambda_3 x_3' + \lambda_1 x_1' - \lambda_2 x_2'),\ \lambda_3 (\lambda_1 x_1' + \lambda_2 x_2' - \lambda_3 x_3').$$

Now if L_x touch the conic, we have

$$\frac{l}{L_1} + \frac{m}{L_2} + \frac{n}{L_3} = 0.$$

Hence the required locus, omitting accents, is the right line

$$\frac{\lambda_1 (\lambda_2 x_2 + \lambda_3 x_3 - \lambda_1 x_1)}{L_1} + \frac{\lambda_2 (\lambda_3 x_3 + \lambda_1 x_1 - \lambda_2 x_2)}{L_2} + \frac{\lambda_3 (\lambda_1 x_1 + \lambda_2 x_2 - \lambda_3 x_3)}{L_3} = 0. \quad (582)$$

6. The triangles formed by three given points, and their polars with respect to any conic, are in perspective.

Dem.—Let y, z, w, be the angular points of the original triangle; their polars, with respect to $a_x^2 = 0$, are $a_x . a_y$, $a_x . a_z$, $a_x . a_w$, respectively; and the equation of the join of y to the intersection of the polars of z and w is

$$(a_x . a_z)(a_y . a_w) - (a_x . a_w)(a_y . a_z) = 0,$$

with two similar equations for the other lines of connexion; and these, when added, vanish identically. Hence, &c.

7. It is required to determine when the general equation

$$a\alpha^2 + b\beta^2 + c\gamma^2 + 2h\alpha\beta + 2f\beta\gamma + 2g\gamma\alpha = 0$$

represents an ellipse, a parabola, or a hyperbola. If we eliminate γ between this and the equation

$$\alpha \sin A + \beta \sin B + \gamma \sin C = 0,$$

which represents the line at infinity, and if the resulting equation in α, β be the product of two real factors, it will be a hyperbola; if the product of two imaginary factors, it will be an ellipse; and if a perfect square, it will be a parabola. In this way we find it to be an ellipse, a parabola, or a hyperbola, according as

$$A \sin^2 A + B \sin^2 B + C \sin^2 C + 2F \sin B \sin C + 2G \sin C \sin A + 2H \sin A \sin B$$

is positive, zero, or negative.

8. If the condition of Ex. 3 be fulfilled, what is the locus of the pole of the join of the points $\psi + \psi'$, $\psi - \psi'$?

Denoting the co-ordinates of the poles by x, y, z, from equation (580), we have

$$lx = \cos\psi, \quad my = \sin\psi, \quad nz = \cos\psi'.$$

Hence, from (581), we get

$$\frac{l^4x^2}{l'^2} + \frac{m^4y^2}{m'^2} + \frac{n^4z^2}{n'^2} = 0. \tag{583}$$

This conic is the polar reciprocal of $l'^2\alpha^2 + m'^2\beta^2 + n'^2\gamma^2 = 0$, with respect to $l^2\alpha^2 + m^2b^2 + n^2y^2 = 0$. Hence the polar reciprocal of $a'\alpha^2 + b'\beta^2 + c'\gamma^2 = 0$, with respect to $a\alpha^2 + b\beta^2 + c\gamma^2 = 0$, is

$$\frac{a^2\alpha^2}{a'} + \frac{b^2\beta^2}{b'} + \frac{c^2\gamma^2}{c'} = 0. \tag{584}$$

9. Find the condition that the line $\lambda\alpha + \mu\beta + \nu\gamma = 0$ will touch the conic $l^2\alpha^2 + m^2\beta^2 - n^2\gamma^2 = 0$.

Comparing $\lambda\alpha + \mu\beta + \nu\gamma = 0$ with equation (579), and eliminating ψ, we get the required condition,

$$\frac{\lambda^2}{l^2} + \frac{\mu^2}{m^2} = \frac{\nu^2}{n^2}. \tag{585}$$

Hence, if one tangent to the conic $l^2\alpha^2 + m^2\beta^2 = n^2\gamma^2$ be given, three others are determined.

10. If the chord in Ex. 3 passes through the point α', β' γ', the locus of its pole is

$$l^2\alpha'\alpha + m^2\beta'\beta + n^2\gamma'\gamma. \tag{586}$$

11. The locus of the pole of any tangent to the conic $a_x{}^2$, with respect to $x_1{}^2 + x_2{}^2 + x_3{}^2 = 0$, is

$$A_x{}^2 = 0. \tag{587}$$

12. Find the equation of the director circle of the conic

$$a\alpha^2 + b\beta^2 + c\gamma^2 = 0.$$

If $\psi + \psi'$, $\psi - \psi'$ be the parametric angles of the points of contact of two rectangular tangents, then the condition of perpendicularity will give us the required result, after eliminating ψ, ψ' by means of the co-ordinates in equation (580), and putting a, b, c for l^2, m^2, $-n^2$; thus we get

$$a(b+c)\alpha^2 + b(c+a)\beta^2 + c(a+b)\gamma^2 + 2bc\cos A\,.\,\beta\gamma + 2ca\cos B\,.\,\gamma\alpha$$
$$+ 2ab\cos C\,.\,\alpha\beta = 0. \tag{588}$$

160. *To discuss the equation* $\alpha\beta = \gamma^2$.

This is the special case of the last proposition, when the coefficients of the products $\beta\gamma$, $\gamma\alpha$ vanish, and also the coefficients of α^2, β^2. The form of equation (Art. 146, 3°) shows that α, β are tangents, and γ their chord of contact. If in the equation $\alpha\beta = \gamma^2$ we put $\alpha = \gamma \tan\phi$, $\beta = \gamma \cot\phi$, the equation is satisfied. Hence the co-ordinates of any point on the curve may be represented by $\tan\phi$, $\cot\phi$, 1. This point will be called the point ϕ.

161. The equation of the join of two points ϕ, ϕ' is the determinant

$$\begin{vmatrix} \alpha, & \beta, & \gamma, \\ \tan\phi, & \cot\phi, & 1, \\ \tan\phi', & \cot\phi', & 1 \end{vmatrix} = 0,$$

or

$$\frac{\alpha}{\tan\phi + \tan\phi'} + \frac{\beta}{\cot\phi + \cot\phi'} = \gamma. \qquad (589)$$

Cor. 1.—If $\tan\phi + \tan\phi'$ be constant, the join of the points ϕ, ϕ'. passes through a given point. For writing the equation (589) in the form $\alpha + \beta \tan\phi \tan\phi' - \gamma(\tan\phi + \tan\phi')$, it represents a line through the intersection of $\alpha - \gamma(\tan\phi + \tan\phi') = 0$ and $\beta = 0$; that is, through a fixed point on β. In like manner, if $\cot\phi + \cot\phi'$ be given, it passes through a fixed point on α; and if the product $\tan\phi \,.\, \tan\phi'$ be given, it passes through a fixed point on γ.

Cor. 2.—The tangent at the point ϕ is

$$\alpha \cot\phi + \beta \tan\phi = 2\gamma. \qquad (590)$$

Cor. 3.—The tangents at ϕ, ϕ' intersect on the line $\alpha - \beta \tan\phi \tan\phi'$, got by eliminating γ between their equations. Hence, if $\tan\phi \,.\, \tan\phi'$ be constant, the tangents at ϕ, ϕ' intersect on a fixed line passing through the point $\alpha\beta$. In like manner, it may be shown that if $\tan\phi + \tan\phi'$ be constant, the tangents meet on a fixed line passing through $\gamma\alpha$, and if $\cot\phi + \cot\phi'$ be constant, on a fixed line through $\beta\gamma$.

Cor. 4.—If the equation (589) be written in the form

$$(\alpha - \gamma \tan \phi) - (\gamma - \beta \tan \phi) \tan \phi' ;$$

or, say $L - M \tan \phi' = 0$; and since (Art. 33) the anharmonic ratio of the pencil of four lines $\alpha - k\beta$, $\alpha - k'\beta$, $\alpha - k''\beta$, $\alpha - k'''\beta$ is

$$(k - k')(k'' - k''') \div (k - k'')(k' - k'''),$$

we infer that the anharmonic ratio of the pencil of lines from any variable point of the conic to the four fixed points ϕ', ϕ'', ϕ''', ϕ'''' is

$$(\tan \phi' - \tan \phi'')(\tan \phi''' - \tan \phi'''') \div (\tan \phi' - \tan \phi''')(\tan \phi'' - \tan \phi''''),$$

or $\quad \sin(\phi' - \phi'') \sin(\phi''' - \phi'''') \div \sin(\phi' - \phi''') \sin(\phi'' - \phi'''')$,

and is therefore constant.

The theorem just proved was discovered by CHASLES, and is the fundamental one in his *Sections Coniques*, Paris, 1865. On account of its great importance we shall give another proof. Let the quadrilateral formed by the four fixed points be $ABCD$, and let O be any variable point; then, if the equations of the sides AB, BC, CD, DA of the quadrilateral be α, β, γ, δ respectively, the equation of the conic (Art. 146, 5°) may be written $\alpha\gamma - k\beta\delta = 0$; but α being the perpendicular from O on AB, we have

$$\alpha = \frac{OA \cdot OB \cdot \sin AOB}{AB},$$

with similar values for β, γ, δ; and these substituted in the equation $\alpha\gamma - k\beta\delta = 0$, give

$$\frac{\sin AOB \cdot \sin COD}{\sin BOC \cdot \sin AOD} = k \cdot \frac{AB \cdot CD}{BC \cdot AD}.$$

The right-hand side of this equation is constant, and the left-hand side is the anharmonic ratio of the pencil $(O \cdot ABCD)$. Hence the proposition is proved. (See SALMON'S *Conics*, p. 240.)

Cor. 5.—The tangent at ϕ intersects the tangent at ϕ' on the line $\alpha \cot \phi - \beta \tan \phi' = 0$. Hence, as in *Cor.* 4, we infer that the anharmonic ratio of the four points, where tangents at four fixed points ϕ', ϕ'', ϕ''', ϕ'''' meet the tangent at any variable point ϕ, is

$$\sin(\phi' - \phi'')\sin(\phi''' - \phi'''') \div \sin(\phi' + \phi''')\sin(\phi'' - \phi''''),$$

and is therefore independent of ϕ.

Cor. 6.—If the line $\lambda\alpha + \mu\beta + \nu\gamma$ touch the conic at the point ϕ, we must have λ, μ, ν proportional to $\cot\phi$, $\tan\phi$, -2. Hence

$$4\lambda\mu = \nu^2, \qquad (591)$$

which is the tangential equation of the conic.

Examples.

1. The co-ordinates of the point of intersection of tangents at ϕ, ϕ' are proportional to $\tan\phi \tan\phi'$, 1, $\frac{1}{2}(\tan\phi + \tan\phi')$.

2. The length of the perpendicular from the intersection of tangents at ϕ', ϕ'' on the tangent at ϕ is, putting t for $\tan\phi$, &c.,

$$(t - t')(t - t'') \div f(t), \qquad (592)$$

when $f(t)$ stands for

$$\sqrt{(t^4 + 4\cos A \,.\, t^3 + 2(2 - \cos C)t^2 + 4\cos B \,.\, t + 1)}.$$

3. If $\alpha\beta = k^2\gamma^2$ be the equation of a conic, the circle of curvature at the point $\beta\gamma$ is

$$\beta^2 + \gamma^2 - 2\beta\gamma\cos A = \frac{c}{k^2}\beta\sin B. \qquad \text{(Crofton.)}$$

4. If ϕ, ϕ' be two points, such that the ratio of $\tan\phi : \tan\phi'$ is constant, the envelope of their join is a conic, having double contact with the given conic.

5. If the points ϕ, ϕ' vary but so as that the ratio of $\tan\phi : \tan\phi'$ be given, they divide the conic homographically (see *Cor.* 4).

Hence, if two conics have double contact, any variable tangent to one divides the other homographically. (Townsend.)

6. If β_{12} denote the perpendicular from the intersection of tangents at ϕ', ϕ'' on the tangent β, and π_{12} the perpendicular on any other tangent; then

$$\frac{\pi_{12} \cdot \pi_{34}}{\beta_{12} \cdot \beta_{34}} = \frac{\pi_{13} \cdot \pi_{24}}{\beta_{13} \cdot \beta_{24}} = \frac{\pi_{14} \cdot \pi_{23}}{\beta_{14} \cdot \beta_{23}}. \qquad (593)$$

7. If a polygon of any number of sides be inscribed in a conic, and if ϕ', ϕ'', &c., be the points of contact, and ϕ any variable point; then, with the notation of Ex. 6, we have

$$\frac{\beta_{12}(t' - t'')}{\pi_{12}} + \frac{\beta_{23} t'' - t''')}{\pi_{23}} + \&c., = 0. \qquad (594)$$

8. Since $\beta_{12}(t' + t'') + 2\gamma_{12}$, and $\beta_{12}(t't'') = \alpha_{12}$ (Ex. 1), it follows that

$$\beta_{12}(t' - t'') = 2\sqrt{\gamma_{12}^2 - \alpha_{12}\beta_{12}} = 2\sqrt{S_{12}}, \ \&c.$$

Hence, from (594), we get

$$\frac{\sqrt{S_{12}}}{\pi_{12}} + \frac{\sqrt{S_{23}}}{\pi_{23}} + \frac{\sqrt{S_{34}}}{\pi_{34}} + \&c., + \frac{\sqrt{S_{n1}}}{\pi_{n1}} = 0. \qquad (595)$$

Section IV.—Theory of Envelopes.

161. We have seen (Chapter II. Section III.) that if the coefficients in the equation of a line be connected by a relation of the first degree, the line passes through a given point—in fact, the relation between the coefficients is the equation of the point (Art. 45); and in the last Section it was shown that, if the coefficients be connected by a relation of the second degree, the line will, in all its positions, be a tangent to a curve of the second degree. From these examples we are led to the following definition:—*When a right line or a curve moves according to any law, the curve which it touches in all its positions is called its envelope.* The following examples afford further illustrations of this theory, one of the most interesting in Analytical Geometry.

EXAMPLES.

1. Let $\lambda x + \mu y + 1 = 0$ be the line, and $(a, b, c, f, g, h)(\lambda, \mu, 1)^2$ the relation among the coefficients; it is required to find the envelope of the line. It appears at once that the required envelope is such that two tangents can be drawn to it from any arbitrary point. For, let $x'y'$ be the point; substitute these co-ordinates in $\lambda x + \mu y + 1$, and eliminate between the result and the equation $(a, b, c, f, g, h)(\lambda, \mu, 1)^2$, and we get a quadratic in λ, corresponding to each root of which can be drawn a tangent to the required envelope. Now, if the quadratic have equal roots, the tangents will coincide, and their point of ultimate intersection will be a point on the curve. Hence, forming the discriminant of the quadratic in λ, and removing the accents from $x'y'$, we get the required envelope, viz.

$$(A, B, C, F, G, H)(x, y, 1)^2 = 0, \tag{596}$$

where A, B, C, &c., have the usual meanings.

2. Find the envelope of $\mu^2 x + \mu y + a = 0$. This is the quadratic that would result if we were solving by the foregoing method the problem of finding the envelope of the line $\lambda x + \mu y + a = 0$; λ, μ being connected by the relation $\lambda = \mu^2$. Hence, forming the discriminant with respect to μ of the equation $\mu^2 x + \mu y + a = 0$, we get the parabola $y^2 = 4ax$.

Similarly, we may solve the more general problem to find the envelope of $\mu^2 P + \mu Q + R$, when P, Q, R denote curves of any degree, viz., we get

$$Q^2 = 4PR. \tag{597}$$

3. Find the envelope of the line $ax \cos\phi + by \sin\phi = ab$.

4. Find the envelope of a line if the product of the perpendiculars on it from two fixed points be given.

5. Find the envelope of a line if the squares of perpendiculars let fall on it from any number of fixed points be constant.

6. Find the envelope of a line if the difference of the squares of perpendiculars let fall on it from two given points be constant.

Ans. A parabola.

7. Find the envelope of

$$\frac{x^2}{\lambda^2} + \frac{y^2}{\mu^2} = 1,$$

$\lambda\mu$ being $= c$. *Ans.* $2xy = c$.

8. Find the envelope of a line which makes on the axes of co-ordinates intercepts whose sum is constant.

9. If two conjugate diameters of an ellipse be given in position, and the sum of the squares of its axes given in magnitude; prove that it is inscribed in a given quadrilateral.

10. Find the envelope of a system of confocal conics. Let

$$\frac{x^2}{a^2+\lambda}+\frac{y^2}{b^2+\lambda}+1$$

be one of the conics. Clearing of fractions, and considering the result as a quadratic in λ, we find, by forming the discriminant, the product of four imaginary lines, viz.

$$h \pm x \pm y\sqrt{-1} = 0, \text{ where } h^2 = a^2 - b^2. \qquad (598)$$

11. The envelope of the polar of a given point, with respect to a system of confocal conics, is a parabola whose directrix is the join of the given point to the centre of the confocals.

12. If A, B, C, A', B', C' be two triads of fixed points, μ, μ' two variable points, one on each line; find the envelope of the join of μ, μ', if the anharmonic ratios $(ABC\mu)$, $(A'B'C'\mu')$ be equal.

13. The vertices of a triangle move along three fixed lines, and two of the sides pass through two fixed points; find the envelope of the third side.

14. If two of the sides of an inscribed triangle of the conic $\alpha^2+\beta^2=\gamma^2$ touch the conic $a\alpha^2+b\beta^2=c\gamma^2$, the envelope of the third side is

$$(ca+ab-bc)^2\alpha^2+(ab+bc-ca)^2\beta^2=(bc+ca-ab)^2. \qquad (599)$$

SECTION V.—THEORY OF PROJECTION.

162. DEF.—*Let O be the origin, OX, OY the axes; BB', II' (called the* BASE *line and the* INFINITE *line respectively) two lines parallel to the axis of Y. Then let P be any point in the plane; join IP, cutting BB' in C; through C draw CP' parallel to OX, meeting OP produced in P'. The point P' is called the projection of P.*

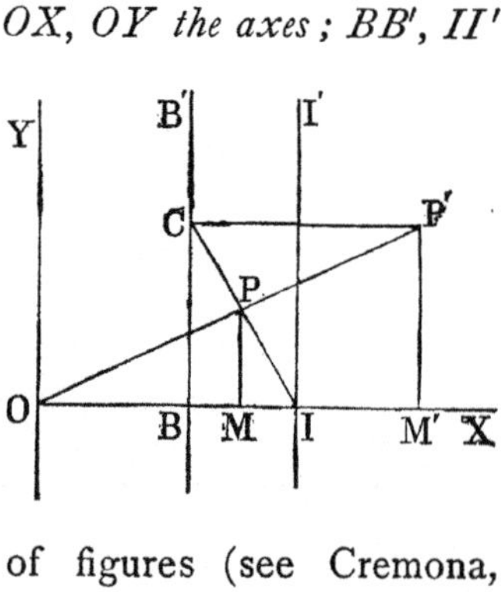

In the ordinary method of treating projective properties of figures (see Cremona,

Elements of Projective Geometry) three planes are required:—(1) A plane passing through the centre of projection. (2) A parallel plane, on which is drawn the projected figure. (3) The plane of the figure to be projected, cutting the former planes in parallel lines. It will be seen that the method which we have adopted is virtually the same, and that while it relieves the student from the embarrassment of having to consider different planes, it has the advantage of admitting the use of analysis.

If the co-ordinates of P be xy, those of P', $x'y'$; then, denoting OI by a and BI by c, we easily get

$$x = \frac{ax'}{c + x'}, \quad y = \frac{ay'}{c + x'}. \tag{600}$$

Cor. 1.—If $x = a$, x' will be infinite. Hence the projection of any point on the line II' will be at infinity.

Cor. 2.—From (600) we get

$$x' = \frac{cx}{a - x}, \quad y' = \frac{cy}{a - x}. \tag{601}$$

163. If any line CD cut the base line and the infinite line in the points C, D respectively, its projection will be a line through C parallel to OD.

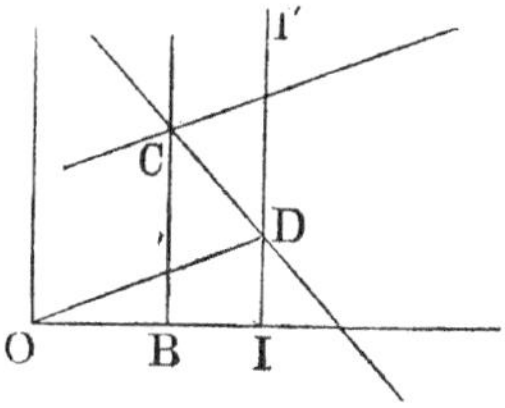

Let the equation of CD be

$$lx + my + n;$$

and since $OI = a$, the equation of II' is $x - a = 0$. Hence the equation of OD is

$$n(x - a) + a(lx + my + n) = 0,$$

or

$$(la + n)x + may = 0.$$

Again, substituting in $lx + my + n$ the values in (600), we get, after omitting accents and clearing of fractions,

$$(la + n)x + may + nc = 0,$$

which is the equation of the projection of CD. Now, since

this differs from the equation of OD only by a constant, it is parallel to it; and since it may be written in the form

$$n(x - a + c) + a(lx + my + n) = 0,$$

it passes through the intersection of the lines

$$x - a + c = 0 \text{ and } lx + my + n = 0;$$

that is, through the point C. Hence the proposition is proved.

Cor. 1.—Any two lines intersecting each other on II' are projected into parallel lines. For, if two lines pass through the point D, the projection of each will be parallel to OD.

Cor. 2.—A line passing through the origin is unaltered by projection.

Cor. 3.—If four lines form a pencil, their projections form an equi-anharmonic pencil. For, if P be the vertex of the pencil, and if its four rays meet the line II' in the points A, B, C, D, their projections will be parallel to OA, OB, OC, OD. Hence the proposition is proved.

Cor. 4.—*Parallel lines are projected into concurrent lines.*

For the projection of $lx + my + n = 0$ is $a(lx + my) + n(c + x) = 0$; if n be variable $(lx + my + n) = 0$ denotes a system of parallel lines, and its projection $a(lx + my) + n(c + x)$ a concurrent system.

164. *A curve of the second degree is projected into another curve of the second degree.*

For, making the substitutions (600) in an equation of any degree, and clearing of fractions, we get an equation of the same degree.

Cor. 1.—The projection of a tangent to a conic is a tangent to its projection.

Cor. 2.—The relations of pole and polar are unaltered by projection.

Cor. 3.—*A system of concentric circles is projected into a system of conics having double contact with each other.*

For, let $x^2 + y^2 = r^2$ be one of the circles: by varying r we get a concentric system; and making the substitutions (600), we get $a(x^2 + y^2) = r^2(c + x)^2$, which, when r varies, denotes a system of conics having double contact with each other.

165. *Any straight line can be projected to infinity, and at the same time any two angles into given angles.*

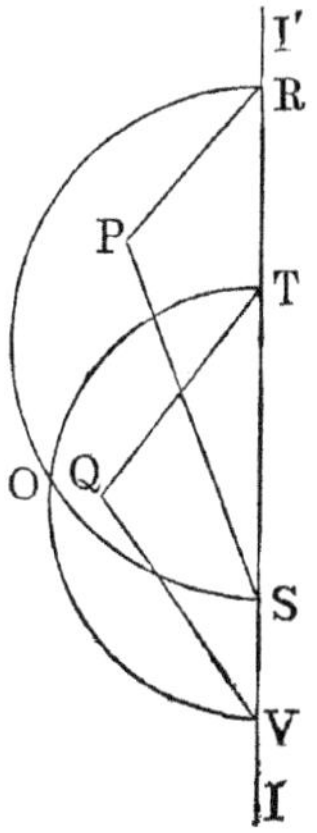

Let II' be the line to be projected to infinity; RPS, TQV the angles to be projected into given angles; say, for example, into right angles. Let II' meet the legs of the angles in the pairs of points R, S; T, V. Upon RS, TV describe semicircles, intersecting in O. Then O will be the required centre of projection, and we can take any line parallel to II' for the base line BB'.

If the circles do not intersect, the point O will be imaginary, in which case imaginary lines in one figure will be projected into real lines in the other. Thus confocal conics, being inscribed in an imaginary quadrilateral, will be projected into conics inscribed in a real quadrilateral.

The substitutions for this case are, for x, y, respectively,

$$\frac{ax}{c+x}, \quad \frac{ay\sqrt{-1}}{c+x}.$$

In this manner we get for the four imaginary lines (598), the four real lines $h(c+x) \pm ax \pm ay = 0$, which are the four sides of the quadrilateral circumscribed to the projection of confocals.

166. *A system of coaxal circles is projected into a system of conics passing through four points.*

Dem.—Let $x^2 + y^2 + 2kx - d^2 = 0$ be a circle, which, by giving k different values, will represent a coaxal system.

Then, making the substitutions (600), we get, after clearing of factors,

$$a^2x^2 + a^2y^2 - d^2(c + x)^2 + 2kax(c + x) = 0,$$

or, say, $$S + 2kLM = 0.$$

Hence the proposition is proved.

This may be shown otherwise, thus: a coaxal system of circles have common the two circular points at infinity, and the two points where they meet the radical axis, and the projections of these points will be common to the projections of the circles.

167. *Any conic S can be projected into a circle having for its centre the projection of any point P in the plane of the conic.*

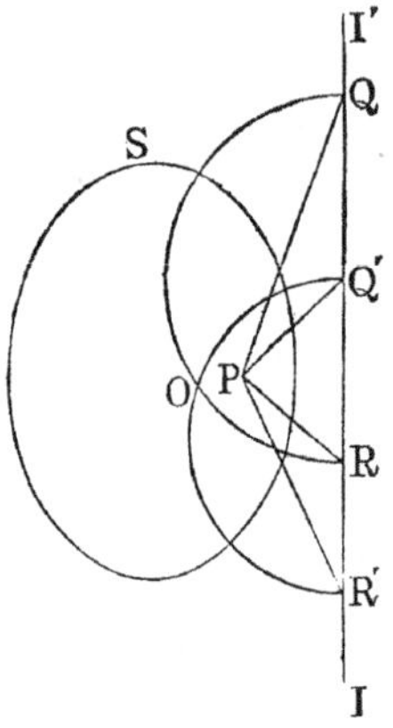

Dem.—Let II' be the polar of P with respect to S; then take this for the *infinite line* (Art. 162), and let Q, R; Q', R' be pairs of conjugate points upon it with respect to S; upon QR, $Q'R'$ describe semicircles, intersecting in O. Now taking O for the centre of projection, and any line parallel to II' for the base line (Art. 162), the lines PQ, PR will be projected into lines parallel to OQ, OR; that is, into rectangular lines. Similarly PQ', PR' will be projected into another pair of rectangular lines. Hence the projection of S will be a conic, having two pairs of rectangular conjugate lines intersecting in the projection of P. In other words, it will be a circle, having the projection of P for centre.

168. *The pencil formed by the two legs of a given angle, and the imaginary lines through its vertex to the circular points at infinity, has a given anharmonic ratio.*

Dem.—Let the given angle be that formed by the axes of co-ordinates, namely, ω. Then the equation of a point circle at the origin is $x^2 + y^2 - 2xy \cos \omega = 0$; and the factors of this, viz., $x - e^{\omega\sqrt{-1}}y = 0$, $x - e^{-\omega\sqrt{-1}}y = 0$, are the lines from the origin to the circular points at infinity. The anharmonic ratio of the pencil, formed by these lines and the axes, is $e^{2\omega\sqrt{-1}}$, and is therefore given. Hence the proposition is proved.

Cor.—If the axes be rectangular the pencil formed by them, and the lines to the circular points at infinity, is a harmonic pencil. For, putting

$$\frac{\pi}{2} \text{ for } \omega, \quad e^{2\omega\sqrt{-1}} = -1.$$

Examples.

1. Any quadrilateral can be projected into a square. For the third diagonal (Art. 165) may be projected to infinity, and the remaining diagonals and a pair of adjacent sides into pairs of rectangular lines.

2. The diagonal triangle of a quadrilateral is self-conjugate with respect to any inconic of the quadrilateral. For projecting the quadrilateral into a square, the intersection of the diagonals of the square will evidently be the centre of the inconic of the square, and will be the pole of the line at infinity with respect to that conic. Hence any diagonal of the quadrilateral is the polar of the intersection of the other two.

3. If four chords of a conic be tangents to an inscribed conic (having double contact), the anharmonic ratio of the points of contact is equal to that of one set of extremities of the chords of the outer conic. For the conics may be projected into concentric circles, and the proposition is evident.

4. Any line passing through a given point in the plane of a conic is cut harmonically by the conic and the polar of the point. For the conic can be projected into a circle and the point into its centre (Art. 167).

5. Any chord of a conic touching an inscribed conic is cut harmonically at the point of contact, and at the point where it meets the chord of contact of the two conics.

6. If two pairs of opposite sides of a hexagon inscribed in a circle be parallel, it is easy to prove that the third pair of opposite sides are parallel. Hence the three pairs of opposite sides intersect on the line at infinity; and, projecting this, we have a proof of PASCAL'S Theorem for any conic.

7. Two tangents to any circle are cut homographically by any variable tangent. For it is easy to see that the pencil formed by joining four points on one tangent to the centre of the circle is equal to the pencil formed by joining their corresponding points to the centre. Hence, by projection, we see that any two fixed tangents to a conic are cut homographically by a variable tangent.

8. If two triangles be such that the intersections of corresponding sides are collinear, the joins of corresponding vertices are concurrent. For, projecting the line of collinearity to infinity, the triangles will be homothetic.

9. If a system of chords of a conic pass through a fixed point P, their extremities divide the conic homographically. Project the conic into a circle, having the projection of P for its centre, and the proposition is evident.

10. Any two conics can be projected into circles. For, project one of them into a circle, and one of their common chords to infinity, then the projection of the other will pass through the circular points at infinity, and therefore it will be a circle.

11. Any two conics can be projected into concentric conics.

12. If a system of conics pass through four points, they cut any transversal in involution.

For the conics can be projected into coaxal circles.

13. If two conics be inscribed in a quadrilateral, their eight points of contact lie on a conic.

Project the quadrilateral into a square, and the proposition is evident.

14–17. What properties of conics are obtained from the following by projection?—If a variable conic pass through four fixed points, the locus of its centre is a conic passing through the middle points of the joins of the four points.

15. If a chord of a given circle pass through a fixed point, the locus of its middle point is a circle.

16. If a variable conic be inscribed in a given quadrilateral, the locus of its centre is a right line bisecting the diagonals of the quadrilateral.

17. The locus of the point, where parallel chords of a given conic are cut in a given ratio, is a conic having double contact with the given conic.

18. If two triangles ABC, $A'B'C'$ be self-conjugate with respect to a conic, their six angular points lie on another conic.

Project the conic into a circle and the line BC to infinity; then A, the pole of BC, will be the centre of the circle; and if, taking the projections of AB, AC as axes, $x'y'$, $x''y''$, $x'''y'''$ be the co-ordinates of the projections of A', B', C', respectively, the equation of a hyperbola passing through the projections of A', B', C', and having its asymptotes parallel to the axes, is—

$$\begin{vmatrix} xy, & x, & y, & 1, \\ x'y', & x', & y', & 1, \\ x''y'', & x'', & y'', & 1, \\ x'''y''', & x''', & y''', & 1 \end{vmatrix} = 0.$$

This hyperbola passes through the projections of the six points. Hence the proposition is proved.

19. In the same case the six lines forming the sides of the two triangles are tangents to a conic.

Project, as in Ex. 18, and it is easy to see that the projections are tangents to a parabola.

169. The projections of focal properties are always imaginary. For the imaginary tangents from a focus are projected into real tangents, and the imaginary circular points at infinity, and the antifoci into real points. It will be seen that all these results follow from the projections of the four lines $\sin h \pm x \pm y\sqrt{-1}$, forming an imaginary circumscribed quadrilateral to a conic, into four real lines.

Examples.

1. If a variable circle touch two fixed lines, the chords of contact are parallel. Hence, by projection, if a variable conic touch two fixed lines, and pass through two fixed points I, J, the chords of contact are concurrent.

2. If a variable circle touch two fixed lines, the locus of its centre is a right line. Hence, if a variable conic touch two fixed lines, and pass

through two fixed points I, J, the locus of the pole of the chord IJ is a right line.

3. If a variable circle pass through a given point and touch a given line, the locus of its centre is a parabola, having the given point as focus. Hence, *if a circumconic of a given triangle touch a given line, the loci of the poles of the sides of the triangle are conics inscribed in it.*

4. Two lines through the focus of a conic are cut by pairs of tangents parallel to them in four concyclic points.

5. The circumcircle of the triangle formed by three tangents to a parabola passes through the focus. Hence *the vertices of two circumtriangles of a conic lie on a conic.*

6. If a circumtriangle to a given circle have two sides fixed, and the third variable, the envelope of its circumcircle is a circle. Hence, if a circumtriangle of a given conic have two sides fixed, and the third variable, the envelope of a conic passing through two fixed points I, J of the former conic, and through the vertices of the triangle, is a conic passing through the two points I, J. (PROF. J. PURSER.)

7. The locus of the centre of a circle touching two given circles is a conic section, having the centres of the given circles as foci. Hence, if a variable conic passing through two given points I, J touch two given conics also passing through I, J, the locus of the pole of the chord IJ with respect to it is a conic inscribed in the quadrilateral formed by the tangents to the fixed conics at the points I, J.

170. In projecting a locus described by the vertex of a constant angle, we consider the pencil formed by its legs and the lines from the vertex to the circular points at infinity; and it follows, from Art. 168, that we get a constant pencil. Again, if the sum or difference of angles be given, we get, by projection, pencils the product or quotient of whose anharmonic ratios is constant. This projection is always imaginary.

Examples.

1. The angle contained in the same segment of a circle is constant. Hence the anharmonic ratio of the pencil formed by lines drawn from any variable point to four fixed points of a conic is constant.

2. If two tangents to a conic be perpendicular to each other they intersect on the director circle. Hence the locus of the point of intersection of tangents to a conic which divide a given line IJ harmonically is a conic through the points I, J, and the envelope of the chord of contact is a conic which touches the tangents to the original conic from I, J.

3. If two tangents to a parabola be at right angles, they intersect on the directrix. Hence the locus of the point of intersection of tangents to a conic which divide harmonically a given line IJ touching the conic is a right line.

4. If from any point on a circle two lines be drawn forming a given angle, the chord joining their other extremities touches a concentric circle. Hence if I, J be two fixed points on a conic; P, Q two variable points, such that the anharmonic ratio of the four points P, Q, I, J is constant, the envelope of PQ is a conic.

5. Project the following properties:—

If two tangents to a parabola include a given angle, the locus of their intersection is a conic.

6. If two circles be such that a quadrilateral can be inscribed in one and circumscribed to another, the chords of contact intersect at right angles.

7. Confocal conics intersect at right angles.

8. If two tangents, one to each of two confocals, lie at right angles, the locus of their intersection is a circle.

9. The circle described about a self-conjugate triangle to another circle cuts it orthogonally.

10. If a variable chord of a conic subtend a right angle at a fixed point not on the conic, the envelope of the chord is a conic.

11. If a variable line, whose extremities rest on the circumferences of two given concentric circles, subtend a right angle at any given fixed point, the locus of its centre is a circle.

Section VI.—Sections of a Cone.

171. A cone *of the second degree* is the surface generated by a variable line passing through the circumference of a fixed circle called *the base*, and through a fixed point not in the plane of the circle. The generating line, in any of its positions, is called an *edge* of the cone, the fixed point its *vertex*, and the line joining the vertex to the centre of the base the *axis* of the cone.

The line generating the cone being produced indefinitely both ways, it is evident that the complete surface consists of two sheets united at the vertex, and the whole is considered only as one cone, of which the vertex is a node or double point.

When the axis of the surface is at right angles to the plane of the base, it is called a *right* cone, in other cases it is *oblique*.

In the following propositions a plane through the axis, perpendicular to the plane of the base, will be the plane of *reference*, and the sections of the cone will be understood to be those made by planes at right angles to the plane of reference.

172. *Sections of a cone made by parallel planes are similar.*

This is evident, for the sections are homothetic with respect to the vertex.

Cor. 1.—Any line drawn through the vertex will meet the planes of two parallel sections in homologous points with respect to those sections.

Cor. 2.—The sections made by planes parallel to the base are circles.

Def.—*A section whose plane intersects the plane of reference in a line antiparallel to the diameter of the base is called a subcontrary section.*

173. *If an oblique cone ABC be cut by a plane ELF in a subcontrary position, the section will be a circle.*

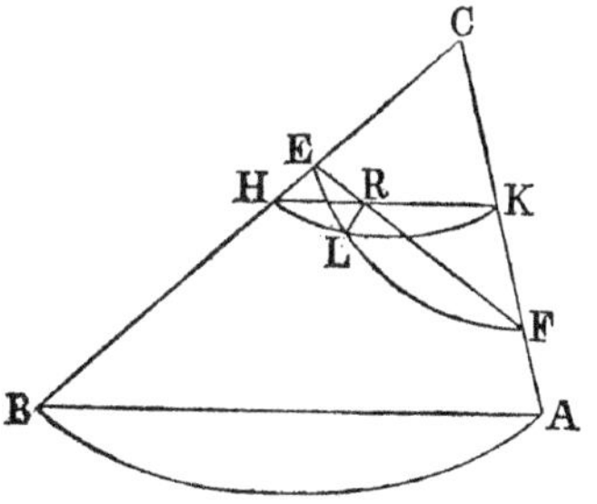

Dem.—Through any point R in EF draw a plane HLK parallel to the base. Then, since the planes ELF, HLK are both normal to the plane ABC, their common section (*Euc.*, XI. xix.), RL, is normal to it. Hence (*Euc.*, III., xxxv.) $RL^2 = HR \cdot RK$. But, from the hypothesis, the four points H, E, K, F are concyclic. Hence $ER \cdot RF = HR \cdot RK$; therefore $ER \cdot RF = RL^2$. Hence the section ELF is a circle.

Cor. 1.—Any sphere passing through the base of a cone will cut the cone again in a subcontrary section.

Cor. 2.—If a sphere be described about a cone, its tangent plane at the vertex is parallel to the plane of subcontrary section.

174. *Any section of an oblique cone which is not subcontrary is either a parabola, an ellipse, or a hyperbola.*

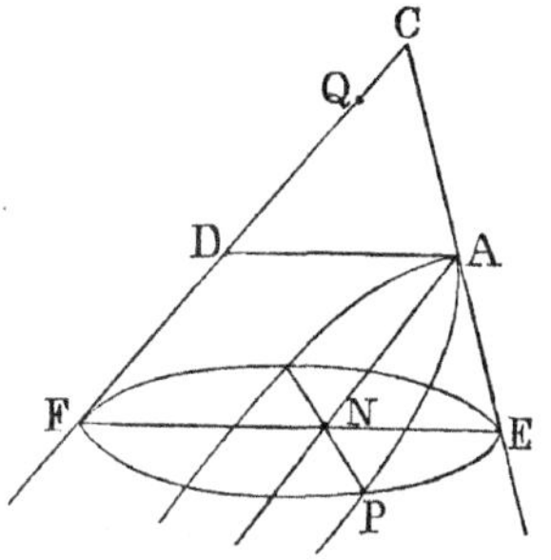

1°. *Let the section be parallel to an edge of the cone.*

Let AN be the intersection of the section with the plane of reference. Then, since AN is parallel to the edge CD, and NE parallel to the diameter of the base, the triangle ANE is given in species. Hence the ratio of $AN : NE$ is given; and since AD is equal to FN, the ratio of the rectangle $AD \cdot AN : FN \cdot NE$ is given; but $FN \cdot NE = NP^2$. Hence the ratio $AD \cdot AN : PN^2$ is given; therefore PN^2 varies as AN. Hence *the section is a parabola.*

Cor.—If the point Q be taken in CD, such that $DC \cdot DQ = DA^2$, then DQ = latus rectum of the section.

2°. *Let the section cut all the edges of one sheet of the cone.*

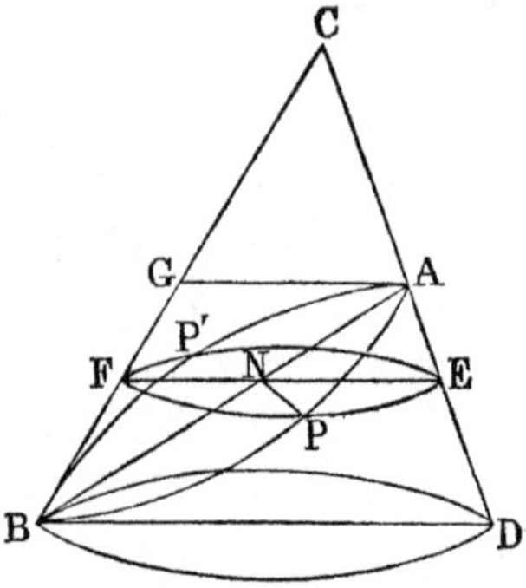

Let A, B be the vertices of the section. Draw any section EF parallel to the base, intersecting the former in the points P, P'. Then, since the planes APB, EPF are both normal to the plane of reference, their common section is normal to it; hence NP is perpendicular to EF. Therefore $PN^2 = EN.NF$.

Again, from the pairs of similar triangles BAG, BNF; ABD, ANE, we get

$$AB^2 : AG.BD :: AN.NB : EN.NF \text{ or } PN^2.$$

Hence the ratio $AN.NB : PN^2$ is given, and therefore the locus of P is an ellipse.

3°. *Let the plane of section meet both sheets of the cone.*

The section in this case will be a hyperbola. The proof is, with slight modification, the same as 2°.

Examples.

1. The square of the conjugate diameter is equal to the rectangle contained by the diameter of the sections through A, B, parallel to the base.

2. The orthogonal projection of the section APB on the base of the cone is a conic having a focus at the centre of the base.

3. If the section of a cone by a plane be a hyperbola; prove that the asymptotes are parallel to the edges in which the cone is cut by a plane parallel to the section. (Make use of Art. 172.)

4. If AB be the diameter of the section of a right cone; C the vertex; F, F' the points of contact of inscribed circle; and the escribed circle of the triangle ABC, touching AC, BC produced; F, F' are the foci of the section.

Dem.—Through A, B draw planes parallel to the base of the cone; then, denoting BC, AC, as in Trigonometry, by a, b; the diameters of the sections made by these planes are, respectively, $2b \sin \frac{1}{2} C$, $2a \sin \frac{1}{2} C$. Hence the square of the conjugate diameter of the section, whose transverse is AB, is (Ex. 1), $4ab \sin^2 \frac{1}{2} C$, or $c^2 - (a - b)^2 = AB^2 - FF'^2$. Hence *$F$, F' are the foci.*

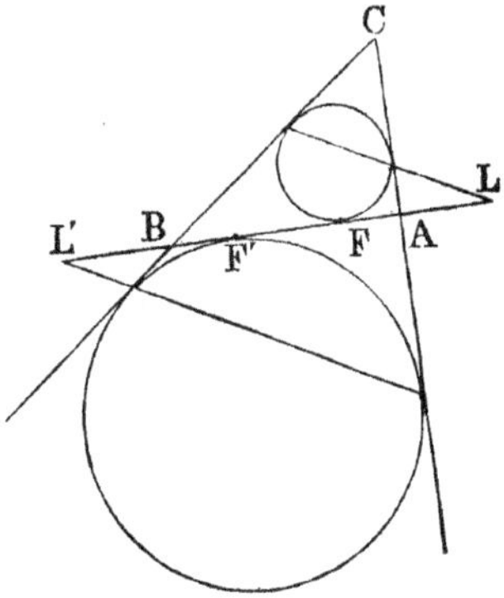

5. If P be any point in the circumference of the section; prove $CP - FP$, and $CP + F'P$, are constants.

6. If the polars of C, with respect to the circles in Ex. 4, meet AB produced in L, L'; prove that the normals to the plane of reference, at the points L, L', are the directrices of the section.

7. The latus rectum of the section is equal to twice the perpendicular from the vertex on the plane, multiplied by the tangent of half the vertical angle.

8. If the section be a hyperbola, state the theorem corresponding to that of Ex. 4.

9. If P be any point in the circumference of the section; prove that the right cone, having $F'P$, PF, PC as edges, has the tangent at P to the curve for its axis.

10. The locus of the vertex of all right cones, out of which a given ellipse can be cut, is a hyperbola, passing through the foci of the ellipse.

11. If through the vertex of an oblique cone standing on a circular base a plane be drawn perpendicular to one of its edges, this plane will cut the base in a line whose envelope is a conic, having the foot of the perpendicular from the vertex on the base as focus.

Section VII.—Theory of Homographic Division.

175. If O be the origin, and the abscissae OA, OB the roots of the equation

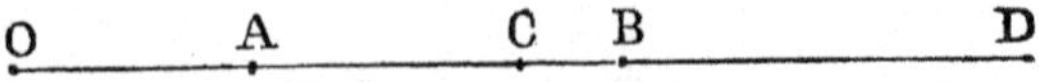

$ax^2 + 2hx + b = 0$, and OC, OD the roots of $a'x^2 + 2h'x + b' = 0$; then, if C, D be harmonic conjugates to A, B,

$$ab' + a'b - 2hh' = 0. \tag{602}$$

Dem.—If the abscissa of C be x', its polar, with respect to $ax^2 + 2hx + b$, is $axx' + h(x + x') + b = 0$; and the points whose abscissae are x, x' will be harmonic conjugates with respect to A, B, and therefore x, x' will be the roots of $a'x^2 + 2h'x + b' = 0$. Hence

$$x + x' = -\frac{2h'}{a'}, \qquad xx' = \frac{b'}{a'};$$

and, substituting in $axx' + h(x + x') + b = 0$, we get

$$ab' + a'b - 2hh' = 0.$$

Cor. 1.—The pair of points denoted by

$$Axx' + B(x + x') + C = 0$$

are harmonic conjugates to the pair

$$Ax^2 + 2Bx + C = 0.$$

Cor. 2.—If the pair of points $ax^2 + 2hx + b = 0$ be harmonic conjugates to

$$U \equiv a'x^2 + 2h'x + b' = 0 \text{ and to } V \equiv a''x^2 + 2h''x + b'' = 0,$$

they are also harmonic conjugates to $U + kV = 0$.

Cor. 3.—If the pair of lines $ax^2 + 2hxy + by^2 = 0$ be harmonic conjugates to the line $a'x^2 + 2h'xy + b'y^2 = 0$, then $ab' + a'b - 2hh' = 0$.

Cor. 4.—The line pairs

$$U \equiv ax^2 + 2hxy + by^2 = 0, \quad V \equiv a'x^2 + 2h'xy + b'y^2 = 0$$

have the line pair

$$(ah' - a'h)x^2 + (ab' - a'b)xy + (hb' - h'b)y^2 = 0$$

as harmonic conjugates. For each of the former line pairs

fulfil with this the condition of harmonicism. The last equation may be written

$$\frac{dU}{dx} \cdot \frac{dV}{dy} - \frac{dU}{dy} \cdot \frac{dV}{dx} = 0. \qquad (603)$$

Cor. 5.—If the line pairs $U = 0$, $V = 0$, be written in ARONHOLD'S notation thus,

$$(a_1x_1 + a_2x_2)^2 = 0, \quad (b_1x_1 + b_2x_2)^2 = 0,$$

the condition that they form a harmonic pencil is

$$(a_1b_2 - a_2b_1)^2 = 0, \qquad (604)$$

where, as usual, a_1 a_2, &c., have no meaning until the multiplication is performed.

176. If $a_x^2 = 0$, $b_x^2 = 0$ be the equations of two conics, it is required to find the locus of a point whence tangents to them form a harmonic pencil. Let x be the point; then if y be a point on a tangent to $a_x^2 = 0$, the equation of a pair of tangents from y to $a_x^2 = 0$ is got by substituting the expressions

$$(x_2y_3 - x_3y_2), \quad (x_3y_1 - x_1y_3), \quad (x_1y_2 - x_2y_1)$$

for λ_1, λ_2, λ_3 in the tangential equation $A_\lambda^2 = 0$ (Art. 158, 6°). Hence the pair of tangents is—

$$\begin{vmatrix} A_1, & A_2, & A_3, \\ x_1, & x_2, & x_3, \\ y_1, & y_2, & y_3 \end{vmatrix}^2 = 0;$$

and putting $y_3 = 0$, the pair of points, where the tangents meet the third side of the triangle of reference, are given by the equation

$$\{(A_2x_3 - A_3x_2)y_1 + (A_3x_1 - A_1x_3)y_2\}^2 = 0;$$

where A_1, A_2, A_3 have no meaning until the multiplication is performed. Similarly we get from the conic, $b_x^2 = 0$,

$$\{(B_2x_3 - B_3x_2)y_1 + (B_3x_1 - B_1x_3)y_2\}^2 = 0.$$

Hence (Art. 175) the condition of harmonicism is—

$$\begin{vmatrix} A_2x_3 - A_3x_2, & A_3x_1 - A_1x_3, \\ B_2x_3 - B_3x_2, & B_3x_1 - B_1x_3 \end{vmatrix}^2 = 0;$$

or,

$$\begin{vmatrix} x_1, & x_2, & x_3, \\ A_1, & A_2, & A_3, \\ B_1, & B_2, & B_3 \end{vmatrix}^2 = 0. \qquad (605)$$

Similarly, the envelope of λ_x, which cuts the conics a_x^2, b_x^2 harmonically, is

$$\begin{vmatrix} \lambda_1, & \lambda_2, & \lambda_3, \\ a_1, & a_2, & a_3, \\ b_1, & b_2, & b_3 \end{vmatrix}^2 = 0. \qquad (606)$$

The two conics (605), (606) may be called, respectively, *the point and line harmonic conics* of $a_x^2 = 0$, $b_x^2 = 0$. Their importance in the theory of a pair of conics was first noticed by DR. SALMON.

Cor.—The point and line harmonic conics of $a_1x_1^2 + a_2x_2^2 + a_3x_3^2 = 0$, and $b_1x_1^2 + b_2x_2^2 + b_3x_3^2 = 0$ are, respectively,

$$a_1b_1(a_2b_3 + a_3b_2)x_1^2 + a_2b_2(a_3b_1 + a_1b_3)x_2^2 + a_3b_3(a_1b_2 + a_2b_1)x_3^2 = 0, \qquad (607)$$

and

$$(a_2b_3 + a_3b_2)\lambda_1^2 + (a_3b_1 + a_1b_3)\lambda_2^2 + (a_1b_2 + a_2b_1)\lambda_3^2 = 0. \qquad (608)$$

177. *If two series of points on the same or on different lines have a* 1 *to* 1 *correspondence; that is, if to a point of either series correspond one, and only one, point of the other, they divide the lines homographically.*

Dem.—From the hypothesis, it is evident that the distances x, x' of corresponding points from two fixed points must be connected by an equation of the form

$$Axx' + Bx + Cx' + D = 0.$$

Now, giving x any four arbitrary values a_1, a_2, a_3, a_4; and to x' the four corresponding values

$$-\frac{Ba_1 + D}{Aa_1 + C}, \quad -\frac{Ba_2 + D}{Aa_2 + C}, \ \&c.,$$

got from $Axx' + Bx + Cx' + D = 0$; we see that the anharmonic ratio

$$(a_1 - a_2)(a_3 - a_4) \div (a_1 - a_3)(a_2 - a_4)$$

of the four points a_1, a_2, a_3, a_4, is equal to that of the four corresponding points. Hence the proposition is proved. *In like manner two pencils with the same or different vertices, and which have a* 1 *to* 1 *correspondence, are homographic.*

Examples.

1. A variable tangent to a conic cuts two fixed tangents homographically. For the points of section have a 1 to 1 correspondence.

2. Two series of rays connecting two fixed points of a conic to a variable point of the same curve are homographic.

3. If S, S', S'', &c., be circumconics of a quadrilateral $ABCD$; through A, B draw two transversals, meeting the conics in the ranges of points a, a', a'', &c.; b, b', b'', &c. These points have a 1 to 1 correspondence. Hence the conics divide the lines homographically.

4. If two systems of points have a 1 to 1 correspondence, and if we take on each system the point which corresponds to infinity on the other, the distances of a pair of corresponding points from these new points have a constant rectangle.

For, making x' infinite, we get $x = -\frac{C}{A}$; and making x infinite, we get $x' = -\frac{B}{A}$. Hence, putting $x - \frac{C}{A}$, $x' - \frac{B}{A}$ for x, x', respectively, in $Axx' + Bx + Cx' + D = 0$, we get $A^2xx' = BC - AD$, which proves the proposition.

5. In forming two homographic systems of points, three points of each system can be taken arbitrarily. For if a, b, c be three points of the first system; a', b', c' three points of the second system; then if x, x' be two variable points fulfilling the conditions of dividing the lines homographically, we get

$$(a - b)(c - x) \div (a - c)(b - x) = (a' - b')(c' - x') \div (a' - c')(b' - x'),$$

which, reduced, gives an equation of the form $Axx' + Bx + Cx' + D = 0$.

178. *If two systems of points which have a* 1 *to* 1 *correspondence be measured on the same line, two points can be formed, each of which coincides with its conjugate.*

For, if in the equation $Axx' + Bx + Cx' + D = 0$ we put $x = x'$, we get a quadratic for determining x'.

Or, geometrically, if A, B, C be three points of one system; A', B', C' the corresponding points of the other, and

O a point which corresponds with its conjugate; then the anharmonic ratios $(OABC)$, $(OA'B'C')$, being equal, we have

$$\frac{OA \cdot BC}{OB \cdot AC} = \frac{OA' \cdot B'C'}{OB' \cdot A'C'}.$$

Hence $$\frac{OA}{OA'} \cdot \frac{OB'}{OB} = \frac{AC \cdot B'C'}{A'C' \cdot BC} = \text{constant, say } k^2.$$

Now $OA \,.\, OB'$, $OA' \,.\, OB$ are the squares of tangents drawn from O to circles described on AB', $A'B$ as diameters; and since the ratio of these tangents is given, the locus of O is a circle, which will intersect the line OX in the two required points.

DEF.—*The two points which coincide with their conjugates are called the double points of the homographic systems.*

If two triads of points A, B, C; A', B', C', be on a conic, the Pascal's line of the hexagon, which they form, will intersect the conic in the double points. For, joining AA', it is evident that $(A' \,.\, OABC) = (A \,.\, OA'B'C')$.

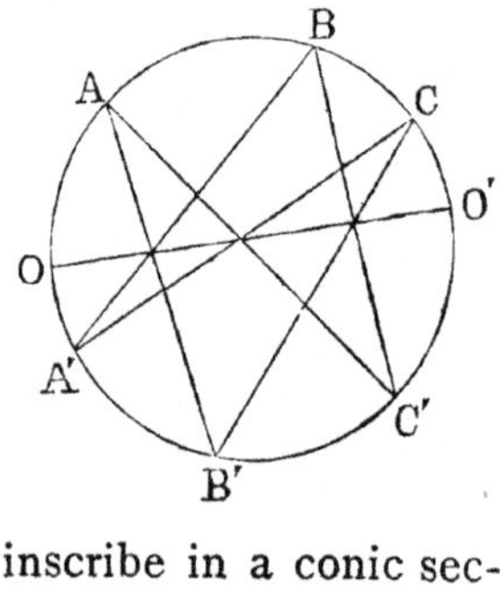

The method of finding double points enables us to solve a number of problems which would be otherwise very difficult. We give the following as an example:—'To inscribe in a conic section a polygon all whose sides pass through fixed points.'

Solution.—Assume any point a at random on the conic for the vertex of the polygon, and form a polygon whose sides pass through the given points: the point a', where the last side meets the conic, will not in general coincide with a. If we make three such attempts, we get three pairs of points a, a'; b, b'; c, c': then a point O, such that the anharmonic ratio $(Oabc) = (Oa'b'c')$, will be the vertex required.

179. If two homographic systems on the same line be determined by the relation $Axx' + Bx + Cx' + D = 0$, the correspondents of a given point will be different, according as it is regarded as belonging to the first or second system. Thus the corespondent of the point, whose abscissa is α in one system, is $-\dfrac{B\alpha + D}{A\alpha + C}$, and in the other system $-\dfrac{C\alpha + D}{A\alpha + B}$; and these in general are different, unless when $B = C$, and then they always coincide. Two systems fulfilling this condition are said to be *in involution*. Thus the systems determined by the equation $Axx' + B(x + x') + D$ are in involution.

Cor. 1.—Three pairs of points, which are harmonic conjugates to a given pair, form a system in involution.

For any pair of points determined by the equation

$$Axx' + B(x + x') + D = 0$$

are harmonic conjugates to the pair of points $Ax^2 + 2Bx + D$.

Cor. 2.—If three pairs of points be in involution, the anharmonic ratio of any four points is equal to that of their four conjugates.

Cor. 3.—If O, O' be harmonic conjugates to any numbers of pairs of points in involution, each pair will be inverse points with respect to the circle described on OO' as diameter.

Def.—The points O, O' are called the *double points* of the involution, and their middle point its centre.

Examples.

1. A system of conics passing through four points cuts any transversal in involution.

2. A system of conics having a common self-conjugate triangle meets every transversal passing through any of its vertices in a system of points in involution.

Section VIII.—Theory of Reciprocal Polars.

180. We have already, in Art. 125, given some propositions connected with reciprocal polars. We propose, in this section, to give a more systematic account of this method of transformation, so important in Modern Geometry.

Def.—*If any figure A be given, by taking the pole of every line and the polar of every point in it with respect to any arbitrary conic S, we construct a new figure B, which is called the polar reciprocal of A with respect to S. The conic S is called the reciprocating conic.*

181. From the definition, we have at once the following results:—

A.	*B.*
1°. For a point on.	1°. A tangent to.
2°. A tangent to.	2°. A point on.
3°. A system of collinear points on.	3°. A pencil of concurrent lines.
4°. A pencil of concurrent lines.	4°. A system of collinear points.
5°. A pair of lines homographically divided.	5°. Two pencils of homographic lines.
6°. The join of two points.	6°. The intersection of two lines.
7°. The locus of a point.	7°. The envelope of a line.

182. The following are a few theorems proved by this method:—

EXAMPLES.

1. *Any two fixed tangents to a conic are cut homographically by any variable tangent.*

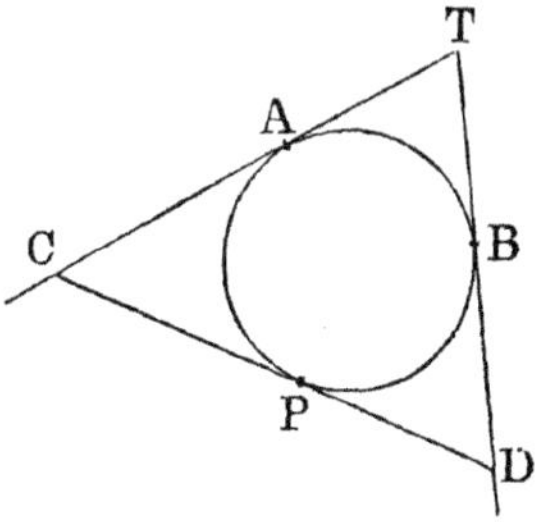

Let AT, BT be two fixed tangents, touching the conic at the points A, B; CD any variable tangent touching it at P. Join AP, BP. Now AP is the polar of C, and BP of D; and if P take four different positions, the point C will take four corresponding positions, and so will D. Then the anharmonic ratio of the four positions of C will be equal to the anharmonic ratio of the pencil from A to the four positions of P. Similarly, the anharmonic ratio of the four positions of D will be equal to the anharmonic ratio of the pencil from B to the same positions of P; but the pencils from A and B are equal. Hence the anharmonic ratio of the four positions of C is equal to the anharmonic ratio of the corresponding positions of D.

From the theorem just proved it follows, *that if two lines be divided equianharmonically by four others, the six lines are tangents to a conic.* And, more generally, *If two lines be divided homographically, the envelope of the join of corresponding points is a conic.*

2. *Any four fixed tangents to a conic are cut by a variable tangent in points whose anharmonic ratio is constant.*

Dem.—The joins of the point of contact of the variable tangent to the points of contact of the fixed tangents are the polars of the points of intersection of the variable tangent with the fixed ones; but these form a constant pencil. Hence the proposition is proved.

3. *If a hexagon be described about a conic, the joins of opposite angular points are concurrent.*

For the circumhexagon is the polar reciprocal of the inhexagon, and the joins of its opposite vertices are the polars of the intersection of opposite sides. Hence the proposition is the reciprocal of PASCAL'S Theorem.

4. The three pairs of points, in which a transversal meets three circumconics of a quadrilateral, are in involution.

5. The common tangent to any two of three circumconics of a quadrilateral is cut harmonically by the third conic. Hence, if three conics

S, S', S'' be inscribed in a quadrilateral; and if from P, a point of intersection of S, S', tangents be drawn to S'', these form a harmonic pencil with the tangents at P to S, S'.

6. From Ex. 2 it follows that the intercepts made on any variable tangent to a parabola made by three fixed tangents have a given ratio.

7. The reciprocal of Ex. 2, Art. 179, is—pairs of tangents to a system of conics having a common self-conjugate triangle, drawn from any point in one of its sides, form a pencil in involution.

8. The six sides of two inscribed triangles of a conic are such that any two are cut equianharmonically by the remaining four. Hence they touch another conic.

Reciprocally, if two triangles circumscribe a conic, the six vertices lie on another conic.

9. The locus of the pole of a given line, with respect to any circum-conic of a quadrilateral, is another conic. Hence the envelope of the polar of a given point, with respect to a conic inscribed in a quadrilateral, is a conic.

183. When the reciprocating conic is a circle, its centre is called the centre of reciprocation. The following results will be evident from a diagram :—

1°. The angle between any two lines is equal or supplemental to the angle at the centre of reciprocation subtended by the join of their poles.

2°. Since the nearer any line is to the centre of reciprocation the more remote its pole, it is evident that the pole of any line passing through the centre must be at infinity, and in the direction perpendicular to the line through the centre. Hence it follows, since two real tangents can be drawn from any external point O to a conic, that the polar reciprocal of that conic with respect to O is a hyperbola. Similarly, the polar reciprocal of any conic with respect to any point on it is a parabola, and its polar reciprocal with respect to any internal point is an ellipse.

3°. If a conic reciprocate into a hyperbola, the asymptotes of the hyperbola are perpendicular to the tangents drawn from the centre of reciprocation to the original curve.

4°. If a conic reciprocate into an equilateral hyperbola, the locus of the centre of reciprocation is the auxiliary circle.

5°. The polar of the centre of reciprocation with respect to any conic will reciprocate into the centre of the reciprocal conic.

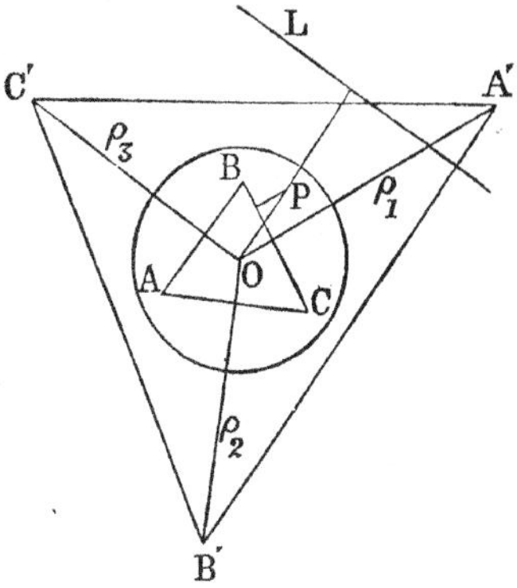

6°. If the original conic be a circle, its centre will reciprocate into the directrix.

184. If O be the centre of reciprocation; ABC the triangle of reference for trilinear co-ordinates; $A'B'C'$ its reciprocal; L the polar of any point P; $\lambda_1, \lambda_2, \lambda_3$ perpendiculars from $A'B'C'$ on L; and $\alpha_1, \alpha_2, \alpha_3$ the trilinear co-ordinates of P; then (Sequel, Book III., Prop. 27), if OA', OB', OC' be denoted by ρ_1, ρ_2, ρ_3, we have

$$\alpha_1 = OP.\frac{\lambda_1}{\rho'}, \text{ \&c.}$$

Hence, if $(a, b, c, f, g, h)(\alpha_1, \alpha_2, \alpha_3)^2 = 0$ be the equation of any conic, the equation of its reciprocal with respect to the circle O will be

$$(a, b, c, f, g, h)\left(\frac{\lambda_1}{\rho_1}, \frac{\lambda_2}{\rho_2}, \frac{\lambda_3}{\rho_3}\right)^2 = 0. \qquad (609)$$

Again, if $(A, B, C, F, G, H)(\lambda_1, \lambda_2, \lambda_3)^2 = 0$ be the tangential equation of a conic, where $\lambda_1, \lambda_2, \lambda_3$ denote perpendiculars from the angles A', B', C' of the triangle of reference on any tangent L to the conic; then if x_1, x_2, x_3 be the trilinear co-ordinates of O with respect to the reciprocal triangle ABC, we have $x_1\rho_1 = r^2$, where r is the radius of reciprocation. Hence, eliminating ρ_1 between this equation

and $$\alpha_1 = \frac{OP.\lambda_1}{\rho_1},$$

we get $$\lambda_1 = \frac{r^2}{OP}\frac{\alpha_1}{x_1},$$

and similar values for λ_2 λ_3. Hence the transformed equation is—

$$(A, B, C, F, G, H)\left(\frac{\alpha_1}{x_1}, \frac{\alpha_2}{x_2}, \frac{\alpha_3}{x_3}\right)^2 = 0. \quad (610)$$

EXAMPLES.

1. The equation of the circumcircle of the triangle of reference is—

$$\frac{\sin A}{\alpha_1} + \frac{\sin B}{\alpha_2} + \frac{\sin C}{\alpha_3} = 0.$$

Now it is easy to see that the angles A, B, C of the old triangle of reference will be the supplements of the angles, which the sides of the new triangle of reference subtend at the centre of reciprocation. Hence, denoting these angles by ψ_1, ψ_2, ψ_3, respectively, the result of reciprocation gives the following theorem:—*Given a focus and a triangle circumscribed to a conic, its tangential equation is*—

$$\sin \psi_1 \cdot \frac{\rho_1}{\lambda_1} + \sin \psi_2 \cdot \frac{\rho_2}{\lambda_2} + \sin \psi_3 \cdot \frac{\rho_3}{\lambda_3} = 0. \quad (611)$$

2. If a polygon of any number of sides be inscribed in a circle, and if the angles which the sides subtend at any point in the circumference be denoted by ψ_1, ψ_2, ψ_3, &c., we have (Art. 79), if $\alpha_1 = 0$, $\alpha_2 = 0$, $\alpha_3 = 0$, &c., be the standard equations of its sides, $\Sigma \dfrac{\sin \psi_1}{\alpha_1} = 0$. Hence, reciprocating with respect to any point in the circumference, we get the following theorem:—*If a polygon of any number of sides circumscribe a parabola, and if* ψ_1, ψ_2, ψ_3, *&c., be the angles subtended at its focus by the sides of the polygon* λ_1, λ_2, λ_3, *&c., perpendiculars from the vertices on any tangent* ρ_1, ρ_2, ρ_3, *&c., the distances of the angular points from the focus, then*

$$\Sigma \frac{\sin \psi_1 \cdot \rho_1}{\lambda_1} = 0. \quad (612)$$

3. In equation (270), if we put $\sin A$, $\sin B$, $\sin C$ for a, b, c, the tangential equation of the circumcircle of the triangle of reference may be written

$$\sin A \sqrt{\lambda_1} + \sin B \sqrt{\lambda_2} + \sin C \sqrt{\lambda_3} = 0.$$

Hence, by the foregoing substitutions, being given a focus and three tangents, the equation of the conic is

$$\sin \psi_1 \sqrt{\frac{\alpha_1}{x_1}} + \sin \psi_2 \sqrt{\frac{\alpha_2}{x_2}} + \sin \psi_3 \sqrt{\frac{\alpha_3}{x_3}} = 0. \quad (613)$$

4. If the focus be one of the Brocard points, viz., the point whose co-ordinates are—

$$\frac{c}{b}, \frac{a}{c}, \frac{b}{a};$$

then the angles ψ_1, ψ_2, ψ_3, which the sides subtend at that point, are the supplements of the angles C, A, B, respectively. Hence the equation of the Brocard ellipse, that is, the inscribed ellipse, whose foci are the Brocard points, is—

$$\sqrt{\frac{\alpha_1}{a}} + \sqrt{\frac{\alpha_2}{b}} + \sqrt{\frac{\alpha_3}{c}} = 0. \qquad (614)$$

5. If the angles of a polygon circumscribed to a circle be denoted by A, B, C, &c., and the perpendiculars from its angular points on any tangent to the circle by λ_1, λ_2, &c., we have

$$\Sigma \left(\frac{\cot \frac{1}{2} A}{\lambda_1}\right) = 0.$$

Hence, if a polygon of any number of sides be inscribed in a conic; and if x_1, x_2, x_3, &c., be the perpendiculars from one of its foci on the sides, and ψ_1, ψ_2, &c., the angles subtended at that focus by the sides, we have

$$\Sigma \left(\frac{x_1 \tan \frac{1}{2} \psi 1}{\alpha_1}\right) = 0. \qquad (615)$$

Section IX.—Invariants and Covariants.

185. Def. i.—*An invariant is a function of the coefficients of the equation of a curve expressed in point co-ordinates, which remains unaltered by linear transformation.*

Def. ii.—*A covariant is a function of both coefficients and variables, which is unaltered by linear transformation.*

Def. iii.—*If the equation of the curve be expressed in line co-ordinates, the functions corresponding to invariants and co-variants are called contravariants.*

186. If $S = 0$, $S' = 0$ be the equations of two conics, and if by linear transformation they become $\overline{S}$, $\overline{S'}$, it is evident that the conic $S + kS' = 0$ will become by the same transformation $\overline{S} + k\overline{S'} = 0$. Hence, if k be determined so as to make $S + kS' = 0$ fulfil some special condition—such, for

instance, as to represent an equilateral hyperbola, or to touch a given line—the same value of k will make $\overline{S} + k\overline{S'} = 0$ fulfil the same condition. Now if in any function of the coefficients expressing a special property of S we substitute $a + ka'$, $b + kb'$, &c., for a, b, &c., the resulting equation in k will represent the same property for the conic $S + kS' = 0$; and since the value of k remains unaltered by transformation of co-ordinates, the ratios of the coefficients of the several powers of k will be unaltered by transformation of co-ordinates. Hence the ratios of the coefficients will be invariants.

EXAMPLES.

1. If $S \equiv (a, b, c, f, g, h)(\alpha, \beta, \gamma)^2 = 0$, $S' \equiv (a', b', c', f', g', h')(\alpha, \beta, \gamma)^2 = 0$, it is required to find the value of k, for which $S + kS' = 0$ represents an equilateral hyperbola.

If $S = 0$ represent an equilateral hyperbola, putting it into Cartesian co-ordinates, and equating to zero the sum of the coefficients of x^2 and y^2, we get

$$a + b + c - 2f \cos A - 2g \cos B - 2h \cos C = 0.$$

Hence the value of k, for which $S + kS' = 0$ represents an equilateral hyperbola, is obtained from the equation

$$(a + b + c - 2f \cos A - 2g \cos B - 2h \cos C)$$
$$+ k(a' + b' + c' - 2f' \cos A - 2g' \cos B - 2h' \cos C) = 0;$$

or, say, $\Theta + k\Theta' = 0$; eliminating k between this and $S + kS' = 0$, we get

$$\Theta' S - \Theta S' = 0. \qquad (616)$$

Cor. 1.—In general, only one equilateral hyperbola can be described through four points.

Cor. 2.—If $S = 0$, $S' = 0$ denote equilateral hyperbolas, $S + kS' = 0$ will be an equilateral hyperbola for every value of k.

2. Find the value of k, for which the conic $S + kS' = 0$ will touch the line $\lambda_x = 0$.

Let $S \equiv a_x^2 = 0$, $S' \equiv b_x^2 = 0$; then the condition that S touch λ_x is—

$$\begin{vmatrix} a_{11}, & a_{12}, & a_{13}, & \lambda_1, \\ a_{21}, & a_{22}, & a_{23}, & \lambda_2, \\ a_{31}, & a_{32}, & a_{33}, & \lambda_3, \\ \lambda_1, & \lambda_2, & \lambda_3, & 0 \end{vmatrix}$$

and substituting $a_{11}+kb_{11}$, $a_{12}+kb_{12}$, &c., for a_{11}, a_{12}, &c., we get a result which may be written

$$A_\lambda{}^2 + k\,\{(a_2b_3 - a_3b_2)\lambda_1 + (a_3b_1 - a_1b_3)\lambda_2 + (a_1b_2 - a_2b_1)\lambda_3\}^2 + k^2B_\lambda{}^2 = 0;$$

or, denoting the middle term by ϕ,

$$A_\lambda{}^2 + k\phi + k^2B_\lambda{}^2 = 0. \qquad (617)$$

Cor. 1.—Two conics of the pencil $S + kS' = 0$ touch $\lambda_2 = 0$.

For the equation (617) in k is of the second degree. The equation of these conics is got by eliminating k between the equation (617) and $S + kS' = 0$.

Cor. 2.—By supposing $\lambda_x = 0$ to be the line at infinity, we see that two conics of the pencil $S + kS' = 0$ denote parabolas.

Cor. 3.—If the line $\lambda_x = 0$ pass through one of the points of intersection of S, S', it is evident only one conic can be drawn to touch λ_x, and the two values of k in equation (617) will be equal.. Hence the discriminant of that equation is the condition that λ_x will pass through one of the four points of intersection of the conics S, S', or, in other words, the tangential equation of the four points of intersection of the conics

$$a_x{}^2 = 0,\ b_x{}^2 = 0 \quad \text{is} \quad 4A_\lambda{}^2 . B_\lambda{}^2 - \phi_2 = 0. \qquad (618)$$

Cor. 4.—The locus of the pole of $\lambda_x = 0$, with respect to all the conics of the pencil $S + kS' = 0$, is a conic.

For since two conics of the pencil touch λ_x, the locus will meet $\lambda_x = 0$ in two points. Hence the locus is a conic.

3. Find the value of k, for which $a_x{}^2 + k(\lambda_x)^2$ represents two right lines. The discriminant is—

$$\begin{vmatrix} a_{11} + k\lambda_1{}^2, & a_{12} + k\lambda_1\lambda_2, & a_{13} + k\lambda_1\lambda_3, \\ a_{21} + k\lambda_2\lambda_1, & a_{22} + k\lambda_2{}^2, & a_{23} + k\lambda_2\lambda_3, \\ a_{31} + k\lambda_3\lambda_1, & a_{32} + k\lambda_3\lambda_2, & a_{33} + k\lambda_3{}^2 \end{vmatrix} = 0;$$

or, denoting the discriminant of $a_x{}^2 = 0$ by Δ,

$$\Delta + kA_\lambda{}^2 = 0;$$

eliminating k between this and the equation $a_x{}^2 + k(\lambda_x)^2$, we get the equation of the pair of tangents drawn to $a_x{}^2$, at the points where it meets $\lambda_x = 0$,

$$A_\lambda{}^2 . a_x{}^2 - \Delta(\lambda_x) = 0. \qquad (619)$$

Cor.—By supposing $\lambda_x = 0$ to be the line at infinity, we get the equation of the asymptotes. (Compare Art. 105, *Cor.* 4.)

4. If $A\lambda^2 = 0$, $B\lambda^2 = 0$ be the tangential equations of two conics, it is required to determine the value of k, in order that the conic $A_\lambda^2 + kB_\lambda^2 = 0$, which is inscribed in the quadrilateral formed by their four common tangents, may pass through the point whose equation $x_\lambda = 0$.

If the equations in point co-ordinates corresponding to $A_\lambda^2 = 0$, $B_\lambda^2 = 0$, be $a_x^2 = 0$, $b_x^2 = 0$; and Δ, Δ' denote their discriminants, the equation corresponding to $A_\lambda^2 + kB_\lambda^2 = 0$ is

$$\Delta a_x^2 + kF + k^2\Delta' b_x^2 = 0. \tag{620}$$

Where F has the value in equation (605), and since this must pass through the point $x_\lambda = 0$, we have a quadratic to determine k. Hence two conics of the system $A_\lambda^2 + kB_\lambda^2 = 0$ will pass through the point; if the point be on one of the four common tangents of $A_\lambda^2 = 0$, $B_\lambda^2 = 0$, the two conics will coincide; and we infer, as in Ex. 2, that the equation of the four common tangents to the two conics $a_x^2 = 0$, $b_x^2 = 0$ is

$$F^2 = 4\Delta\Delta'\, a_x^2 b_x^2. \tag{621}$$

Cor. 1.—The envelope of the polar of a given point with respect to a system of conics inscribed in a given quadrilateral is a conic.

For, since the equation (620) is of the second degree, two conics of the system pass through the given point. Hence two tangents to the envelope can be drawn through it. Therefore the envelope is a conic.

187. *Three conics of the pencil* $S + kS' = 0$ *represent line-pairs.*

Dem.—Let $S \equiv a_x^2 = 0$, $S' \equiv b_x^2 = 0$; then the discriminant of $S + kS'$ is—

$$\begin{vmatrix} a_{11} + kb_{11}, & a_{12} + kb_{12}, & a_{13} + kb_{13} \\ a_{21} + kb_{21}, & a_{22} + kb_{22}, & a_{23} + kb_{23} \\ a_{31} + kb_{31}, & a_{32} + kb_{32}, & a_{33} + kb_{33} \end{vmatrix};$$

or $\Delta + \Theta k + \Theta' k^3 + \Delta' k_3$, where Δ, Δ' are the discriminants of a_x^2, b_x^2, respectively, and $\Theta \equiv A_b^2$, $\Theta' \equiv B_a^2$. Hence the condition required is

$$\Delta + \Theta k + \Theta' k^2 + \Delta' k^3 = 0; \tag{622}$$

giving three values for k, which proves the proposition.

Cor.—By eliminating k between $S + kS' = 0$ and the equation (622), we get the equation of the three line-pairs, viz.—

$$\Delta S'^3 - \Theta S'^2 S + \Theta' S' S^2 - \Delta' S^3 = 0. \qquad (623)$$

188. The equation (622) is called the *invariant equation* of the pencil of conics $S + kS' = 0$. It will be found that every relation which is independent of the axes can be expressed in terms of its coefficients. We shall now examine the geometrical interpretation of the vanishing of some of these coefficients.

1°. If $\Delta' = 0$, the equation (622) reduces to a quadratic: this happens when one of the conics S, S' denotes a line-pair.

Examples.

1. Find the equation of the bisectors of the angles of the line-pair $ax^2 + 2hxy + by^2 = 0$, the axes being oblique. The equation $x^2 + y^2 + 2xy \cos\omega - r^2 = 0$ represents a circle. Hence the quadratic in k, which is the discriminant of

$$ax^2 + 2hxy + by^2 + k(x^2 + y^2 + 2xy\cos\omega - r^2) = 0,$$

or of

$$(a + k)x^2 + (b + k)y^2 + 2(h + k\cos\omega)xy - kr^2 = 0,$$

will evidently give us two line-pairs, which, from the property of the circle, will be such that each pair denotes parallel lines, and such that one pair is perpendicular to the other. Now, make $r = 0$ in the equation of the circle, and each line-pair becomes a perfect square; hence, making $r = 0$, the discriminant of

$$(a + k)x^2 + (b + k)y^2 + 2(h + k\cos\omega)xy$$

is

$$(a + k)(b + k) - (h + k\cos\omega)^2 = 0;$$

and, eliminating k between these equations, we get

$$\{(a\cos\omega - h)x^2 + (a - b)xy + (h - b\cos\omega)y^2\}^2 = 0. \qquad (624)$$

Hence the equation of the bisectors is

$$(a\cos\omega - h)x^2 + (a - b)xy + (h - b\cos\omega)y^2 = 0.$$

The same result may be obtained by differentiating the equation

$$(a + k)x^2 + (b + k)y^2 + 2(h + k\cos\omega)xy$$

with respect to x and y, respectively, and eliminating k between the differentials.

2. When $\Delta' = 0$. If the quadratic $\Delta + \Theta k + \Theta' k^2 = 0$, to which the invariant equation reduces, has equal roots, only one line-pair distinct from S' can be drawn through the intersection of S and S'. This can happen only when S touches one of the lines of S', as is evident from a figure. Hence the condition that S touch one of the lines of S' is,

$$\Theta^2 = 4\Delta\Theta'. \tag{625}$$

2°. When $\Theta = 0$.

The full value of Θ for the equations

$$(a, b, c, f, g, h)(\alpha, \beta, \gamma)^2 = 0, \quad (a', b', c', f', g', h')(\alpha, \beta, \gamma)^2 = 0,$$

is $$(bc - f^2)\,a' + (ca - g^2)\,b' + (ab - h^2)\,c' + 2\,(gh - af)\,f'$$

$$+ 2\,(fg - ch)\,g' + 2\,(fg - ch)\,h' = 0. \tag{626}$$

Hence, if a', b', c', f, g, h, each separately vanishes, Θ vanishes; that is, if the equations of the two conics be of the forms

$$a\alpha^2 + b\beta^2 + c\gamma^2 = 0, \quad f'\beta\gamma + g'\gamma\alpha + h'\alpha\beta = 0\,;$$

or, in other words, Θ *vanishes when S' is a conic described about a triangle which is self-conjugate with respect to* S. Again, Θ vanishes: if f', g', h', $bc - f^2$, $ca - g^2$, $ab - h^2$, each vanish, which will happen, if the equations of the conics can be written in the forms

$$\sqrt{l\alpha} + \sqrt{m\beta} + \sqrt{n\gamma} = 0, \quad a'\alpha^2 + b'\beta^2 + c'\gamma^2 = 0\,;$$

that is, if S can be inscribed in a triangle, which is self-conjugate with respect to S'.

Similar results may be inferred from the vanishing of Θ'.

EXAMPLES.

1. If four equilateral hyperbolas have a common point, and be connected with the conic S' by the relation $\Theta = 0$, the points of intersection of any pair and those of the remaining pair lie on an equilateral hyperbola.

(PROF. CURTIS, S.J.)

For, taking the common point as origin of co-ordinates, and the four hyperbolas as S_1, S_2, S_3, S_4, where

$$S \equiv a_1(x^2 - y^2) + 2h_1(xy) + 2g_1x + 2f_1y = 0, \quad S_2 \equiv a_2(x^2 - y^2) + \&c.,$$

we have from the given conditions four equations of the form

$$a_1(A - B) + 2h_1H + 2g_1G + 2f_1F = 0;$$

therefore,

$$\begin{vmatrix} a_1, & h_1, & g_1, & f_1, \\ a_2, & h_2, & g_2, & f_2, \\ a_3, & h_3, & g_3, & f_3, \\ a_4, & h_4, & g_4, & f_4 \end{vmatrix} = 0.$$

Hence, multiplying the first column by $x^2 - y^2$, the second by $2xy$, the third by $2x$, the fourth by $2y$, and adding the other columns to the first, we get—

$$\begin{vmatrix} S_1, & h_1, & g_1, & f_1, \\ S_2, & h_2, & g_2, & f_2, \\ S_3, & h_3, & g_3, & f_3, \\ S_4, & h_4, & g_4, & f_4 \end{vmatrix} = 0;$$

or, as it may be written, $lS_1 - mS_2 + nS_3 - pS_4 \equiv 0$. Therefore the equilateral hyperbola $lS_1 - mS_2 = 0$, passing through the intersection of S_1, S_2, is identical with the equilateral hyperbola $pS_4 - nS_3 = 0$, passing through the intersection of S_3, S_4.

2. If two conics S_1, S_2 be homothetic, and connected with S' by the relation Θ', their common chord passes through the centre of S'. (*Ibid.*)

From the hypothesis, we have

$$\frac{a_1}{a_2} = \frac{h_1}{h_2} = \frac{b_1}{b_2};$$

and

$$a_1A' + 2h_1H' + b_1B' + 2f_1F' + 2g_1G' + c_1C' = 0;$$

$$a_2A' + 2h_2H' + b_2B' + 2f_2F' + 2g_2G' + c_2C' = 0;$$

therefore $$2(f_1a_2 - f_2a_1)\frac{F'}{C'} + 2(g_1a_2 - g_2a_1)\frac{G'}{C'} + c_1a_2 - c_2a_1 = 0;$$

but $$\frac{G'}{C'}, \quad \frac{F'}{C'}$$

are the co-ordinates of the centre of S'. Hence the proposition is proved.

3. If a variable conic S be connected with four fixed conics S_1, S_2, S_3, S_4 by the relation $\Theta' = 0$, the locus of its centre is a right line. (*Ibid.*)

From the hypothesis, we have four relations of the form

$$\frac{A}{C}a_1 + \frac{B}{C}b_1 + \frac{H}{C}h_1 + \left(\frac{2F}{C}f_1 + \frac{2G}{C}g_1 + c_1\right) = 0;$$

and eliminating $$\frac{A}{C'} \quad \frac{B}{C'} \quad \frac{H}{C'}$$

we get a linear relation between $\frac{F}{C'}$, $\frac{G}{C'}$

which are the co-ordinates of the centre of S.

Cor. 1.—If S_1, S_2, &c., break up into line-pairs, we have the theorem that the locus of the centre of a conic, which has four given pairs of conjugate lines, is a right line.

Cor. 2.—If S_1, S_2, &c., become perfect squares, they must be the squares of lines touching S. Hence the locus of the centre of a conic inscribed in a given quadrilateral is a right line.

189.— EXAMPLES OF INVARIANTS.

1. Calculate the invariant equation for the conics

$$a\alpha^2 + b\beta^2 + c\gamma^2 = 0, \quad \alpha^2 + \beta^2 + \gamma^2 = 0.$$

Ans. $(k+a)(k+b)(k+c)$. (627)

2. Calculate the invariant equation for the conic

$$\frac{x^2}{a^2} + \frac{y^2}{b^2} - 1 = 0,$$

and the circle $$(x - x')^2 + (y - y')^2 = r^2.$$

Ans. $$\frac{x'^2}{a^2 - k} + \frac{y'^2}{b^2 - k} + \frac{r^2}{k} - 1 = 0. \quad (628)$$

Hence $\Theta = \frac{1}{a^2b^2}(x'^2 + y'^2 - a^2 - b^2 - r^2)$; and, therefore, if Θ vanish, $x'^2 + y'^2 = a^2 + b^2 + r^2$. Hence we have the theorem that a circumcircle of a triangle self-conjugate with respect to a central conic cuts its director circle orthogonally. (FAURE.)

3. Calculate the invariants for the conics

$$\sqrt{\alpha} + \sqrt{\beta} + \sqrt{\gamma} = 0 \quad \text{and} \quad 2f\beta\gamma + 2g\gamma\alpha + 2h\alpha\beta = 0.$$

Ans. $\Delta = -4$, $\Theta = 4(f+g+h)$, $\Theta' = -(f+g+h)^2$, $\Delta' = 2fgh$.

Hence $$\Theta^2 = 4\Delta\Theta'; \quad (629)$$

which is therefore the condition that a triangle may be inscribed in one conic, and circumscribed to another. (CAYLEY.)

4. If a variable triangle inscribed in one conic be circumscribed to another, it is self-conjugate with respect to a fixed conic.

(R. A. ROBERTS.)

For, forming the equation of the conic F, which is the locus of points whence tangents to the conics S, S' (Ex. 3) form a harmonic pencil, we have

$$F \equiv 4(f+g+h)(f\beta\gamma + g\gamma\alpha + h\alpha\beta) - 4(gh\alpha^2 + hf\beta^2 + fg\gamma^2) = 0,$$

it is evident that $gh\alpha^2 + hf\beta^2 + fg\gamma^2$ is expressible in the form $2F - \Theta S = 0$, which proves the proposition.

5. Calculate the invariants for the Brocard circle

$$abc\,(\alpha^2 + \beta^2 + \gamma^2) - (a^3\beta\gamma + b^3\gamma\alpha + c^3\alpha\beta) = 0;$$

and the Brocard ellipse

$$\sqrt{\frac{\alpha}{a}} + \sqrt{\frac{\beta}{b}} + \sqrt{\frac{\gamma}{c}} = 0;$$

where a, b, c denote the lengths of the sides of the triangle of reference.

Ans. $\Delta = -\dfrac{4}{a^2b^2c^2}$, $\Theta = -\left(\dfrac{a^2+b^2+c^2}{abc}\right)$, $\Theta' = -\frac{1}{4}\{(a^2+b^2+c^2)^2 - 6(a^4+b^4+c^4)\}$,

$$\Delta' = -\frac{abc}{4}(a^6 + b^6 + c^6 - 3a^2b^2c^2).$$

In terms of these and of the radius of the circumcircle can be expressed several metrical relations in the recent geometry of the triangle. Thus, if ρ, ρ' denote the radii of the Lemoine and Brocard circles,

$$3\rho^2 = R^2\left(\frac{\Theta^3 + \Delta\Delta'}{\Theta^3}\right), \quad \rho'^2 = -\frac{\Delta\Delta'}{\Theta^3}R^2. \qquad (630)$$

190. If the conics S, S' touch, it is evident from a diagram that instead of three distinct line-pairs, passing through their intersection, we shall have only two: this will happen because one of the three line-pairs coincides with another. Hence the invariant equation (622) will have two equal roots. Thus the condition for the contact of two conics, called their TACT-INVARIANT, is the vanishing of the discriminant of the equation (622), viz.,

$$\Theta^2\Theta'^2 + 18\Delta\Delta'\Theta\Theta'^2 - 27\Delta^2\Delta'^2 - 4(\Delta\Theta'^3 + \Delta'\Theta^3) = 0. \qquad (631)$$

The Tact-invariant just written is the product of six anharmonic ratios.

Dem.—Let the conics be referred to their common self-conjugate triangle; then their equations may be written

$$a\alpha^2 + b\beta^2 + c\gamma^2 = 0, \quad \alpha^2 + \beta^2 + \gamma^2 = 0,$$

and the cubic in k will be (Art. 189, Ex. 1),

$$(k + a)(k + b)(k + c) = 0.$$

Again, if we denote by θ', θ'', θ''', the angles of the anharmonic ratios (Chap. II., Ex. 32), in which the sides of the triangle of reference are intersected by the two conics; then, to determine θ', we must find the anharmonic ratio in which the side $\alpha = 0$ is intersected by the two conics. After an easy calculation, we get

$$\sin^2 \tfrac{1}{2}\theta' = -\frac{(b^{\frac{1}{2}} - c^{\frac{1}{2}})^2}{4b^{\frac{1}{2}}c^{\frac{1}{2}}}, \qquad \cos^2 \tfrac{1}{2}\theta' = \frac{(b^{\frac{1}{2}} + c^{\frac{1}{2}})^2}{4b^{\frac{1}{2}}c^{\frac{1}{2}}}.$$

Hence $$\sin^2\theta = -\frac{(b-c)^2}{4bc}.$$

Now, if we denote the roots of the invariant equation by k', k'', k''', since these roots are $-a$, $-b$, $-c$, we have the following system of equations:—

$$\sin^2\theta' = -(k'' - k''')^2 \div 4k''k''',$$
$$\sin^2\theta'' = -(k''' - k')^2 \div 4k'''k',$$
$$\sin^2\theta''' = -(k' - k'')^2 \div 4k'k''.$$

Hence the condition that the discriminant of $S + kS' = 0$ will give two equal values of k is

$$\sin^2\theta' . \sin^2\theta'' . \sin^2\theta''' = 0;$$

and as $\sin^2\theta'$, &c., are each the product of two anharmonic ratios, the proposition is proved.

191. *To find the Tact-invariant of two conics $S - L^2$, $S - M^2$, both inscribed in the same conic $S \equiv \alpha^2 + \beta^2 + \gamma^2 = 0$.*

If we write the equations in the forms $S^{\frac{1}{2}} - L = 0$, $S^{\frac{1}{2}} - M = 0$, $S^{\frac{1}{2}} - L + k(S^{\frac{1}{2}} - M) = 0$ denotes a conic passing through the two points, in which the common chord $L - M$ meets the conics $S - L^2$, $S - M^2$; and forming the discriminant of $S^{\frac{1}{2}} - L + k(S^{\frac{1}{2}} - M)$, after clearing of radicals, we get

$$(1 - S'')k^2 + 2(1 - R)k + (1 - S') = 0, \qquad (632)$$

where S', S'' denote the powers of the poles of L and M with respect to S, and R the power of the pole of L with respect to M. Now, since the equation (632) is of the second degree in k, two line-pairs can be drawn through the intersection of the conics $S - L^2$, $S - M^2$ with the common chord $L - M$, each having double contact with S, as is geometrically evident. Hence the condition of contact of $S - L^2$, $S - M^2$ is, that these line-pairs coincide; that is, the vanishing of the discriminant of (632) with respect to k. Therefore the Tact-invariant of the conics is

$$(1 - R)^2 - (1 - S')(1 - S'') = 0. \qquad (633)$$

We should have got the same result if we had worked with the equation $S^{\frac{1}{2}} + L + k(S^{\frac{1}{2}} + M) = 0$; but if we had used the equation $S^{\frac{1}{2}} \mp L + k(S^{\frac{1}{2}} \pm M)$, we should have got

$$(1 + R)^2 - (1 - S')(1 - S'') = 0. \qquad (634)$$

Hence there are two Tact-invariants.

Cor. 1.—If we put $1 - R = \sqrt{(1 - S')(1 - S'')} \cos\theta$, we get, denoting the roots of (632) by k_1, k_2,

$$e^{2\theta\sqrt{-1}} = \frac{k_1}{k_2}; \qquad (635)$$

and if we had formed the discriminant of $S^{\frac{1}{2}} \mp L + k(S^{\frac{1}{2}} \pm M) = 0$, and denoted the roots of the resulting equation in k by k_3, k_4, we should get, putting $1 + R = \sqrt{(1 - S')(1 - S'')} \cos\phi$,

$$e^{2\phi\sqrt{-1}} = \frac{k_3}{k_4}.$$

Hence the Tact-invariants of the conics $S - L^2$, $S - M^2$ are equivalent to the equations $\theta = 0$, $\phi = 0$.

Cor. 2.—If $\theta = \frac{\pi}{2}$, $\frac{k_1}{k_2} = -1$, and the pencil is harmonic which is formed of the lines L, M and the chords of contact of the two line-pairs, which can be drawn to touch S through the intersection of $L - M$ with $S - L^2$; hence, what corresponds, in the geometry of conics inscribed in a given conic, to two circles cutting orthogonally, are two conics whose angle θ is right, we shall by an extension of the term say that the conics cut *orthogonally*. In general, to the angle of intersection of two conics corresponds the angle θ of the conics.

EXAMPLES.

1. If Σ, Σ' be the tangential equations of S, S', the discriminant of $\Sigma + k\Sigma'$ is

$$\Delta^2 + k\Delta\Theta' + k^2\Delta'\Theta + k^3\Delta'^2. \tag{636}$$

2. If the conic S' be the product of two lines, $\Theta = 0$ is the condition that they should be conjugate with respect to S; for $\Theta = 0$ is the condition that a triangle self-conjugate with respect to S can be inscribed in S'; and when S' denotes a pair of lines, this implies that they must be conjugate lines with respect to S.

3. If Σ, Σ' be tangential equations, and if Σ' denote two points, $\Theta' = 0$ is the condition that their join should be cut harmonically by Σ. Hence, if Σ' represent the circular points at infinity, $\Theta' = 0$ is the condition that Σ shall be an equilateral hyperbola.

4. If a system of conics touch two rectangular lines OX, OY in two fixed points, the normals in these points intersect in a point P. Prove that the line joining the feet of the two other normals drawn from P to each conic passes through a fixed point.

5. If $\Theta = 0$, the centre of perspective of any triangle inscribed in S', and its reciprocal with respect to S, is a point on S'. (SALMON.)

6. In the same case, the axis of perspective of any triangle circumscribed to S, and its reciprocal with respect to S', is a tangent to S. (*Ibid.*)

7. The polar reciprocal of S with respect to S' is

$$\Theta S' - F. \tag{637}$$

8. If the polar reciprocal of S with respect to S' be denoted by S''; prove that the invariant angles (see Art. 190) are $2\theta'$, $2\theta''$, $2\theta'''$.

9. The envelope of the line λ_a, cut harmonically by the conics S, S (Art. 190), is

$$\left(\frac{\cos\theta'}{\sqrt{k'}}\right)\lambda_1^2 + \left(\frac{\cos\theta''}{\sqrt{k''}}\right)\lambda_2^2 + \left(\frac{\cos\theta'''}{\sqrt{k'''}}\right)\lambda_3^2 = 0. \qquad (638)$$

10. Prove that the conic F, which is the locus of points whence tangents to S, S' form a harmonic pencil, may be written in the form

$$\left(\sqrt{k'}\cos\theta'\right)\alpha_1^2 + \left(\sqrt{k''}\cos\theta''\right)\alpha_2^2 + \left(\sqrt{k'''}\cos\theta'''\right)\alpha_3^2 = 0. \qquad (639)$$

11. If the relation $1 - R - \sqrt{(1-S')(1-S'')}$ be denoted by (12), &c.; rove that the invariants of four conics inscribed in a given conic S, and tangential to a fifth conic, also inscribed in S, are connected by the relation

$$\sqrt{(12)(34)} \pm \sqrt{(23)(14)} \pm \sqrt{(31)(24)} = 0. \qquad (640)$$

This theorem was discovered by me in 1867, and published in my *Bicircular Quartics*, which was read before the Royal Irish Academy in that year. The following proof is due to Dr. Salmon:—

Let the conic S be $x_1^2 + x_2^2 + x_3^2$, and let L be λ_x, $M = \mu_x$, &c.; then

$$S' = \lambda_1^2 + \lambda_2^2 + \lambda_3^2,\quad S'' = \mu_1^2 + \mu_2^2 + \mu_3^2,\quad R = \lambda_1\mu_2 + \lambda_2\mu_2 + \lambda_3\mu_3, \text{ or } \lambda_\mu;$$

then the Tact-Invariant $S - L^2$, $S - M^2$ (see Art. 191) is (12).

Now, let us multiply the two matrices, each containing five columns and six rows—

$$\begin{vmatrix} 1, & 0, & 0, & 0, & 0 \\ 1, & \lambda_1, & \lambda_2, & \lambda_3, & \sqrt{(1-S')}, \\ 1, & \mu_1, & \mu_2, & \mu_3, & \sqrt{(1-S'')}, \\ 1, & \nu_1, & \nu_2, & \nu_3, & \sqrt{(1-S''')}, \\ 1, & \rho_1, & \rho_2, & \rho_3, & \sqrt{(1-S_4)}, \\ 1, & \pi_1, & \pi_2, & \pi_3, & \sqrt{(1-S_5)}, \end{vmatrix} \quad \begin{vmatrix} 0, & 0, & 0, & 0, & 1, \\ -1, & \lambda_1, & \lambda_2, & \lambda_3, & \sqrt{(1-S')}, \\ -1, & \mu_1, & \mu_2, & \mu_3, & \sqrt{(1-S'')}, \\ -1, & \nu_1, & \nu_2, & \nu_3, & \sqrt{(1-S''')}, \\ -1, & \rho_1, & \rho_2, & \rho_3, & \sqrt{(1-S_4)}, \\ -1, & \pi_1, & \pi_1, & \pi_3, & \sqrt{(1-S_5)} \end{vmatrix}$$

The result must be equal zero, since there are more rows than columns.

Hence

$$\begin{vmatrix} 0, & 1, & 1, & 1, & 1, & 1, \\ \sqrt{(1-S')}, & 0, & (12), & (13), & (14), & (15), \\ \sqrt{(1-S'')}, & (12), & 0, & (23), & (24), & (25), \\ \sqrt{(1-S''')}, & (13), & (23), & 0, & (34), & (35), \\ \sqrt{(1-S_4)}, & (14), & (24), & (34), & 0, & (45), \\ \sqrt{(1-S_5)}, & (15), & (25), & (35), & (45), & 0 \end{vmatrix} = 0.$$

A relation between the invariants of five conics inscribed in the same conic S. Suppose now that the conic (5) touches the other four, then (15), &c., vanish, and we have the theorem that the invariants of four circles, all inscribed in the same conic S, and tangential to the same fifth, are connected by the relation

$$\begin{vmatrix} 0, & (12), & (13), & (14), \\ (12), & 0, & (23), & (24), \\ (13), & (23), & 0, & (34), \\ (14), & (24), & (34), & 0 \end{vmatrix} = 0,$$

or

$$\sqrt{(12)(34)} \pm \sqrt{(13)(24)} \pm \sqrt{(14)(23)} = 0.$$

Cor.—This expressed in terms of the invariant angles θ, &c., is identical in form with the corresponding theorem for circles.

12. If corresponding lines of two figures directly similar be conjugate to each other with respect to a given conic, the envelope of each line is a conic.

13. In the same case, the join of the points, where each line touches its envelope, is a conic.

14. If the two conics $S^{\frac{1}{2}} - L$, $S^{\frac{1}{2}} - M$ be connected by the relation $1 - R = 0$, the pole of L with respect to S is also the pole of $L - M$ with respect to $S - M^2$.

15. The equation of a conic touching the three conics $S^{\frac{1}{2}} - L$, $S^{\frac{1}{2}} - M$, $S^{\frac{1}{2}} - N$ is

$$\sqrt{(23)(S^{\frac{1}{2}} - L)} \pm \sqrt{(31)(S^{\frac{1}{2}} - M)} \pm \sqrt{(12)(S^{\frac{1}{2}} - N)} = 0).$$

(*Proc*. R. I. A., 1866.)

Dem.—From equation (632) we see, taking $k = 0$, that the discriminant of $S^{\frac{1}{2}} - L$ is $1 - S'$. Hence, when $S^{\frac{1}{2}} - L = 0$ denote sone of a line-pair, the Tact-invariant of $S^{\frac{1}{2}} - L$ and $S^{\frac{1}{2}} - M$ reduces to $R = 1$. Again, if $\alpha_1, \alpha_2, \alpha_3$ be co-ordinates of any point on the conic $S^{\frac{1}{2}} - L = 0$, or

$$(\alpha_1^2 + \alpha_2^2 + \alpha_3^2)^{\frac{1}{2}} - (\lambda_1\alpha_1 + \lambda_2\alpha_2 + \lambda_3\alpha_3),$$

then
$$(x_1^2 + x_2^2 + x_3^2)^{\frac{1}{2}} - \frac{(\alpha_1 x_1 + \alpha_2 x_2 + \alpha_3 x_3)}{(\alpha_1^2 + \alpha_2^2 + \alpha_3^2)^{\frac{1}{2}}} = 0$$

denotes a conic whose discriminant vanishes, and which touches $S^{\frac{1}{2}} - L$.

Now if $\alpha_1, \alpha_2, \alpha_3$ be the co-ordinates of any point on the conic which touches the three conics $S^{\frac{1}{2}} - L = 0$, $S^{\frac{1}{2}} - M = 0$, $S^{\frac{1}{2}} - N = 0$, and take the conic

$$(x_1^2 + x_2^2 + x_3^2)^{\frac{1}{2}} - \frac{(\alpha_1 x_1 + \alpha_2 x_2 + \alpha_3 x_3)}{(\alpha_1^2 + \alpha_2^2 + \alpha_3^2)^{\frac{1}{2}}}$$

for a fourth; then the functions (14), (24), (34) are respcctively

$$1 - \frac{L}{S^{\frac{1}{2}}}, \quad 1 - \frac{M}{S^{\frac{1}{2}}}, \quad 1 - \frac{N}{S^{\frac{1}{2}}}.$$

Hence, from equation (640), we have

$$\sqrt{(23)\,(S^{\frac{1}{2}} - L)} \pm \sqrt{(31)\,(S^{\frac{1}{2}} - M)} \pm \sqrt{(12)\,(S^{\frac{1}{2}} - N)} = 0.$$

For another proof see *Bicircular Quartics*, p. 70.

16. The operation $l_1 \dfrac{d}{d\alpha_1} + l_2 \dfrac{d}{d\alpha_2} + l_3 \dfrac{d}{d\alpha_3}$ performed on the conic $S^{\frac{1}{2}} - l_\alpha = 0$, where $S \equiv \alpha_1^2 + \alpha_2^2 + \alpha_3^2 = 0$, gives a conic cutting $S^{\frac{1}{2}} - l_\alpha$ orthogonally.

17. The conic

$$\begin{vmatrix} S^{\frac{1}{2}}, & \alpha_1, & \alpha_2, & \alpha_3, \\ 1, & l_1, & l_2, & l_3, \\ 1, & m_1, & m_2, & m_3, \\ 1, & n_1, & n_2, & n_3, \end{vmatrix} = 0 \qquad (641)$$

cuts orthogonally the conics $S^{\frac{1}{2}} - l_\alpha$, $S^{\frac{1}{2}} - m_\alpha$, $S^{\frac{1}{2}} - n_\alpha = 0$; and three others are got by changing the signs of l_1, l_2, l_3 in the second column; of m_1, m_2, m_3 in the third; and of n_1, n_2, n_3 in the fourth.

18. If the four conics of Ex. 17 be denoted by J_1, J_2, J_3; prove that the poles of their chords of contact with S are the four radical centres of the conics $S - L^2$, $S - M^2$, $S - N^2$.

Miscellaneous Exercises.

1. The two lines forming any of the three line-pairs, joining four concyclic points on a conic, are equally inclined to either axis.

2. The axes of all conics passing through four concyclic points are parallel.

3. Find the equation of the circle whose diameter is the normal at the origin to the conic $ax^2 + 2hxy + by^2 + 2fy = 0$.

Ans. $b(x^2 + y^2) + 2fy = 0$.

4. Find the locus of a variable point, if the perpendicular from a fixed point on its polar with respect to $(a, b, c, f, g, h)(x, y, 1)^2 = 0$, be constant.

5. If two triangles be self-conjugate with respect to any conic, their six vertices lie on one conic, and their six sides are tangential to another.

6. If two lines be at right angles to each other, the diameters with respect to them of the triangle of reference meet on the line

$$\alpha \cos A + \beta \cos B + \gamma \cos C. \qquad \text{(M'Cay.)}$$

7. If ω be the Brocard angle of the triangle of reference, prove that

$$(\alpha^2 + \beta^2 + \gamma^2) \sin \omega - \{\alpha\beta \sin (C - \omega) + \beta\gamma \sin (A - \omega) + \gamma\alpha \sin (B - \omega)\} = 0$$

is the equation of its Brocard circle.

8. The locus of the point of intersection of the polars of any point, with respect to two conics, is a circumconic of the common self-conjugate triangle.

9. The locus of the pole of the line $\lambda_a = 0$, with respect to a system of confocal conics given by their general equation, is, if $\Sigma = 0$ be the tangential equation of one of them, and

$$\Omega \equiv \lambda_1^2 + \lambda_2^2 + \lambda_3^2 - 2\lambda_2\lambda_3 \cos A - 2\lambda_3\lambda_1 \cos B - 2\lambda_1\lambda_2 \cos C,$$

and Ω_1, Ω_2, Ω_3, the differential coefficients,

$$\begin{vmatrix} a_1, & a_2, & a_3, \\ \Sigma_1, & \Sigma_2, & \Sigma_3, \\ \Omega_1, & \Omega_2, & \Omega_3 \end{vmatrix} = 0.$$

10. If $S = 0$, $S' = 0$, be two circles in trilinear co-ordinates, their radical axis $\Theta' S + \Theta S' = 0$.

11. Find the locus of a point from which tangents to two given conics are proportional to their parallel semidiameters.

12. If two figures be directly similar, and if corresponding points be conjugate with respect to a given circle, the locus of each is a circle, and the envelope of their line of connection is a conic.

13. Show that the normal to an ellipse, which cuts the curve most obliquely at its second intersection with the curve, is parallel to one of the equiconjugate diameters. (PROF. J. PURSER.)

14. The directrix of a conic, and any two rectangular lines through the focus, form a self-conjugate triangle with respect to the conic.

15. If $y = x \tan \phi$, the equation of a tangent to a conic may be written $x \cos \phi + y \sin \phi - e\gamma = 0$ where $\gamma = 0$ is a directrix.

16. If two points on a conic subtend a given angle at a focus, the locus of the intersection of the tangent at these points is a conic, having the same focus and directrix; and so also is the envelope of their chord.

17. If two semidiameters of an ellipse make a given angle, the line joining their extremities meets its envelope at the point in which it meets a symmedian of the triangle formed by it and the semidiameters. (D'OCAGNE.)

18. If two tangents to an ellipse intersect at a given angle, their chord of contact meets its envelope at the point in which it meets a symmedian of the triangle formed by it and the tangents. (*Ibid.*)

19. Given the base and area of a triangle, prove that the locus of its symmedian point is a hyperbola.

20. A circle S passes through a fixed point O, and intersects a fixed circle in a varying chord L. Show that if L envelops any curve given by its polar equation, with O as the origin, the polar equation of the envelope of S may be at once written down; and hence show—1°. If S envelop a conic concentric with O, L will envelop a conic, having O as focus. 2°. If S touch a line, L will envelop a conic. (MR. F. PURSER, F.T.C.D.)

21. Two conics U, V are taken; U inscribed in a triangle ABC; V touching the sides AC, BC in A, B. Prove that the poles, with respect to U of a common chord of U, V, lies on V. (*Ibid.*)

22. If from any point on a given normal to a conic the three other normals be drawn; prove that the circle through their feet belongs to a fixed coaxal system. (*Ibid.*)

23. If from a point O, whose distances are ρ, ρ' from the foci of an ellipse (whose major axis is $2a$), two tangents be drawn making an angle θ; prove

$$\cos\theta = \frac{\rho^2 + \rho'^2 - 4a^2}{2\rho\rho'}.$$

Dem.—If F, F' be the foci; T one of the points of contact; join FT, $F'T$. Produce FT to S, making $TS = TF'$. Join OS; then the sides of the triangle OFS are equal to ρ, $2a$, ρ', respectively; and the angle $FOS = \theta$. Hence the proposition is proved.

24. B', C' are variable points on the sides AC, AB of a fixed triangle, such that $AB' : B'C :: BC' : C'A$. Prove that the envelope of B_1C_1 is a parabola.

25. If F be the focus of the parabola in Ex. 24, M the circumcentre of the triangle ABC, the angle AFM is right.

26. The directrix of the parabola bisects the portion of the perpendicular between vertex and orthocentre.

27. If a variable conic S' be connected with two fixed conics S_1, S_2 by the relation $\Theta = 0$, the locus of the centre of perspective of the triangle of reference, and its polar reciprocal with regard to S', is a right line.

(PROF. CURTIS, S.J.)

28. Two concentric and coaxal conics U, V are such that a triangle can be inscribed in U, and circumscribed to V. Show that the normals to U at the vertices are concurrent, and that the locus of their centre of concurrence is a coaxal conic. (MR. F. PURSER, F.T.C.D.)

29. If a self-conjugate triangle, with respect to a conic section, be indefinitely small, the radius of its circumcircle is half the corresponding radius of curvature.

30. If a triangle be formed by three consecutive tangents to a conic section, the radius of its circumcircle is one-fourth the corresponding radius of curvature.

31. If α, β, γ be the trilinear co-ordinates of a point in the plane of a triangle, through which are drawn parallels to the sides meeting them respectively in the points 1. 4; 2, 5; 3, 6; prove that the trilinear co-ordinates of the centre of the conic inscribed in the hexagon 123456 are

$$\tfrac{1}{4}(\alpha + b\sin C),\quad \tfrac{1}{4}(\beta + c\sin A,\quad \tfrac{1}{4}(\gamma + a\sin B).$$

32. If for α, β, γ of the last exercise we substitute successively the coordinates of the points Θ, Θ_1, Θ_2, Θ_3, of Ex. 58, chap. ii.; prove that the resulting conics will be the inscribed and escribed circles of the triangle of reference. (LEMOINE.)

33. The locus of the points of contact of tangents from the point $\alpha'\beta'\gamma'$ to the system of conics $\alpha\beta = k\gamma^2$ when k varies, is the conic

$$\frac{\alpha'}{\alpha} + \frac{\beta'}{\beta} = \frac{2\gamma'}{\gamma}.$$

34. If e vary, the locus of the points of contact of tangents from $x'y'$ to $x^2 + y^2 = e^2\gamma^2$ is $(xx' + yy') \div (x^2 + y^2) = \gamma' \div \gamma$.

35. The locus of a point, whose polars with respect to two circles meet on a given line, is a hyperbola.

36. The equation $\sqrt{\alpha \sin A} + \sqrt{\beta \sin B} + \sqrt{-\gamma \sin C} = 0$ denotes a hyperbola whose asymptotes are parallel to the lines α, β.

37. If a circle whose diameter is d passes through the origin and intersects the conic $(a, b, c, f, g, h)(x, y, 1)^2$ in four points, whose radii vectors are ρ_1, ρ_2, ρ_3, ρ_4; prove that

$$\rho_1\rho_2\rho_3\rho_4 \{4h^2 + (a - b)^2\}^{\frac{1}{2}} = cd^2.$$

38. The lines through the origin, and the intersection of

$$(a, b, c, f, g, h)(x, y, 1)^2 = 0, \text{ with } \lambda x + \mu y + \nu = 0,$$

are at right angles if

$$c(\lambda^2 + \mu^2) - 2(f\mu + g\lambda)\nu + (a + b)\nu^2 = 0.$$

39. In the same case, the locus of the foot of the perpendicular from the origin on $\lambda x + \mu y + \nu = 0$ is the circle $(a + b)(x^2 + y^2) + 2gx + 2fy + c = 0$, and the envelope of $\lambda x + \mu y + \nu = 0$ is the conic

$$c\{a + b)(x^2 + y^2) + 2gx + 2fy + c\} = (fx - gy)^2.$$

40. If the axes be oblique, find the equation of the rectangular hyperbola, making intercepts λ, λ'; μ, μ' on them.

41. Find the condition that $\lambda x + \mu y + \nu = 0$ should be normal to

$$\frac{x^2}{a^2} + \frac{y^2}{b^2} - 1 = 0.$$

Ans. $\dfrac{a^2}{\lambda^2} + \dfrac{b^2}{\mu^2} = \dfrac{c^4}{\nu^2}$.

42. Find the equation of the locus of the centre of a conic touching the four right lines

$$\alpha \equiv x \cos\alpha + y \sin\alpha - p_1 = 0, \quad \beta \equiv x \cos\beta + y \sin\beta - p_2 = 0, \text{ \&c.}$$

PROF. CURTIS, S.J.

As in Ex. 3, Art. 188, from the given conditions we have four equations of the form

$$\frac{A}{C}\cos^2\alpha + \frac{B}{C}\sin^2\alpha + \frac{2H}{C}\sin\alpha\cos\alpha = p_1(2\alpha + p_1).$$

Hence, by elimination,

$$L \equiv \begin{vmatrix} \cos^2\alpha, & \sin^2\alpha, & \sin\alpha\cos\alpha, & p_1(2\alpha+p_1), \\ \cos^2\beta, & \sin^2\beta, & \sin\beta\cos\beta, & p_2(2\beta+p_2), \\ \cos^2\gamma, & \sin^2\gamma, & \sin\gamma\cos\gamma, & p_3(2\gamma+p_3), \\ \cos^2\delta, & \sin^2\delta, & \sin\delta\cos\delta, & p_4(2\delta+p_4) \end{vmatrix} = 0,$$

which is the required equation. If the determinant be expanded, and putting $l = \sin\widehat{\beta\gamma}\,.\,\sin\widehat{\gamma\delta}\,.\,\sin\widehat{\delta\beta}$, &c., we get

$$L \equiv lp_1(2\alpha+p_1) - mp_2(2\beta+p_2) + np_3(2\gamma+p_3) - rp_4(2\delta+p_4) = 0,$$

and the origin being transferred to any point of the locus, by putting $p_1 = \alpha$, $p_2 = \beta$, &c., this becomes $L \equiv l\alpha^2 - m\beta^2 + n\gamma^2 - r\delta^2 = 0$, which, though apparently of the second degree, is only of the first; for, on substituting $x\cos\alpha + y\sin\alpha - p$, for α, &c., the coefficients of x^2, xy, y^2 vanish identically.

43. If the equation in Ex. 42 be written in the form

$$l\alpha^2 - m\beta^2 + n\gamma^2 \equiv r\delta^2 + L,$$

we infer that a parabola may be described, having the triangle $\alpha\beta\gamma$ as self-conjugate, and touching L at the point where it meets δ. (*Ibid.*)

44. In the same case, prove that $l\alpha^2 - m\beta^2 = 0$ is a pair of common tangents to the parabolae $r\delta^2 + L = 0$, $n\gamma^2 - L = 0$, and $n\gamma^2 - r\delta^2 = 0$, a pair of common tangents to the parabolae $m\beta^2 + L = 0$, $l\alpha^2 - L = 0$, and that the former pair intersects the latter on L.

45. If α vary in position while β, γ, δ remain fixed; then, if α touches a fixed conic to which β and γ are tangents, the envelope of L is a conic. (*Ibid.*)

46. Given three tangents to a conic, and the sum of the squares of its axes, the locus of its centre is a circle.

47. The director circles of all conics inscribed in a given quadrilatera are coaxal.

48. The covariant F of two conics S, S' passes through their eight points of contact with their common tangents.

49. Find, by means of the covariant F, the equation of the director circle of the conic $(a, b, c, f, g, h)(x, y, 1)^2$.

50. The envelope of the line cutting the conics S, S' harmonically touches the eight tangents drawn to S, S' at their four points of intersection.

51. If two of the vertices of a self-conjugate triangle with respect to S lie on S', the locus of the third vertex is $\Theta' S - \Delta S'$.

52. If the joins of the points in which $(a, b, c, f, g, h)(\alpha, \beta, \gamma)^2$ meets the sides of the triangle of reference to the opposite vertices form two triads of concurrent lines; prove $abc - 2fgh - af^2 - bg^2 - eh^2 = 0$.

Compare the equation with

$$ll'\alpha^2 + mm'\beta^2 + nn'\gamma^2 - (mn' + m'n)\beta\gamma - (nl' + n'l)\gamma\alpha - (lm' + l'm)\alpha\beta = 0,$$

which meets the sides in points which connect with the opposite verti ces by the two triads $l\alpha = m\beta = n\gamma$, $l'\alpha = m'\beta = n'\gamma$.

53. Find, in this manner, the equation of the nine-points circle, the Lemoine circle, the inscribed conic, and the inscribed circle, &c.

54. If λ, μ, ν denote the perpendiculars from the angular points on a tangent; prove that $\lambda^2 \tan A + \mu^2 \tan B + \nu^2 \tan C = 0$ denotes a circle.

55. From last Example prove by reciprocation, if $l\alpha^2 + m\beta^2 + n\gamma^2 = 0$ denote a circle, that

$$l : m : n :: \frac{\tan \psi_1}{\alpha'^2} : \frac{\tan \psi_2}{\beta'^2} : \frac{\tan \psi_3}{\gamma'^2},$$

where α', β', γ' denote the co-ordinates of the centre, and ψ_1, ψ_2, ψ_3 the angles subtended by the sides at the centre.

56. Four concentric equilateral hyperbolas can be described, having the four triangles formed by any four arbitrary lines as self-conjugate.

57. Prove that the polars of the four radical centres of $S - L^2$, $S - M^2$, $S - N^2$, with respect to $S - L^2$, are

$$\begin{vmatrix} L, & a_1, & a_2, & a_3, \\ 1, & l_1, & l_2, & l_3, \\ 1, & m_1, & m_2, & m_3, \\ 1, & n_1, & n_2, & n_3 \end{vmatrix} = 0,$$

and three others got by changing the sign of l_1, l_2, l_3 in the second row; of m_1, m_2, m_3 in the third; and of n_1 n_2, n_3 in the fourth.

58. Prove the following method of constructing the conics J_1, J_2, J_3, J_4, cutting the three conics $S^{\frac{1}{2}} - L$, $S^{\frac{1}{2}} - M$, $S^{\frac{1}{2}} - N$ orthogonally.

Draw tangents from the radical centres to the three conics, and describe a conic through the four systems of six points of contact corresponding to the four radical centres.

59. The conic J_1 is the locus of all the double points of

$$\lambda_1 (S^{\frac{1}{2}} - L) + L_2 (S^{\frac{1}{2}} - M) + L_3 (S^{\frac{1}{2}} - N) = 0.$$

60. If a triangle be turned round the centre of its inscribed circle through two right angles, the triangle in its new position, and that formed by the points of contact of its sides in its original position with its inscribed circle, are in perspective: the centre of perspective is the point Θ. (See Chap. II., Ex. 58.)

61. If through any point in the axis of perspective of a triangle and its orthocentric triangle parallels be drawn to the three sides, these parallels meet the sides in six points which are on an equilateral hyperbola.

62. Parallels to the sides of the triangle of reference through any of the points ω, ω_1, ω_2, ω_3 of Chap. II., Ex. 59, are in each case equally distant from the vertices of the triangle; the distances being in the respective cases the diameters of the inscribed and escribed circles.

63. In a given conic inscribe a triangle whose sides shall pass through given points.

Let the given conic be $\alpha\beta = \gamma^2$, the given points abc, $a'b'c'$, $a''b''c''$, and the perametric angles of the angular points of the inscribed triangle θ, θ', θ''; then, putting $t = \tan\theta$, &c., we have (Art. 160) the three equations

$$a + btt' - c(t+t') = 0, \quad a' + b't't'' - c'(t'+t'') = 0, \quad a'' + b''t''t - c''(t''+t) = 0.$$

Hence, eliminating t', t'', we get a quadratic in t, viz.

$$(a'bb'' + b'cc'' - cc'b'' - c'c''b(t^2 + \{2c(c'c'' - a''b') - \Delta\}t$$
$$+ (a'cc'' + b'aa'' - c'ac'' - c'a''c) = 0,$$

where Δ denotes the determinant $(ab'c'')$. Hence, in general, two triangles can be inscribed: the condition for only one is the equation in t, having equal roots. Hence, if two of the points be given, and the third variable, its locus, so that only one triangle can be described, is a conic.

64. The conics

$$\frac{1}{\alpha} + \frac{1}{\beta} + \frac{1}{\gamma} = 0, \quad \sin\tfrac{1}{2}A\sqrt{\alpha} + \sin\tfrac{1}{2}B\sqrt{\beta} + \sin\tfrac{1}{2}C\sqrt{\gamma} = 0,$$

are confocal. (LEMOINE.)

65. In the same case, the symmedian point of the triangle formed by the centres of the escribed circles of the triangle of reference is the common centre of the conics. (LEMOINE.)

66. A triangle is inscribed in $x^2+y^2-z^2=0$, and two of the sides touch $ax^2+by^2-cz^2=0$; find the envelope of the third side. (SALMON.)

The condition that $\lambda x+\mu y+\nu z$ shall touch $ax^2+by^2-cz^2$ is

$$\frac{\lambda^2}{a}+\frac{\mu^2}{b}-\frac{\nu^2}{c}=0;$$

and denoting (Art. 159, *Cor.* 2) the parametric angles of the vertices of the triangle inscribed in $x^2+y^2-z^2$ by θ, θ', θ'', the equation of the join θ, θ'' is

$$x\cos\tfrac{1}{2}(\theta+\theta'')+y\sin\tfrac{1}{2}(\theta+\theta'')-z\cos\tfrac{1}{2}(\theta-\theta'')=0.$$

Hence the condition for this touching $ax^2+by^2-cz^2=0$ is

$$\frac{\cos^2\frac{1}{2}(\theta+\theta'')}{a}+\frac{\sin^2\frac{1}{2}(\theta+\theta'')}{b}-\frac{\cos^2\frac{1}{2}(\theta-\theta'')}{c}=0;$$

that is,

$$\left(\frac{1}{a}+\frac{1}{b}+\frac{1}{c}\right)+\left(\frac{1}{a}-\frac{1}{b}+\frac{1}{c}\right)\cos\theta\cos\theta''\left(\frac{1}{b}-\frac{1}{c}-\frac{1}{a}\right)\sin\theta\sin\theta''=0;$$

or, say, $$l+m\cos\theta'\cos\theta''+n\sin\theta\sin\theta''=0.$$

In like manner, we get

$$l+m\cos\theta'\cos\theta''+n\sin\theta'\sin\theta''=0.$$

Hence $$\frac{m\cos\theta''}{l}=-\frac{\cos\frac{1}{2}(\theta+\theta')}{\cos\frac{1}{2}(\theta-\theta')},\quad \frac{n\sin\theta''}{l}=-\frac{\sin\frac{1}{2}(\theta+\theta')}{\cos\frac{1}{2}(\theta-\theta')}.$$

Now the chord of $x^2+y^2-z^2$, which is the join of the points θ, θ', is

$$x\cos\tfrac{1}{2}(\theta+\theta')+y\sin\tfrac{1}{2}(\theta+\theta')-z\cos\tfrac{1}{2}(\theta-\theta')=0.$$

Hence $$mx\cos\theta''+ny\sin\theta''+lz=0,$$

and the envelope is $$m^2x^2+n^2y^2-l^2z^2=0.$$

67. The equation of a conic confocal with

$$S\equiv(a,b,c,f,g,h)(\alpha,\beta,\gamma)^2=0,$$

and touching $\lambda\alpha+\mu\beta+\nu\gamma=0$, is

$$\Omega^2S-\Omega\Sigma F+\Sigma^2M^2=0\ [M\equiv\alpha\sin A+\beta\sin B+\gamma\sin C].$$

68. PT, QT are tangents to a conic at the points P, Q; from the centres of curvature at P, Q perpendiculars are drawn to the chord of contact

PQ; prove that the parallels to PT, PQ drawn through the feet of the perpendiculars meet on the symmedian line of the triangle PQT drawn through T. (D'OCAGNE.)

69. The hyperbola

$$\frac{1}{\alpha} + \frac{1}{\beta} - \frac{1}{\gamma} = 0,$$

and the hyperbola

$$(\cos^2 \tfrac{1}{2} A \,.\, \alpha + \cos^2 \tfrac{1}{2} B \,.\, \beta + \sin^2 \tfrac{1}{2} C \,.\, \gamma)^2 - 4 \sec^2 \tfrac{1}{2} A \,.\, \sec^2 \tfrac{1}{2} B \,.\, \alpha\beta = 0$$

are confocal, and their common centre is the symmedian point of one of the triangles formed by the incentre and the centres of two of the escribed circles. (LEMOINE.)

70. A system of four conics having two points common, and each connected with a fifth conic by the relation $\Theta' = 0$, are such that their points of intersection, six by six, lie on three conics.

For, taking the common points as vertices of the triangle of reference, the equation wil be of the form

$$S \equiv a_1 x^2 + 2 f_1 y + 2 g_1 z x + 2 h_1 x y = 0, \text{ \&c.};$$

and there are four relations,

$$a_1 A' + 2 f_1 F' + 2 g_1 G' + 2 h_1 H' = 0, \text{ \&c.}$$

Hence, as in Art. 188, 2°, Ex. 1,

$$l S_1 - m S_2 + n S_3 - p S_4 \equiv 0, \text{ \&c.}$$

71. The condition that the line $(y - y') = m (x - x')$ should be normal to

$$\frac{x^2}{a^2} + \frac{y^2}{b^2} - 1 = 0$$

is

$$m'' (b^2 x'^2) - 2m^3 (b^2 x' y') + m^2 (b^2 x'^2 + a^2 y'^2 - c^4) - 2m (a^2 x' y') + a^2 y'^2 = 0.$$

Hence, find an expression for the sum of the angles which the four normals from any point make with the axis of x.

72. The sum of the angles made with a given line by the four normals from any point to a series of confocal conics is constant.

73. The locus of points having the same eccentric angle on a series of confocal ellipses is a confocal hyperbola.

74. Through a given point A on a conic two rectangular lines are drawn, meeting the conic again in the points B, C; BC meets the normal at A in O. Prove, if A move along the conic, that the locus of O is a homothetic conic.

75. A circle passing through three points on any one of a series of confocal ellipses, the points always lying on fixed confocal hyperbolae, meets the ellipse again, where it is met by another of the confocal hyperbola.

76. In the last question, supposing the three points to coincide, we have a theorem for the circle's curvature of a series of confocal ellipses.

77. The locus of the centres of curvature at points on confocal ellipses where a confocal hyperbola meets them is

$$\frac{\cos^6 \phi}{x^2} - \frac{\sin^6 \phi}{y^2} = \frac{1}{c^2}.$$

78. The centres of the six circles of curvature which can be drawn through any point to a given conic lie on a conic; find its equation.

79. If four normals OA, OB, OC, OD be drawn to a conic from the point $x'y'$; prove that the tangents at the points A, B, C, D, and the axis of the conic, all touch the parabola

$$(xx' + yy' + c^2)^2 = 4c^2x'x.$$

80. Prove that the directrix of the parabola in Ex. 79 is the join of the given point $x'y'$ to the centre.

81. If on perpendiculars erected at the middle points of the sides of a triangle portions be taken (measured either inwards or outwards) proportional to its sides, the triangle formed by the points thus constructed is in perspective with the original triangle. (KIEPERT.)

It is easy to see that the co-ordinates of the corresponding points may be denoted by $\sin\theta$, $\sin(C-\theta)$, $\sin(B-\theta)$; $\sin(C-\theta)$, $\sin\theta$, $\sin(A-\theta)$; $\sin(B-\theta)$, $\sin(A-\theta)$, $\sin\theta$.

Hence the lines of connection of the vertices of the triangles are

$$\alpha\sin(A-\theta) = \beta\sin(B-\theta) = \gamma\sin(C-\theta),$$

which proves the proposition.

82. The envelope of the axis of perspective of the two triangles is the parabola

$$\sqrt{a(b^2-c^2)\alpha} + \sqrt{b(c^2-a^2)\beta} + \sqrt{c(a^2-b^2)\gamma} \mp 0.$$

83. The envelope of the polar line of the centre of perspective with respect to the triangle is the conic

$$\sqrt{\sin(B-C)\alpha} + \sqrt{\sin(C-A)\beta} + \sqrt{\sin(A-B)\gamma} = 0.$$

DEF.—*If the co-ordinates of one of the foci of a conic inscribed in the triangle of reference be* α, β, γ, *the co-ordinates of the other focus will be* $\frac{1}{\alpha}, \frac{1}{\beta}, \frac{1}{\gamma}$. *Hence, being given the locus described by one focus, we can write down the locus described by the other. This transformation, which is of considerable importance, we shall, after* NEUBERG (*Mathesis*, tom. i., p. 184), *call the* ISOGONAL TRANSFORMATION; *and two points related as* $\alpha\beta\gamma$, $\left(\frac{1}{\alpha}\frac{1}{\beta}\frac{1}{\gamma}\right)$, ISOGONAL POINTS. For example, the orthocentre and circumcentre of a triangle are isogonal points, and the centroid and the symmedian point.

84. Prove that the isogonal transformation of the circumcircle is the line at infinity.

85. Prove that the isogonal transformation of a line is an ellipse, a parabola, or a hyperbola, according as it is exterior to, tangential to, or a secant of, the circumcircle of the triangle of reference. (BROCARD.)

86. The isogonal transformation of any diameter of the circumcircle is an equilateral hyperbola circumscribed to the triangle of reference. (*Ibid.*)

87. The locus of the centres of the isogonal transformations of all the diameters of the circumcircle is the nine-points circle. (*Ibid.*)

88. The isogonal transformation of the line joining the symmedian point to the circumcentre is the locus of the centre of perspective in KIEPERT'S theorem 81.

89. The isogonal transformations of the four lines $l\alpha \pm m\beta \pm n\gamma$ are the four conics

$$\frac{l}{\alpha} \pm \frac{m}{\beta} \pm \frac{n}{\gamma} = 0,$$

which, being four circumconics to the triangle of reference, correspond to the four in-conics.

90. Prove that the envelope of Tucker's circles is the Brocard ellipse.

91. Given four tangents to a conic, viz., $\alpha = 0$, $\beta = 0$, $\gamma = 0$, $\delta = 0$; find the locus of the foci. Let $a\alpha + b\beta + c\gamma + d\delta \equiv 0$ be an identical relation; then

$$\frac{a}{\alpha} + \frac{b}{\beta} + \frac{c}{\gamma} + \frac{d}{\delta} = 0 \qquad \text{(SALMON.)}$$

is the locus of the foci.

92. If a variable conic pass through two given points, and have double contact with a given conic, the chord of contact passes through one or other of two given points: prove this, and thence infer that four circumconics of a given triangle can be described, each having double contact with a given conic.

93. If $S \equiv (1, 1, 1 - \cos A, - \cos B, - \cos C)(x, y, z)^2 = 0$; prove that each of the four conics $S - (x \pm y \pm z)^2 = 0$ touches the four conics

$$S - \{x \cos (B \pm C) + y \cos (C \pm A) + Z \cos (A \pm B)\}^2 = 0,$$

where the choice of sign is such that there must be an odd number of negatives.

94. If $a_x^2 = 0$, $b_x^2 = 0$, $c_x^2 = 0$ be three conics, fulfilling the condition that each shall circumscribe a triangle self-conjugate with respect to the other two; prove that

$$\begin{vmatrix} a_1, & a_2, & a_3, \\ b_1, & b_2, & b_3, \\ c_1, & c_2, & c_3 \end{vmatrix}^2 = 0.$$

95. Given a tangent to a variable conic, its eccentricity, and one of the foci, prove that the locus of the other focus is a circle.

96. *M'Laurin's Method of describing Conics.*—The locus of the vertex of a variable triangle whose sides pass through three fixed points, and whose base angles move on two fixed lines, is a conic section.

97. The circumcentre of any triangle self-conjugate with respect to a parabola is on the directrix.

98. If a quadrilateral be described about a parabola, the three circles described on the diagonals of the quadrilateral as diameters have the directrix for their common radical axis.

99. A, B, C; A', B', C' are two triads of points on two lines L, M. Three homothetic conics through ABC', BCA', CAB' meet M again in the points P', Q', R'; and three other homothetic conics through $AB'C'$, $BC'A'$, $CA'B'$ meet L again in P, Q, R; prove that the lines PP', QQ, RR' are parallel. (Mr. F. Purser, F.T.C.D.)

100. *Newton's Method of Generating Conics.*—Two angles of given magnitude turn about two fixed vertices; then, if two of their legs intersect on a fixed line, the locus of the intersection of the other legs is a conic passing through the vertices of the angles.

101. Given two conjugate semi-diameters CP, CQ of a hyperbola, construct the axis.

102. If X, Y be the co-ordinates of a focus of $ax^2 + 2hxy + by^2 + c = 0$, prove that

$$\frac{X^2 - Y^2}{a - b} = \frac{XY}{h} = \frac{c}{ab - h^2};$$

and if μ denote the product of the perpendiculars from the foci on any tangent, prove that

$$\left(\frac{c}{\mu} + a\right)\left(\frac{c}{\mu} + b\right) = h^2.$$

103. Prove that the equation of the pair of tangents from $x'y'$ to $\frac{x^2}{a^2} + \frac{y^2}{b^2} - 1$ may be written $(x - x')^2 a^2 + (y - y')^2 b^2 = (xy' - x'y)^2$.

(PROF. CROFTON.)

104. If R be the circumradius of the triangle of reference, ω the Brocard angle, prove that $R \sin \omega$ is the radius of the auxiliary circle of the Brocard ellipse.

105. Prove that the eccentricity of the conic given by the general equation in terms of its invariants I_1, I_2 of the first and second degree in the coefficients is given by the equation

$$\frac{e^4}{1 - e^2} = \frac{I_1^2 - 4I_2}{I_2}.$$

106. If from the points 1, 2, 3, 4 perpendiculars be drawn to the four lines $\alpha = 0$, $\beta = 0$, $\gamma = 0$, $\delta = 0$; then

$$\begin{vmatrix} \alpha_1, & \beta_1, & \gamma_1, & \delta_1, \\ \alpha_2, & \beta_2, & \gamma_2, & \delta_2, \\ \alpha_3, & \beta_3, & \gamma_3, & \delta_3, \\ \alpha_4, & \beta_4, & \gamma_4, & \delta_4 \end{vmatrix} \equiv 0. \quad \text{Also} \quad \begin{vmatrix} \alpha_1, & \beta_1, & \gamma_1, & 1, \\ \alpha_2, & \beta_2, & \gamma_2, & 1, \\ \alpha_3, & \beta_3, & \gamma_3, & 1, \\ \alpha_4, & \beta_4, & \gamma_4, & 1 \end{vmatrix} \equiv 0.$$

(PROF. CURTIS, S.J.)

107. Hence infer that if p', p'', p''' be the perpendiculars of a triangle r, r', r'', r''', the radii of its inscribed and escribed circles

$$\frac{1}{r} = \frac{1}{p'} + \frac{1}{p''} + \frac{1}{p'''}; \quad \frac{1}{r'} = \frac{1}{p''} + \frac{1}{p'''} - \frac{1}{p'}, \text{ \&c.}$$

Also, if λ', λ'', λ''' denote perpendiculars from the vertices of any triangle on any line through the centre of the in-circle, prove that

$$\frac{\lambda'}{p'} + \frac{\lambda''}{p''} + \frac{\gamma'''}{p'''} = 0. \qquad (\textit{Ibid.})$$

108. If L_1, L_2, L_3, L_4 be perpendiculars from four points A, B, C, D, to a line L; then $L_1(BCD) - L_2(CDA) + L_3(DAB) - L_4(ABC) = 0$. (Compare equation (216).) (*Ibid.*)

109. Given three tangents to a conic, and the length of the minor axis b, to find the focus. Let the co-ordinates of the foci $\alpha\beta\gamma$, $\alpha'\beta'\gamma'$; and the perpendiculars of the triangle of reference p', p'', p'''; then, from (106), we get

$$\begin{vmatrix} \alpha', & \beta', & \gamma', & 1, \\ p', & 0, & 0, & 1, \\ 0, & p'', & 0, & 1, \\ 0, & 0, & p''', & 1 \end{vmatrix} = 0; \quad \therefore \begin{vmatrix} \alpha\alpha', & \beta\beta', & \gamma\gamma', & 1, \\ p'\alpha, & 0, & 0, & 1, \\ 0, & p''\beta, & 0, & 1, \\ 0, & 0, & p'''\gamma, & 1 \end{vmatrix} = 0; \quad \therefore \begin{vmatrix} b^2, & b^2, & b^2, & 1, \\ p'\alpha, & 0, & 0, & 1, \\ 0, & p''\beta, & 0, & 1, \\ 0, & 0, & p'''\gamma, & 1 \end{vmatrix} = 0;$$

or
$$\frac{1}{p'\alpha} + \frac{1}{p''\beta} + \frac{1}{p'''\gamma} = \frac{1}{b^2}; \text{ or } S = \frac{\alpha\beta\gamma}{b^2},$$

where S denotes the circumcircle of the triangle of reference. When the conic is a parabola, b is infinite, and the equation reduces to $S = 0$.

(*Ibid.*)

110. If ABC be a triangle self-conjugate to a conic; λ, μ, ν perpendiculars from A, B, C on the tangent at any variable point D on the curve; prove that

$$\lambda(BCD) + \mu(CAD) + \nu(ABD) = 0. \qquad (Ibid.)$$

111. The circumcircles of the triangles formed by four right lines α, β, γ, δ meet in a point O; tangents at the vertices of the triangle $\beta\gamma\delta$ to its circumcircle meet α in the points A, A', A''. Similarly are found, on the lines β, γ, δ, the triads B, B', B''; C, C', C''; D, D', D''. These points lie four by four on three circles, each passing through O, and through the extremities of a diagonal of the quadrilateral $\alpha\beta\gamma\delta$.

(*Ibid.*)

112. If Σ be the circle through the circumcentres of the triangles $\alpha\beta\gamma$, $\alpha\beta\delta$, $\alpha\gamma\delta$, $\beta\gamma\delta$, the diameters of the circumcircles of the triangles $\alpha\beta\gamma$, $\alpha\beta\delta$, $\alpha\gamma$, passing through the vertices opposite the common base α, concur in Σ.

113. Being given a self-conjugate triangle and a tangent to a conic, the locus of its centre is a right line. (See Art. 188, Ex. 3.)

114. If one of four sides of a quadrilateral envelop a conic, the other three being fixed, the line through the middle points of the diagonals will also envelop a conic. (PROF. CURTIS, S.J.)

115. If six line-pairs xx', yy', zz', uu', vv', ww' be conjugate pairs to the same conic, they are connected by a linear relation

$$lxx' + myy' + nzz' + puu' + fvv' + rww' \equiv 0. \qquad (Ibid.)$$

116. Hence, if two triangles be self-conjugate to the same conic, they are both inscribed in another conic.

For, if $x' = y$, $y' = z$, $z' = x$, $u' = v$, $v' = w$, $w' = u$, we have

$$lxy + myz + nzx = -(puv + gvw + rwu). \qquad (Ibid.)$$

117. Also, in the same case, if

$$u = u', \quad v = v', \quad w = w', \quad x = y', \quad y = z', \quad z = x',$$

the triangle xyz is self-conjugate to the conic, and uvw circumscribes it; and, from (114), we get $lxz + myx + nzy \equiv -(pu^2 + gv^2 + rw^2)$; or a conic circumscribes the triangle xyz, and has uvw self-conjugate.

118. Tangents drawn to a parabola, from the centre of a circumconic of a self-conjugate triangle of the parabola, are conjugate diameters of the conic.

For if x, y, z be the sides of the triangle; u, v the tangents, and taking w to denote the line at infinity by (116), we get the conic

$$lxz + myx + nzy = -(pu^2 + gx^2 + \text{constant}).$$

119. If a circle be described about a triangle self-conjugate with respect to a parabola, its centre is on the directrix.

For the tangents from the centre, being conjugate diameters, must be at right angles.

120. If the centre of the conic be a point on the parabola, an asymptote of the conic is a tangent to the parabola.

121. If corresponding points of similar figures, similarly described on two sides of a triangle, be the poles with respect to a circle of corresponding lines of the same figures; prove that the points are equally distant from the centre of the circle.

122. The four conics which touch three given lines, and have double contact with a given conic Σ, are all touched by four other conics, each having double contact with Σ. (See Ex. 93.)

123. Given $S \equiv ax^2 + 2hxy + by^2 + c = 0$; prove that the equation of any pair of conjugate diameters is

$$lx\frac{ds}{dy} + my\frac{ds}{dx} = 0;$$

and if the diameters be equiconjugate, their equation is

$$\frac{S}{ab - h^2} = \frac{2(x^2 + y^2)}{a + b}.$$

124. The equation of the reciprocal of the parabola at the distance r with respect to the circle $x^2 + y^2 = k^2$ is

$$(k^2x^2 - a^2y^2)^2 = r^2x^2(x^2 + y^2).$$

125. The reciprocal of the parallel to an ellipse at the distance r with respect to the circle $x^2 + y^2 = k^2$ is

$$4k^4r^2(x^2 + y^2) = \{(a^2 - r^2)x^2 + (b^2 - r^2)y^2 - k^4\}.$$

126. The circumcentre of a triangle, its symmedian point, and the orthocentre of its pedal triangle, are collinear. (TUCKER.)

127. The orthocentre of a triangle, its symmedian point, and the symmedian of its pedal triangle, are collinear. (E. VAN AUBEL.)

128. If L, M, N be three collinear points, L', M', N' their corresponding isogonal points (Ex. 84); prove that if the triads L', M, N'; M', L, N' be respectively collinear, the points L', M', N are collinear.

129. Hence show if L', M', N' be points on Kiepert's hyperbola, and if N' be the fourth point where the hyperbola meets the circumcircle of the triangle ABC, that the chord LM is parallel to the Brocard line OK.

130. In the same case, if N' be either of the points where the hyperbola meets infinity, N will be one of the points where the Brocard line cuts the circumcircle.

131. The asymptotes of Kiepert's hyperbola are the Simson's lines of the points where the Brocard line meets the circumcircle. (BROCARD.)

132. The trilinear equation of Neuberg's circle, page 120, is

$$(\alpha\beta \sin C + \beta\gamma \sin A + \gamma\alpha \sin B)$$
$$= \sin A (\beta \operatorname{cosec} C + \gamma \operatorname{cosec} B)(\alpha \sin A + \beta \sin B + \gamma \sin C).$$

(NEUBERG.)

133. The trilinear equation of M'Cay's circle (α), page 253, is

$$3(\alpha\beta \sin C + \beta\gamma \sin A + \gamma\alpha \sin B)$$
$$= \sin A (\beta \operatorname{cosec} C + \gamma \operatorname{cosec} B + 2\alpha \cot A)(\alpha \sin A + \beta \sin B + \gamma \sin C).$$

134. If the base and the Brocard angle of a triangle be given, the locus of the centre of its Brocard circle is an ellipse. (NEUBERG.)

135. If a variable conic S, passing through two fixed points I, J, touch a fixed conic S' at a fixed point; prove that the locus of the point of intersection of a pair of common tangents to S, S' is a conic inscribed in the quadrilateral formed by the tangents from the points I, J to S'.

136. If the axes and a tangent to a conic be given in position; prove that the locus of the centre of the circle osculating it at the point where it touches the tangent is a parabola.

137. If the extremities of the base of a triangle be given in position, and also the symmedian passing through one of these extremities, the locus of the vertex is a circle. (NEUBERG.)

138. In the same case, the envelope of the symmedian passing through the vertex is a conic.

139. The extremities B, C of a triangle are given in position, and the vertex moves on a given conic, passing through the points B, C; prove, if BA, AC pass through corresponding points C', B' of two similar figures, that the loci of the points C', B' are conics. (NEUBERG.)

140. The base BC of a triangle is given in position, and the angle B in magnitude; prove, if $A'B'C'$ be the triangle formed by the tangents to the circumcircle at A, B, C, that the following loci are conics:—1°. Of the point C'; 2°. of the symmedian point of ABC; 3°. of the point of intersection of BB' and AC. (*Ibid.*)

141. In the same case, prove that the envelopes of the lines $B'C'$, AA', and the join of the circumcentre and orthocentre are conics. (*Ibid.*)

142. If from a point P perpendiculars be drawn to the sides of the triangle ABC, and produced, such that

the perpendicular on	a	meets	a in A_1,	b in A_2,	c in A_3;
,,	b	,,	b in B_1,	c in B_2,	a in B_3;
,,	c	,,	c in C_1,	a in C_2,	b in C_3;

then denoting by T_1, T_2, T_3, the areas of the triangles $A_1B_1C_1$, $A_2B_2C_2$, $A_3B_3C_3$, the locus of points for which $T_2 = T_3$ is Kiepert's hyperbola; and for every point in the plane the ratio of $T_1 : T_2 + T_3$ is constant. (*Ibid.*)

143. Prove that the equations of the three axes of perspective of the triangle ABC and Brocard's first triangle are—

$$1°.\quad \frac{\sin^2 A \,.\, \alpha}{\sin(A - 2\omega)} + \frac{\sin^2 B \,.\, \beta}{\sin(B - 2\omega)} + \frac{\sin^2 C \,.\, \gamma}{\sin(C - 2\omega)} = 0;$$

$$2°.\quad \frac{\alpha}{\sin B \,.\, \sin(C - 2\omega)} + \frac{\beta}{\sin C \,.\, \sin(A - 2\omega)} + \frac{\gamma}{\sin A \,.\, \sin(B - 2\omega)} = 0;$$

$$3°.\quad \frac{\alpha}{\sin(B - 2\omega) \,.\, \sin C} + \frac{\beta}{\sin(C - 2\omega) \,.\, \sin A} + \frac{\gamma}{\sin(A - 2\omega) \,.\, \sin B} = 0.$$

INDEX.

THE END.

SECOND EDITION, Price 4s. 6d., post free; or in Two Parts, each Half-a-Crown.

THE

ELEMENTS OF EUCLID,

BOOKS I.–VI., and Propositions I.–XXI. of BOOK XI.;

TOGETHER WITH

An APPENDIX on the CYLINDER, SPHERE, CONE, &c.

With Copious Annotations and Numerous Exercises.

BY

JOHN CASEY, LL.D., F.R.S.,

Fellow of the Royal University of Ireland; Vice-President, Royal Irish Academy; &c. &c.

DUBLIN: HODGES, FIGGIS, & CO. LONDON: LONGMANS, GREEN, & CO.

OPINIONS OF THE WORK.

From the late Rev. Prof. TOWNSEND, F.T.C.D., F.R.S., &c.

"I have no doubt whatever of the general adoption of your work through all the schools of Ireland immediately, and of England also before very long."

From MRS. BRYANT, F.C.P., D.Sc., Principal of the North London Collegiate School for Girls.

"I am heartily glad to welcome this work as a substitute for the much less elegant text-books in vogue here. I have begun to use it already with some of my classes, and find that the arrangement of exercises after each proposition works admirably."

"NATURE," *October* 23, 1884.

"This is the second edition of a work which so accomplished a geometer as Prof. Henrici has pronounced in these columns to be in many respects an 'excellent' book. As the first edition contained 254 pages, and this one reaches 312 pages, it is manifest that the work has grown—and with its growth we find that it has acquired an accession of strength. We will indicate in what directions it has increased. First, and foremost, is the addition of the propositions of Euclid's Eleventh Book, which are generally read by junior students, and an Appendix (*well suited for candidates for the London Intermediate Examination*) on the properties of the Prism, Pyramids, Cylinder, Sphere, and Cone. There is also now given an explanation of the ratio of incommensurable quantities, and a still greater number than in the first edition of alternative proofs. Further, we can testify, by a careful perusal of the text, that the work has been 'thoroughly revised as well as greatly enlarged.' . . . A very large and well selected collection of exercises (upwards of 800), with the addition (now) of numerous examination questions, complete a work every way worthy of the reputation of the great Irish geometer."

THE "NOTTINGHAM GUARDIAN."

"The edition of the *First Six Books of Euclid*, by Dr. John Casey, is a particularly useful and able work. . . . The illustrative exercises and problems are exceedingly numerous, and have been selected with great care. Dr. Casey has done an undoubted service to teachers in preparing an edition of Euclid adapted to the development of the Geometry of the present day."

BY THE SAME AUTHOR.

A SEQUEL TO THE FIRST SIX BOOKS OF THE ELEMENTS OF EUCLID.

THIRD EDITION, *Revised and Enlarged, Price 3s. 6d., cloth.*

DUBLIN: HODGES, FIGGIS, & CO. LONDON: LONGMANS, GREEN, & CO.

EXTRACTS FROM CRITICAL NOTICES.

"NATURE," *April* 17, 1884.

"We have noticed two previous editions of this book, and are glad to find that our favourable opinion of it has been so convincingly endorsed by teachers and students in general. The novelty of this edition is a Supplement of additional Propositions and Exercises. This contains an elegant mode of obtaining the Circle tangential to three given Circles by the methods of false positions, constructions for a quadrilateral, and a full account—for the first time in a text-book—of the Brocard, triplicate ratio, and (what the author proposes to call) the Cosine Circles. Dr. Casey has collected together very many properties of these Circles, and, as usual with him, has added several beautiful results of his own. He has done excellent service in introducing the Circles to the notice of English students. . . .

THE "MATHEMATICAL MAGAZINE," ERIE, PENNSYLVANIA.

"Dr. Casey, an eminent Professor of the Higher Mathematics and Mathematical Physics in the Catholic University of Ireland, has just brought out a second edition of his unique 'Sequel to the First Six Books of Euclid,' in which he has contrived to arrange and to pack more geometrical gems than have appeared in any single text-book since the days of the self-taught Thomas Simpson. The principles of Modern Geometry contained in the work are, in the present state of Science, indispensable in Pure and Applied Mathematics, and in Mathematical Physics; and it is important that the student should become early acquainted with them.

"Eleven of the sixteen sections into which the work is divided exhibit most excellent specimens of geometrical reasoning and research. These will be found to furnish very neat models for systematic methods of study. The other five sections contain 261 choice problems for solution. Here the earnest student will find all that he needs to bring himself abreast with the amazing developments that are being made almost daily in the vast regions of Pure and Applied Geometry. On pp. 152 and 153 there is an elegant solution of the celebrated Malfatti's Problem.

"As our space is limited, we earnestly advise every lover of the 'Bright Seraphic Truth' and every friend of the 'Mathemtical Magazine' to procure this invaluable book without delay."

THE "EDUCATIONAL TIMES."

"We have certainly seen nowhere so good an introduction to Modern Geometry, or so copious a collection of those elementary propositions not given by Euclid, but which are absolutely indispensable for every student who intends to proceed to the study of the Higher Mathematics. The style and general get up of the book are, in every way, worthy of the 'Dublin University Press Series,' to which it belongs."

THE "SCHOOL GUARDIAN."

"This book is a well-devised and useful work. It consists of propositions supplementary to those of the first six books of Euclid, and a series of carefully arranged exercises which follow each section. More than half the book is devoted to the Sixth Book of Euclid, the chapters on the 'Theory of Inversion' and on the 'Poles and Polars' being especially good. Its method skilfully combines the methods of old and modern Geometry; and a student, well acquainted with its subject-matter, would be fairly equipped with the geometrical knowledge he would require for the study of any branch of physical science."

THE "PRACTICAL TEACHER."

"Professor Casey's aim has been to collect within reasonable compass all those propositions of Modern Geometry to which reference is often made, but which are as yet embodied nowhere. . . . We can unreservedly give the highest praise to the matter of the book. In most cases the proofs are extraordinarily neat. . . . The Notes to the Sixth Book are the most satisfactory. Feuerbach's Theorem (the nine-points circle touches inscribed and escribed circles) is favoured with two or three proofs, all of which are elegant. Dr. Hart's extension of it is extremely well proved. . . . We shall have given sufficient commendation to the book when we say, that the proofs of these (Malfatti's Problem, and Miquel's Theorem), and equally complex problems, which we used to shudder to attack, even by the powerful weapons of analysis, are easily and triumphantly accomplished by Pure Geometry.

"After showing what great results this book has accomplished in the minimum of space, it is almost superfluous to say more. Our author is almost alone in the field, and for the present need scarcely fear rivals."

THE "ACADEMY."

"Dr. Casey is an accomplished geometer, and this little book is worthy of his reputation. It is well adapted for use in the higher forms of our schools. It is a good introduction to the larger works of Chasles, Salmon, and Townsend. It contains both a text and also numerous examples."

"JOURNAL OF EDUCATION."

"Dr. Casey's 'Sequel to Euclid' will be found a most valuable work to any student who has thoroughly mastered Euclid, and imbibed a real taste for geometrical reasoning. . . . The higher methods of pure geometrical demonstration, which form by far the larger and more important portion, are admirable; the propositions are for the most part extremely well given, and will amply repay a careful perusal to advanced students."

"MATHESIS," *April*, 1885.

"*A Sequel to Euclid* de M. J. Casey est un de ces livres classiques dont le succès n'est plus à faire. La première édition a paru en 1881, la seconde en 1882, la troisième en 1884, et l'on peut prédire sans crainte de se tromper, qu'elle sera suivie de beaucoup d'autres. C'est un ouvrage analogue aux *Théorèmes et Problèmes de Géométrie* de M. Catalan, et il a les mêmes qualités: il est clair, concis, et renferme beaucoup de matières, sous un petit volume. . . . . . Le sixième livre de l'ouvrage de Casey est aussi étendu à lui seul que les quatre premiers. Il occupe les pages 67 à 158, c'est-à-dire la moitié du volume. C'est en réalité une *Introduction à la Géométrie Supérieure.* Dans cette partie de l'ouvrage, on rencontre des demonstrations d'une rare élégance dues à M. Casey lui-même.

"Le livre se termine par des propositions et exercices supplémentaires. Nous avons remarqué avec plaisir que cet appendice est consacré surtout à ces propriétès du triangle et de certains cercles spéciaux trouvées par nos collaborateurs MM. Brocard et Lemoine, et par M. Tucker. Ces propriétés, d'ont l'étude est, continuée par divers géomètres, ont été le point de départ d'un dévellopement inattendu de la géométrie du triangle. . . .

"Nous recommandons vivement le livre de M. Casey aux professeurs et aux élèves de nos collèges. . . .

"Les nombreux corollaires que M. Casey ajoute à la plupart des théorèmes ne peuvent manquer d'éveiller l'esprit d'observation chez ceux qui pratiqueront son livre; les exercices qui en suivent chacune des grandes divisions permettront au lecteur de se familiariser avec les propositions et les théories nouvelles qui y sont exposées."

Printed in Dunstable, United Kingdom

91220103R00201